T0255341

A Closer Examination of Applicant Faking Behavior

a volume in
Research in Organizational Analysis

Series Editor:
Daniel J. Svyantek, Auburn University

A Closer Examination of Applicant Faking Behavior

edited by

Richard L. Griffith
Florida Institute of Technology

and

Mitchell H. Peterson
Florida Institute of Technology

INFORMATION AGE
PUBLISHING

Greenwich, Connecticut • www.infoagepub.com

Library of Congress Cataloging-in-Publication Data

A closer examination of applicant faking behavior / edited by Richard L. Griffith.
    p. cm. -- (Research in organizational science)
 Includes bibliographical references.
 ISBN 1-59311-513-X (pbk.) -- ISBN 1-59311-514-8 (hardcover)
 1. Applications for positions. 2. Truthfulness and falsehood. I.
Griffith, Richard L. II. Series.
 HF5383.C58 2006
 658.3'112--dc22

                                      2006007218

ISBN 13: 978-1-59311-513-5 (pbk.)
        978-1-59311-514-2 (hardcover)
ISBN 10: 1-59311-513-X (pbk.)
        1-59311-514-8 (hardcover)

Printed in the United States of America

CONTENTS

FOREWORD

There is a tendency for the field of psychology, surprisingly often, to ignore obvious and important topics. The "faking" of personality tests by applicants seeking employment is just one example. It is obvious to anyone who has taken a personality test, or tried to use the results of a personality test for decision making, that people might try to alter their responses in order to receive the scores they desire. And it is equally obvious that if faking were to occur, it might be important. The measurements might become invalid, causing decisions based upon those measurements to be flawed. Yet the topic of faking has not only failed to receive the attention it deserves, some prominent researchers, as Griffith and McDaniel observe in their opening chapter, have actually argued that the topic does not deserve research!

Why would a psychologist argue that *any* topic should not be researched? One reason, as Griffith and McDaniel suggest, may be that the possibility that faking, and the probability that the techniques psychologists use to correct for faking do not really work, are "family secrets" that industrial psychologists are not eager to advertise to their potential client base. Another likely possible reason is that the topic of faking is not quite as obvious as it seems at first glance. A reader of this book will quickly discover that everything about applicant faking, beginning with the very definition of the concept, is complex. Faking is not only hard to define precisely, it is also difficult to detect because by its very nature the behavior is hidden. So it is perhaps understandable when some researchers are tempted by the "drunkard's search"—looking for the house key where the light is best instead of where it was hidden—choose to deal with seemingly more tractable topics instead of grappling with the difficult definitional and empirical issues that the study of faking necessarily entails.

The present volume is a much needed, timely corrective to this attitude. In a wide range of chapters representing different philosophical and empirical approaches, the assembled authors demonstrate the courage to tackle this important and difficult topic head-on, as it deserves to be. The writers of these chapters identify two critical concerns with faking. First, if people fake their responses to personality tests, the resulting scores and the inferences drawn from them might become invalid. For example, people who fake their responses by describing themselves as diligent and prompt might earn better conscientiousness scores, and therefore be hired for jobs requiring this trait that in fact they might not perform satisfactorily. Second, the dishonesty of the faker might itself be a problem, separate from its effect on a particular score. Someone who lies on a pre-employment test might also lie about the hours he or she works, or how much cash is in the till at the end of the shift. Worse, these two problems might exacerbate each other: a dishonest applicant might get higher scores on the traits the employer desires through his or her lying, whereas the compulsively honest applicant might get low scores as an ironic penalty for being honest. Outcomes like these harm employers and applicants alike.

The more one delves into the complexities of faking, as the authors of the chapters in this volume do so thoroughly and so well, the more one will recognize that this seemingly specialized topic ties directly to more general issues in psychology. One of these is test validity. The bottom-line question about any test score, faked or not, is whether it will predict the behaviors and outcomes that it is designed to predict. As Johnson and Hogan point out in their chapter, the behavior of someone faking a test is a subset of the behavior of the person in his or her entire life, and the critical research question concerns the degree to which and manner in which behavior in one domain generalizes to behavior in other domains. This observation illuminates the fact that the topic of faking is also a key part of understanding the relationship between personality and behavior. The central goal of theoretical psychology is to understand why people do the things they do. The central goal of applied psychology is to predict what someone will do in the future. Both of these goals come together in the study of applicant faking.

David C. Funder
Riverside, California
November, 2005

CHAPTER 1

THE NATURE OF DECEPTION AND APPLICANT FAKING BEHAVIOR

Richard Griffith and Michael McDaniel

The use of personality based employment tests has made the intersection of the applicant and the organization much more interesting. While the interpretation of cognitive ability tests is straightforward, noncognitive selections tools offer applicants another opportunity to put their best foot forward, albeit in a fashion not intended by the creators or users of the instruments. Applicants have a great deal of latitude in how they choose to respond to noncognitive employment tests, and may choose to do so in a fashion that does not reflect their true level of the trait.

In fact, between 30 and 50% of applicants elevate their scores (Donovan, Dwight, & Hurtz, 2003; Griffith, Chmielowski, & Yoshita, in press). Not to be outsmarted, Industrial Organizational (I/O) psychologists have armed themselves with more difficult formats and often warn the applicant that if they misrepresent themselves they will be detected (likely a lie on our part). Some I/O professionals find the discussion of applicant faking distasteful and refuse to acknowledge it. Other I/O psychologists try to

A Closer Examination of Applicant Faking Behavior, 1–19
Copyright © 2006 by Information Age Publishing
All rights of reproduction in any form reserved.

hush any conversation regarding faking so we can keep this little family secret to ourselves.

However, the news is out. There are several popular "how to" guides concerning methods for optimizing one's score on personality tests. These helpful hints to applicants date back to the 1950s (Whyte, 1956). In a recent *Time* magazine article titled "What Are They Probing For?" Ehrenreich (2001) described her experience of applying for low paying jobs and faking personality-based selection measures. She stated, "The tests are easy to ace" and offered several faking strategies. Applicants can now purchase the book *Ace the Corporate Personality Test* (Hoffman, 2000). The book offers a tutorial on how applicants can manipulate noncognitive employment tests to maximize their chance of getting hired. The author describes the dimensions of personality, scoring procedures, and lie scales. Hoffman provides example questions and provides advice on how to frame responses that fit specific positions (sales, management, etc.). The Internet provides a wealth of information about faking and personality measures, with stories appearing in widely accessed Web sites such as CNN.com and ABCnews.com. Several Web sites discuss personality assessment, the faking process, and how to watch for lie scales and response patterns that might end in detection (Butcher, 2003; Connolly, 2005; Song, 2005; Swartz, 2003). In some ways these popular applicant resources may be helpful to our science. Instead of ignoring a *potential* weakness in noncognitive assessment, our field may now be compelled to more closely examine the faking issue.

In this chapter, we will discuss the use of personality-based selection measures, and introduce the topic of applicant faking. We will then review literature examining the nature of deception, and its prevalence in society. While the discussion of the animal kingdom and Greek philosophy is not often covered in the applicant faking literature, we use examples from these diverse fields of study to demonstrate the robustness of deceptive behavior across settings. In addition, we briefly discuss the foundations of our current assumptions regarding deception, and how those assumptions may color our understanding of faking behavior. Finally, we will pose some questions regarding the state of our knowledge about applicant faking behavior and introduce the remaining chapters of this text.

PERSONALITY MEASUREMENT AND FAKING BEHAVIOR

The use of personality measures for employee selection has increased greatly in the last 20 years, and has been accompanied by a tremendous surge in research. This resurgence followed a relative lull in the literature, largely due to the influence of authors such as Guion and Gottier (1965) and Mischel (1968). Previously, researchers have claimed that personality measures did not add significantly to the prediction of job performance over and above other selection techniques (Schmidt, Gooding, Noe, & Kirsch, 1984). However due to the wide-ranging support of the five factor model, a generally accepted taxonomy of personality traits (Costa & McCrae, 1992; Costa, 1996; Digman, 1990), and meta-analytic support for the validity of the scales, personality measures have made a strong comeback. Self-report measures are favored by organizations for several reasons. First, many personality measures have demonstrated useful validity with respect to personnel-related decisions (Barrick & Mount, 1991, 1996; Hurtz & Donovan, 2000; Tett, Jackson, & Rothstein, 1991). Not only do they demonstrate criterion validity, but they are also essentially unrelated to ability measures and therefore add incremental variance when included in a selection battery (Salgado & de Fruyt, 2005; Schmidt & Hunter, 1998). Second, they are easy and inexpensive to administer to large groups of applicants. Third, self-report measures, such as personality measures and integrity inventories, exhibit less adverse impact than alternative selection devices such as cognitive ability tests (Hough, Oswald, & Ployhart, 2001; Sackett & Wilk, 1994).

Although personality measures are generally supported as effective tools for personnel selection, they have been criticized because they are easily faked by job applicants (Douglas, McDaniel, & Snell, 1996; Hough & Oswald, 2000; McFarland & Ryan, 2000). This phenomenon has been studied under a variety of names including response distortion, impression management, social desirability, unlikely virtues, self-enhancement and self-presentation (Hogan, 1991; Hough, Eaton, Dunnette, Kamp, & McCloy, 1990; Hough & Paullin, 1994; Rosse, Stecher, Miller, & Levin, 1998; Viswesvaran & Ones, 1999). While many of these terms are conceptually distinct, they all pertain to the elevation of scores on a personality inventory under motivated applicant conditions. Under these circumstances, respondents often raise their scores on positive attributes (such as conscientiousness) and lower their scores on undesirable traits (such as neuroticism). While the general concept of trait elevation is understood, little consensus on the operational definition of faking has been reached (Zickar, Gibby, & Robie, 2004). In addition, we know little about the outcomes associated with faking behavior. Thus we have few answers to the elusive question of what constitutes faking behavior, and *the extent* to which faking matters.

Research has suggested that practitioners are aware of applicant faking and are interested in efforts to reduce it. In a study conducted by Rees and Metcalfe (2003), 39% of managers reported that personality measures were easy to fake, and that they believed over half of all applicants engaged in faking behavior. Goffin and Christiansen (2003) suggested that 69% of practitioners either use, or would like to use, corrections for socially desirable responding on noncognitive selection measures.

Why the concern? Perhaps employers harbor the fear that if an applicant has been dishonest on a selection measure, that applicant may choose to be dishonest on the job. Previous research has suggested that honesty and integrity are rated as the most important characteristics required in applicants (Coyne & Bartram, 2000; Bartram, Lindley, Marshall, & Foster, 1995). This research demonstrated that these characteristics were considered more essential than ability, work experience, and academic qualifications. Thus, as far as employers are concerned, fakers may already have one strike against them. Where do these concerns come from? Why do fakers garner so much attention? Many of these concerns stem from our assumptions about deceptive behavior in general. Therefore to understand where the attitudes towards faking behavior originate, it may be useful to examine the broader literature surrounding deception. The deception literature ranges across a number of diverse disciplines including philosophy (Fingarette, 1998; Frankfurt, 2005; Mele, 1999; Solomon, 1996, 2003), law (Alexander & Sherwin, 2003; Beahrs, 1991; Meissner & Kassin, 2002), anthropology (Dongen, 2002; Smith, 1987; Whiten & Byrne, 1988), business ethics (Cramton & Dees, 1993; Friedman, 2000; Strudler, 1995) and psychology (DePaulo, Lindsay, Malone, Muhlenbruck, Charlton, & Cooper, 2003; Ekman, 1985; Hyman, 1989; Lykken, 1979). It is our hope that briefly introducing a broader perspective on deception may challenge previously held notions regarding the prevalence and pervasiveness of dishonesty, and that these new assumptions may help reframe the assumptions surrounding applicant faking behavior.

THE NATURE OF DECEPTION

Deception is defined as "the deliberate attempt to conceal, fabricate, and/ or manipulate factual or emotional information in order to create and maintain in another a belief that the communicator considers false" (Masip, Garrido, & Herrero, 2004). Intention is a key component in this definition, and it is a common element in other definitions (Bok, 1978; DePaulo & Depaulo, 1989; Ekman, 1985; Miller & Stiff, 1993).

It has been suggested that deception has a basis in evolution, and is an adaptive characteristic (Smith, 2004). Organisms that are able to gain a competitive advantage by deceiving predator or prey have a better chance to survive and reproduce (Bond & Robinson, 1988). Thus, deception is all around us. At first glance the king snake looks remarkably similar to their poisonous cohort the coral snake, which deters predators from dining on them (Greene & McDiarmid, 1981). The opossum deceives potential predators by "playing dead" (Bracha, 2004). Species select for any advantage that will aid in survival against enormous odds. Surviving on this planet is statistically challenging, and deception may provide an edge to those species that effectively use it (Bond & Robinson, 1988). Animals that blend into background with camouflage are less likely to be eaten, and are more likely to eat. The same is true for species that mimic their surroundings, or other species. However, these types of deception are different from our current focus in that they are passive, and outside of the organism's control.

Evolutionary psychology suggests that more active forms of deception also play an adaptive role. In a study of chimpanzees, researchers have found that these primates utilize deception to achieve their goals (Scott, 2001). For instance chimps have been observed faking a limp to garner attention, and ignoring food found while other chimps were present to retrieve it later so the food would not have to be shared (de Waal, 1982; Goodall, 1988). Some researchers have suggested that social complexity, and the deception that accompanies it, may in part be responsible for the advanced development of the human brain (Adenzato & Ardito, 1999). Unlike other mammalian species, the determining characteristic of primate dominance is not size alone. Primate species achieve dominance through the development and maintenance of social networks, and deception is a key tool in this coalition building (Adenzato & Ardito, 1999).

Thus, it should not be surprising that deceptive behavior is part of the competition for employment. Rather, we should be shocked if it were not. Nature provides an abundance of examples of how deception is used to gain a competitive advantage. The employment setting can be viewed as an extension (albeit a socially constructed one) of the competitive forces of nature.

Across many disciplines and literature sources, deception has a strong negative connotation. Deception is often associated with a lack of morals or integrity (Bok, 1978). Western society is based on the notion that truth is a virtue, and that deception is a vice. In fact, the beginning of Judeo Christian ethics commences with the tale of a lie. In the Garden of Eden, Eve told God "The serpent deceived me, and I ate." Western philosophy almost universally portrays deception as a breach of character. In a discussion regarding the inflation of one's accomplishments, Aristotle suggested that deception reflects on the character of an individual, and ultimately on one's virtues (Aristotle, 1991). Kant (1787) describes truthfulness as a moral imperative, and deception a disruptive social force that can erode trust in communication. Nietzsche (1989) stated that deception inevitably ends in compromised relationships. While some philosophical positions have less stringent views of deception, in only a few instances is it portrayed in a positive manner. History also reveals a consistent public attitude toward deception. While George Washington is revered for his inability to tell a lie, Bill Clinton was impeached for it.

Despite our strong social norms against deception, it is a pervasive element in society. We all engage in deception, and we do it quite frequently. On average, people tell three lies for every 10 minutes of conversation (Smith, 2004). While most individuals view themselves as moral (Taylor & Brown, 1988), most lie every day (DePaulo, Kashy, Kirkendol, Dwyer, & Epstein, 1996). Some researchers suggest that deception is a part of our social fabric and that successful functioning of society would be hampered without it (Nyberg, 1993; Scheibe, 1980; Smith, 2004). In some ways we have not been socialized to be completely honest, but to know when and how to use deception so that it is socially appropriate. Anybody who has ever had a significant other ask the question "Do you like my new haircut?" knows that there is only one right answer to that question! While we have come to associate deception with a lack of integrity and unethical behavior, it is often a socially acceptable alternative to completely truthful communication. In fact, we seldom like people to be *totally* honest. Those people who do not understand the rules of acceptable deception, or choose to ignore them are often seen as grating, and are socially scorned (Nyberg, 1993). While we scold others for being deceptive, we are likely engaging in the same behavior to achieve a desired goal of our own. Therefore "fakers," who have been somewhat stigmatized in the I/O literature, may not be behaving much differently than most other people are at any given moment.

Deception is a persuasion strategy. Few individuals deceive for the sake of dishonesty; most use deception as a means towards achieving a desired goal (Miller & Stiff, 1983). During this process the deceiver alters his or her communication in an attempt to influence the beliefs of the target. Altering the beliefs of the target is only one step in the intended action of achieving a predetermined end state that benefits the deceiver. To successfully complete this communication strategy, the deceiver must assess the situation, the current belief state of the target, and the target's ability to facilitate goal achievement. The deceiver then can modify his or her message in a fashion that will maximize the chances of obtaining the goal, and minimize the consequences of deception (Buller & Burgoon, 1994). This description of deception seems fairly cognitive and suggests decision-making components, but deception is not an entirely cognitive phenomenon. Emotions are closely associated with deception (Ekman, 1985; Ekman & Frank, 1993).

One such emotion is the fear of getting caught. If the magnitude of the fear outweighs the valence of the desired goal, individuals are less likely to engage in deception. Ekman and Frank (1993) suggested that fear will be greater if the deceiver perceives the target to be difficult to mislead and if the consequences associated with getting caught are severe. Another emotion that is associated with deception is guilt. Our socialization regarding deception is strong and we consistently receive messages reinforcing the notion that deceptive actions are wrong. Depending on the strength of the socialization of the deceiver and the situation, guilt may be strong enough to interfere with the attempts at deception. In some cases this guilt may be very strong! In a research study we conducted at the University of Akron, we instructed students to respond to a personality questionnaire in such a fashion as to make the respondent look desirable to a potential employer. One subject approached us in the course of the study and pointed out that we were asking her to lie, and that she would not do it! The subject's socialization against deception was so strong that even responding to an instructional set caused her great distress. Situational factors that are associated with guilt and deception include the relationship between the deceiver and target, and the social sanctions associated with deception in that context. Eckman and Frank (1993) suggested that liars feel less guilty when their targets are impersonal or totally anonymous. In addition, if the deceiver feels that "everyone is doing it," they may not be deterred by the general social sanctions that apply to dishonesty.

Extrapolating these individual differences and situational cues to the applicant situation does not paint an encouraging picture in terms of discouraging deception. Some applicants might believe that their attempts at faking can be detected (if they actually believe our warnings, which again are likely fabrications on our part). However the consequences for getting caught are minimal. Perhaps they will not get the job, but their name will not appear on a national registry of known deceivers and application fakers. They will simply move to the next available job and complete another measure (unless conditions are such that other jobs are not available, in which case the punitive consequences of being detected may be higher). Other factors are likely to make the selection setting prone to deception on the part of the applicant. The applicant is not likely to have a relationship with the target (in this case the organization). In an increasingly corporate service economy, the target may be perceived as a distant disembodied giant, not a person. Applicants may also feel that *not faking* may leave them at a competitive disadvantage if they believe the behavior is widespread. In a recent study, 74% of applicants believed that other applicants were engaging in faking behavior (English, Griffith, Graseck, & Steelman, 2005). Thus, the conditions surrounding the application process seem to be fertile ground for deception.

Much of the deception literature refers to a central concept called Theory of Mind (Premack & Woodruff, 1978). This theory suggests that to effectively deceive an individual, the deceiver must understand what the target knows and how that information can be altered in a way to achieve a goal and evade detection. Individuals must develop this "mind reading" skill to effectively utilize deception. Small children, while often willing to deceive, are not very good at it. Research has suggested that the inability to place themselves in the mind of the target may be responsible for this failure to effectively lie (Wimmer & Perner, 1983). This same inability has been posited to explain why individuals with autism may be less effective at deception (Baron-Cohen, Leslie, & Frith, 1985).

It may be surprising to researchers and practitioners, but many applicants who fake also have the inability to effectively deceive on noncognitive employment instruments. Perhaps as many as 20% of applicants fake in the wrong direction (Christiansen & Montgomery, 2005; Griffith, Yoshita, Gujar, Malm, & Socin, 2005; R. L. Hogan, personal communication, October 11, 2005), and much of this dysfunctional faking may be due to the theory of mind explanation. Applicants who do not understand the job or the organizational requirements actually reduce their score, and may fake themselves out of a job. Vasilopoulos, Reilly, and Leaman (2000) suggested that applicants who fake effectively are better able to develop an adopted schema (or implicit job theory) detailing the traits of a successful employee. Deceiving another individual requires a considerable amount of information processing. Not only must the deceiver understand the relationship of the distorted information with the desired goal, they must also understand the target's construal of the situation, and the response that will be most effective with each target.

Simple mathematics tells us that *some* process must be occurring when applicants respond to a measure. Completing the questionnaire takes considerably more time than reading the sentences that comprise the items. Reading a 20-item scale takes about 1.5 minutes, however the average administration time for a scale of that length takes about 10-15 minutes. Something is going on during that time. While time is necessary to access self-relevant information, ample time is still left to consider the impact of this self-reported information, and to consider different reporting strategies that might be more successful. If a decision is made to adopt a less than truthful response, additional time may be necessary for the successful completion of the faking process to occur. Previous research has suggested that there is some response latency when faking occurs (Dwight & Alliger, 1997).

When examining the employment testing setting, it becomes clear that deceptive behavior on the part of the applicant is likely, and in some ways a normal response to the situation. First, evolutionary psychology would suggest that we might be born to deceive. Deception is an adaptive mechanism selected by nature, and thus may be a hard-wired component of human behavior. Second, the situational demands of the applicant setting contain the elements that are likely to elicit deceptive behavior. Applying for a job is a goal driven competitive setting, and few emotional barriers are present to suppress deception. Given this combination of determinants, deception may be the rule rather than the exception in applicant settings. Combined with the robust phenomenon of self-enhancing bias, deception in the applicant setting is likely to produce a significant number of individuals who elevate their scores on personality instruments. In light of our knowledge regarding deception, the research question of "do applicants fake" is silly. The better question might be "why wouldn't they fake?"

Faking and Future Research

Despite the almost universal view of deception as a negative behavior associated with a lack of character, we all engage in deception many times in any given day. However, the use of deception in the employment context in the form of applicant faking still has a stigma associated with it, and often leaves future employers uneasy about their selection choices. Because deception varies in form and severity, we would expect a variation in the outcomes associated with faking. In the end, it may be that "fakers" will engage in behavior counterproductive to the organization, but the truth is we do not have evidence to support that position. Nor do we have evidence to support the counter position. At this present time, little is known about the characteristics of individuals who choose to elevate their scores.

To date, our science has been ineffective at building a basic understanding of the faking phenomenon. While most practitioners support the notion that applicants elevate their scores in an applicant setting, researchers have struggled to adequately define and explain this elevation. Furthermore, research has yet to scratch the surface of the possible associated outcomes of this variety of response distortion. Part of our lack of success in uncovering the nature and consequences of applicant faking is the lack of tested models of faking, or even a theoretical backbone to support such a structure. In recent research history a straw man argument divided researchers instead of facilitating collaboration. Researchers quickly took a side on the issue of "does faking matter?" and often, deep lines were drawn in the sand between the two positions.

The idea for this book was to bring some of those divergent opinions on the nature of faking together, and to address questions regarding the phenomenon of faking and our research efforts to untangle it from the personality measurement process. The question "Does faking matter?" is a rather blunt dichotomous question; however, we aim to address that question, perhaps indirectly, through a closer examination of applicant faking behavior. The questions that are the primary focus of our discussion will center on the complexity of faking behavior and attempt to come to some agreement on the basic assumptions that may underlie our research. It is the hope that a more common set of assumptions will allow future research to become integrated so as to provide more complete understanding to guide our practice. To that end, the chapters of this book discuss the following questions:

1. What is "faking"? This phenomenon has gone by many names, but has generally been based on a few common assumptions. Much of the applicant faking literature has equated faking with deception. Deception in turn is often equated with a lack of morals, and those attempting it are cast in a negative light. However the base rate of deception in society is not comprised of a few morally bereft liars who are shunned, but of almost the entire population. Deception in its varying forms is extremely prevalent in our society, with some claiming that it is the very basis of social interaction. So, is applicant faking a form of deception? Can other variables explain the elevation in scores? If so, how do those mechanisms operate?

2. What is the overarching goal of faking research? To some extent, the progress of this research has been slowed by what appears to be several goals of the research that focus too closely on the phenomenon of faking without keeping the larger goals of I/O psychology in mind. In the case of faking research, what is the "end" to our means? Is this goal (or set of goals) appropriately targeted to reach a meaningful resolution to the issue? Are our research questions and designs currently congruent with these goals?

3. What is occurring when an applicant fakes? What is the process of faking, and what variables (individual differences, situational, etc.) interact to create favorable/unfavorable faking conditions? Much of our research has examined the outcomes of faking, but little has examined the process. How can knowledge of this process contribute to our overarching goals?

4. What are the characteristics of the "typical" faker? Is there a stable set of characteristics that are associated with this type of behavior? If so are these characteristics associated with negative performance (as is assumed by many researchers), better performance, or performance equivalent to a nonfaking sample? What kinds of perfor-

mance (if any) may be most impacted by hiring fakers? Are these characteristics and situational influences stable across cultures?

5. How do we assess whether an individual has faked in an applied setting? Much of what we have learned (or have not learned) about faking stems from studies using instructional sets. These studies tell us something about how much an applicant fakes, but do not tell us much about the applicant setting. Social desirability (SD) measures have been widely used to assess faking behavior. How sensitive are these measures to faking? What, if anything, are they assessing? How can this information be used to improve the predictive validity of personality measures?

6. What methods can be used to reduce the impact of applicant faking, and how effective are these techniques? Warnings and alternative items formats have long been used to deter fakers. How well do these techniques work, and what are some of the consequences of their use? How can these methods be improved to help researchers and practitioners meet the overarching goals of personality measurement? What has our research history taught us, and how do we apply those lessons?

7. There have been several prominent calls for the cessation of faking research (most notably Kevin Murphy's discussion in a 1999 SIOP symposium where he stated he would no longer accept faking papers at the *Journal of Applied Psychology*). Why has there been resistance to this kind of research? Why are so many people interested in it? Are some research designs and analytic techniques more promising than others in our efforts to understand the phenomenon? How have the unique artifactual characteristics of the designs and analytic techniques colored our understanding of applicant faking behavior?

These questions are addressed in the remaining chapters of this book, and additional questions are posed by a group of researchers with extensive research expertise in the measurement of personality and applicant faking behavior. There are several consistent themes that run through the chapters of this text. The first is that faking research has largely been atheoretical, and that future research should be based on sound theoretical models of applicant responding. The second theme is that applicant faking behavior is complex, and that "faking" is not likely to be a unitary construct. Finally the third, and perhaps most important, theme that runs through the book is that personality measures are useful predictors of important organizational outcomes, and that although they are flawed by faking, they continue to be beneficial in selection batteries. To this group of researchers, addressing the faking issue only improves the usefulness of the measures.

In chapter 2, Zickar and Gibby present a historical review of faking research. While the topic of faking has received much recent attention, the issue has been discussed for almost as long as personality measures have existed. In this chapter the authors detail the development of the mechanisms used to combat faking, such as lie scales and warnings, which are discussed in later chapters. Much of the emphasis of the chapter is on research conducted prior to 1970, however after decades of research many of the same research questions remain unanswered. The chapter concludes with a brief review of contemporary research, and suggestions for future study.

In chapter 3, Tett and colleagues present seven nested questions regarding the viability of faking research, and propose an interactionist model of faking. While faking researchers are familiar with the traditional 3 questions surrounding faking (can applicants fake? do they? does it matter?), the authors expand this line of questioning and use this framework to address the recent calls for the cessation of faking research. The second portion of this chapter proposes a model of faking that integrates both individual differences and situational factors in a trait activation structure. The authors then apply this framework to explain and reconcile previous research results.

Mesmer-Magnus and Viswesvaran discuss the strengths and weaknesses of various research designs and analytical methods that have been used to examine faking in chapter 4. In addition, the authors discuss the kind of research questions that can be addressed by each of these designs, and how the inappropriate use of design can cloud research results. In the latter portion of the chapter, Mesmer-Magnus and Viswesvaran discuss the extent to which individuals differ on faking behavior, and whether the lie scales and SD measures used to capture faking are beneficial to organizations.

Burns and Christiansen discuss the development of social desirability (SD) measures, and their role in statistical corrections for faking in chapter 5. SD is discussed both as a trait, and as a response set. In addition, the relationship of SD to other substantive constructs is discussed. Measures of socially desirable responding are the most common method of controlling faking, but the authors of this chapter suggest that these proxy measures may not actually be tapping faking behavior. Burns and Christiansen advise caution when drawing conclusions from SD scores, and suggest further refinement of the construct and scales.

In chapter 6, Griffith and colleagues review potential antecedents of applicant faking behavior, including situational influences, cognitive biases, and individual difference variables. The authors then summarize the results of a program of research that tested these antecedents empirically. Their research suggests that a portion of faking behavior can be explained by nonvolitional components such as trait activation and temporal biases. In addition, the individual differences of integrity and locus of control were significantly related to the amount of faking. Perhaps of most interest is the finding that measures of SD were not related to actual levels of faking behavior.

In chapter 7, Snell and Fluckinger reframe the faking issue by focusing on the validity of personality measures and presenting an applicant response model based on James's theory of conditional reasoning. The authors point out that most faking research has been conducted in an atheoretical fashion, and has unsuccessfully searched for a "silver bullet" to put our concerns regarding faking at rest. The authors suggest that a theoretically sound applicant response model will lead to more informative research, and ultimately generate better strategies for improving the validities of noncognitive measures.

Johnson and Hogan present a socioanalytic view of applicant responses in chapter 8 that is based on self-presentation theory. The authors suggest that previous models of applicant responding, which portray applicants as detached individuals who provide accurate account of their reflections, are unrealistic. Instead they suggest that test-takers use the personality assessment as an opportunity to present themselves in a manner that will further their agendas. The model of responding presented by the authors in chapter 8 has interesting implications for the construct validity of noncognitive measures. This model implies that we are not assessing the person's personality per se, but their projected personality.

In chapter 9, Peterson and Griffith examine the faking/job performance relationship. The authors break this issue into two research questions. First they examine the performance of personality measures under faking conditions by asking the question, "Are personality measures useful when a substantial portion of applicants fake?" In addition the authors discuss the job performance of those individuals who choose to fake, and are hired by the organization. The chapter uses several hypothetical scenarios to explore the effects on criterion-related validity and the quality of hiring decisions.

Converse and colleagues review the research surrounding the use of forced choice measures to reduce the impact of faking in chapter 10. Forced choice measures, attempt to reduce faking a priori, versus attempting to make faking corrections post hoc. The authors discuss recent studies that have demonstrated reduced fakability, and criterion-related validities comparable to Likert scales. In addition, the authors address key questions regarding the ipsative nature of the measures, and the actual use of forced choice measures in selection settings.

Chapter 11, authored by Pace and Borman, discusses the use of warnings to reduce the effects of faking on personality measures. Traditionally warnings have been constructed so that applicants are informed that their attempts at faking can be detected, and, that punitive action will be taken if they are detected. Pace and Borman suggest an alternative format of warnings that may have less of an effect on applicant reactions. These warnings emphasize the notion that faking is not in the best interest of the applicant, and therefore reduce the motivation to fake. The authors recommend the development of additional warning approaches grounded in theory.

Chapter 12 details the interaction of cognitive ability with several methods used to deter applicant faking. Vasilopoulos and Cucina suggest that these deterrents (warnings, forced choice, subtle items) can have the unintentional consequence of increasing the cognitive load of the personality item response process. Therefore to effectively fake under these conditions cognitive ability may play a role. The authors also state that the attempts at reducing faking may affect the adverse impact and incremental validity of noncognitive measures. In addition, they point out that the interaction of cognitive ability may affect the construct validity of personality measures.

Frei, Yoshita, and Isaacson examine the phenomenon of faking across cultures in chapter 13. More specifically, the authors suggest that the meaning of faking may differ across cultures, such that elevated scores in one cultural setting may not have the same meaning as another. Frei et al. summarize the studies on response distortion in international employment testing and highlight potential cultural variables that might influence applicant faking. The latter portion of the chapter provides an example of how indigenous cultural variables in Japan might impact applicant faking in such a way that it would be hard to label it as faking.

Finally in chapter 14, Ryan and Boyce address the motivation of researchers who choose to examine applicant faking. They suggest that the desire to answer practitioner's questions about faking has largely driven the research. Ryan and Boyce suggest that faking should be addressed within the broader question of what affects the ability and motivation to respond to items in an accurate manner, and that asking the question of "Does faking matter?" may be too imprecise. These authors correctly point out that being more specific in our research questions and basing research on sound theory is likely to lead to better solutions for the users of personality tests.

CONCLUSION

Our primary goal for this book is to challenge the assumptions that have served as the basis of contemporary faking research. To that end this group of authors was asked to wipe the slate clean and try to get past the "does faking matter" question, and closely examine the complexity of the faking phenomenon. If successful, this book will raise more questions than it answers and the questions will lead to theoretically sound tests regarding the assumptions regarding applicant faking behavior.

AUTHOR'S NOTE

Correspondence on this paper should be sent to Richard L. Griffith, Department of Psychology, Florida Institute of Psychology, 150 W. University Blvd. Melbourne, FL.
E-mail: griffith@fit.edu

REFERENCES

Adenzato, M., & Ardito, R. B. (1999). The role of theory of mind and deontic reasoning in the evolution of deception. In M. Hahn & S. C. Stoness (Eds.), *Proceedings of the twenty-first conference of the cognitive science society* (pp. 7-12). Mahwah, NJ: Erlbaum.

Alexander, L., & Sherwin, E. (2003). Deception in morality and law. *Law and Philosophy, 22*(5), 393-450.

Aristotle. (1991). *On rhetoric: A theory of civil discourse* (G. A. Kennedy, Tran.). Oxford, England: Oxford University Press.

Barrick, M. R., & Mount, M. K. (1991). The Big Five personality dimensions and job performance: A meta-analysis. *Personnel Psychology, 44*(1), 1-26.

Barrick, M. R., & Mount, M. K. (1996). Effects of impression management and self deception on the predictive validity of personality constructs. *Journal of Applied Psychology, 81*(3), 261-272.

Baron-Cohen S., Leslie A. M., & Frith U. (1985). Does the autistic child have a "theory of mind"? *Cognition, 21*, 37- 46.

Bartram, D, Lindley, P. A, Marshall, L., & Foster, J. (1995). The recruitment and selection of young people by small businesses. *Journal of Occupational and Organizational Psychology, 68*(4), 339-358.

Beahrs, J. O. (1991). Volition, deception, and the evolution of justice. *The Bulletin of the American Academy of Psychiatry and the Law, 19*(1), 81-93.

Bok, S. (1978). *Lying: Moral choice in public and private life.* New York: Vintage Books.

Bond, C. F., & Robinson, M. (1988). The evolution of deception. *Journal of Nonverbal Behavior, 12*(4), 295-307.

Bracha, H. S. (2004). Freeze, flight, fight, fright, faint: Adaptationist perspectives on the acute stress response spectrum. *CNS Spectrums, 9*(9), 679-685.

Buller, D. B., & Burgoon, J. K. (1994). Deception: Strategic and nonstrategic communication. In J. A. Daly & J. M. Wiemann (Eds.), *Strategic interpersonal communication* (pp. 191-223). Hillsdale, NJ: Erlbaum.

Butcher, S. (2003, December). How to handle psychometric tests. *eFinancial Careers.* Retrieved November, 2005, from http://news.efinancialcareers.com/ HOW_TO_ITEM/newsItemId-3158

Christiansen, N., & Montgomery, G. E. (2005). *Individual differences in ability to fake forced-choice personality measures.* Unpublished manuscript.

Connolly, K. G. (2005, May). Faking it: Can job applicants "outsmart" personality tests? *Development Resource Group.* Retrieved November, 2005 from http:// www.drgnyc.com/List_Serve/May%2011_2005.htm

Costa, P. T. (1996). Work and Personality: Use of the NEO-PI-R in Industrial/Organizational Psychology. *Applied Psychology: An International Review, 45*, 225-241.

Costa, P. T., & McCrae, R. R. (1992). Four ways five factors are basic. *Personality and Individual Differences, 13*, 653-665.

Coyne, I., & Bartram, D. (2000). Personnel managers' perceptions of dishonesty in the workplace. *Human Resource Management Journal, 10*, 38-45.

Cramton, P. C., & Dees, J. G. (1993). Promoting honesty in negotiation: an exercise in practical ethics. *Business Ethics Quarterly, 3*(4), 359-394.

DePaulo, P. J., & DePaulo, B. M. (1989). Can deception by salespersons and customers be detected through non-verbal cues? *Journal of Applied Social Psychology, 19*(18), 1552-1577.

DePaulo, B. M., Kashy, D. A., Kirkendol, S. E., Dwyer, M. M., & Epstein, J. A. (1996). Lying in everyday life. *Journal of Personality and Social Psychology, 70*(5), 979-995.

DePaulo, B. M., Lindsay, J. J., Malone, B. E., Muhlenbruck, L., Charlton, K., & Cooper, H. (2003). Cues to deception. *Psychological Bulletin, 129*(1), 74-118.

de Waal, F. (1982). *Chimpanzee politics: Power and sex among apes.* London: Jonathan Cape.

Digman, J. M. (1990). Personality structure: Emergence of the five factor model. *Annual Review of Psychology, 41*, 417-440.

Dongen, E. V. (2002). Theatres of the lie: "Crazy" deception and lying as drama. *Anthropology & Medicine, 9*(2), 135–151.

Donovan, J. J., Dwight, S. A., & Hurtz, G. M. (2003). An assessment of the prevalence, severity, and verifiability of entry-level applicant faking using the randomized response technique. *Human Performance, 16*, 81-106.

Douglas, E. F., McDaniel, M. A., & Snell, A. F. (1996, August). *The validity of non-cognitive measures decays when applicants fake.* Paper presented at the meeting of the Academy of Management, Nashville, TN.

Dwight, S. A., & Alliger, G. M. (1997, April). *Using response latencies to identify overt integrity test dissimulation.* Paper presented at the 12th annual conference of the Society for Industrial Organizational Psychology, St. Louis, MS.

Ehrenreich, B. (2001). What are they probing for? *Time, 157*, 22-86.

Ekman, P. (1985). *Telling lies: Clues to deceit in the marketplace, politics, and marriage.* New York: W. W. Norton.

Ekman, P., & Frank, M. G. (1993). Lies that fail. In M. Lewis & C. Saarni (Eds.), *Lying and deception in everyday life*. London: The Guilford Press.

English, A., Griffith, R. L., Graseck, M., & Steelman, L. A. (2005). *Frame of reference, applicant faking and the predictive validity of non-cognitive measures: A matter of context*. Manuscript submitted for publication.

Fingarette, H. (1998). Self-deception needs no explaining. *The Philosophical Quarterly, 48*(192), 289.

Frankfurt, H. G. (2005). *On bullshit*. Princeton, NJ: Princeton University Press.

Friedman, H. H. (2000). Biblical foundations of business ethics. *Journal of Markets & Morality, 3*(1), 43-57.

Goffin, R. D., & Christiansen, N. D. (2003). Correcting personality tests for faking: A review of popular personality tests and an initial survey of researchers. *International Journal of Selection and Assessment, 11*, 340-344.

Goodall, J. (1988). *In the shadow of man* (Rev ed.). London: Phoenix Giants.

Greene, H. W., & McDiarmid, R. W. (1981). Coral snake mimicry: Does it occur? *Science, 213*(45), 1207-1212.

Griffith, R. L., Chmielowski, T. S., & Yoshita, Y. (in press). Do applicants fake? An examination of the frequency of applicant faking behavior. *Personnel Review*.

Griffith, R. L., Yoshita, Y., Gujar, A. Malm, T., & Socin, R. (2005). *The impact of warnings on applicant faking behavior*. Manuscript submitted for publication.

Guion, R. M., & Gottier, R. F. (1965). Validity of personality measures in personnel selection. *Personnel Psychology, 18*, 135-164.

Hoffman, E. (2000). *Ace the corporate personality test*. New York: McGraw-Hill.

Hogan, R. (1991). Personality and personality measurement. In M. D. Dunnette & L. M. Hough (Eds.), *Handbook of industrial and organizational psychology* (Vol. 2, 2nd ed., pp. 873-919). Palo Alto, CA: Consulting Psychologists Press.

Hough, L. M., Eaton, N. K., Dunnette, M. D., Kamp, J. D., & McCloy, R. A. (1990). Criterion-related validities of personality constructs and the effect of response distortion on those validities. *Journal of Applied Psychology, 75*, 581-595.

Hough, L., & Oswald, F. (2000). Personnel selection: Looking toward the future—remembering the past. *Annual Review of Psychology, 51*, 631-664.

Hough, L. M., Oswald, F. L., & Ployhart, R. E. (2001). Determinants, detection and amelioration of adverse impact in personnel selection procedures: Issues, evidence and lessons learned. *International Journal of Selection and Assessment, 9(1-2)*, 152-194.

Hough, L. M., & Paullin, C. (1994). Construct-oriented scale construction: The rational approach. In G. S. Stokes, M. D. Mumford, & W. A. Owens (Eds.), *Biodata handbook: Theory, research, and use of biographical information in selection and performance prediction*. Palo Alto: Consulting Psychologist Press.

Hurtz, G. M., & Donovan, J. J. (2000). Personality and job performance: The Big Five revisited. *Journal of Applied Psychology, 85*, 869-879.

Hyman, R. (1989). The psychology of deception. *Annual Review of Psychology, 40*, 133-154.

Kant, I. (1787). *On a supposed right to lie from benevolent motives* (Lewis White Beck, Ed. & Trans.) Chicago: University of Chicago Press.

Lykken, D. T. (1979). The detection of deception. *Psychological Bulletin, 86*(1), 47-53.

Masip, J., Garrido, E., & Herrero, C. (2004). Defining deception. *Anales de Psicologia, 20*, 147-171.

McFarland, L. A., & Ryan, A. M. (2000). Variance in faking across non-cognitive measures. *Journal of Applied Psychology, 85*, 812-821.

Mele, A. R. (1999). Twisted self-deception. *Philosophical Psychology, 12*(2), 117 -137.

Meissner, C. A., & Kassin, S. M., (2002). He's guilty!: Investigator bias in judgments of truth and deception. *Law and Human Behavior, 26*(5), 469-480.

Miller, G. R., & Stiff, J. B. (1993). *Deceptive communication.* Newbury Park, CA: Sage.

Mischel, W. (1968). *Personality and assessment.* New York: Wiley.

Murphy, K. (1999, April). *Discussant in New Empirical Research on Social Desirability in Personality Measurement.* Symposium presented at the 14th annual meeting of the Society of Industrial and Organizational Psychologists, Atlanta, GA.

Nietzsche, F. (1989). Beyond good and evil (H. Zimmern, Trans.). New York: Prometheus Books.

Nyberg, D. (1993) *The varnished truth: Truth telling and deceiving in ordinary life.* Chicago: University of Chicago Press.

Premack, D., & Woodruff, G. (1978). Does the chimpanzee have a theory of mind? *The Behavioral and Brain Sciences, 4*, 515-526.

Rees, C. J., & Metcalfe, B. (2003). The faking of personality questionnaire results: Who's kidding whom? *Journal of Managerial Psychology, 18*, 156-165.

Rosse, J. G., Stecher, M. D., Miller, J. L., & Levin, R. L. (1998). The impact of response distortion on pre-employment personality testing and hiring decisions. *Journal of Applied Psychology, 83*(4), 634.

Sackett, P. R., & Wilk, S. (1994). Within-group norming and other forms of score adjustment in pre-employment testing. *American Psychologist. 49*(11), 929-954.

Salgado, J. E., & de Fruyt, F. (2005) Personality in personnel selection. In A. Evers, N. Anderson, & O. Voskuijl (Eds.), *Handbook of personnel selection.* Oxford, England: Blackwell.

Scheibe, K. E. (1980). In defense of lying: on the moral neutrality of misrepresentation. *Berkshire Review, 15*, 15-24.

Scott, S. (2001). *Chimpanzee theory of mind: A proposal from the armchair.* Carleton University Cognitive Science Technical Report 2001-06.

Schmidt, N. W., Gooding, R. Z., Noe, R. A., & Kirsch, M. (1984). Meta-analysis of validity studies published between 1964 and 1982 and the investigation of study characteristics. *Personnel Psychology, 37*, 407-422.

Schmidt, F. L., & Hunter, J. E. (1998). The validity and utility of selection methods in personnel psychology: practical and theoretical implications of 85 years of research findings. *Psychological Bulletin, 124*(2), 262-274.

Smith, E. O. (1987). Deception and evolutionary biology. *Cultural Anthropology, 2*(1), 50-64.

Smith, D. L. (2004). *Why we lie. The evolutionary roots of deception and the unconscious mind.* New York: St. Martin's Press.

Solomon, R. C. (1996). *Self, deception and self-deception in philosophy. A cross-cultural philosophical enquiry.* Albany: State University of New York Press.

Solomon, R. C. (2003). Deception, self, and self–deception in philosophy. *The Joy of Philosophy, 21,* 198-218.

Song, K. M., (2005, June). Faking your type to "pass" a personality test. *The Seattle Times.* Retrieved November, 2005, from http://seattletimes.nwsource.com/html/health/2002343492_healthpersonality22.html

Strudler, A. (1995). On the ethics of deception in negotiation. *Business Ethics Quarterly, 5*(4), 804-813.

Swartz, M. (2003). *Should you cheat on pre-employment tests? It depends on your ultimate goals.* Retrieved November 2005, from http://www.careeractivist.com/my-articles/cheat-on-emplyment-tests.htm

Taylor, S. E., & Brown, J. D. (1988). Illusion and well being: A social-psychological perspective on mental health. *Psychological Bulletin, 103,* 299-314.

Tett, R. P., Jackson, D. N., & Rothstein, M. (1991). Meta-analysis of personality-job performance relationships. *Personnel Psychology, 47*(1), 157-172.

Vasilopoulos, N. L., Reilly, R. R., & Leaman, J. A. (2000). The influence of job familiarity and impression management on self-report measure response latencies and scale scores. *Journal of Applied Psychology, 85,* 50-64.

Viswesvaran, C., & Ones, D. S. (1999). Meta-analysis of fakability estimates: Implications for personality measurement. *Educational and Psychological Measurement, 59,* 197-210.

Whiten, A., & Byrne, R. W. (1988). Tactical deception in primates. *Behavioral and Brain Sciences, 11,* 233-244.

Whyte, W. (1956). How to cheat on personality tests. In *The Organizational Man* (pp. 405-410). New York: Simon & Schuster/Touchstone.

Wimmer H., & Perner J. (1983). Belief about beliefs: Representation and constraining functions of wrong beliefs in young children's understanding of deception. *Cognition, 13,* 103-128.

Zickar, M. J., Gibby, R. E., & Robie, C. (2004). Uncovering faking samples in applicant, incumbent, and experimental data sets: An application of mixed-model item response theory. *Organizational Research Methods, 7*(2), 168-190.

CHAPTER 2

A HISTORY OF FAKING AND SOCIALLY DESIRABLE RESPONDING ON PERSONALITY TESTS

Michael J. Zickar and Robert E. Gibby

In recent years, there has been a heightened concern among psychologists about the threat of misrepresentation on personality tests; with over 900 articles, book chapters, and dissertations published since 2000 alone that have investigated misrepresentation in some form.[1] This edited volume of chapters is a culmination of the recent research attention of this important topic in the field of industrial-organizational (I-O) psychology. As will be detailed in this chapter, there have been periodic waves of concern about response distortion throughout the history of personality testing. In fact, concerns about response distortion on personality inventories are nearly as old as the personality tests themselves. In this chapter we review the earliest response distortion research including topics of faking and socially desirable responding. At places throughout the chapter, the concepts of faking and socially desirable responding will seem intertwined with each other. This is important to note because early researchers often failed to distinguish between con-

A Closer Examination of Applicant Faking Behavior, 21–42
Copyright © 2006 by Information Age Publishing
All rights of reproduction in any form reserved.

scious faking and unconscious socially desirable responding. The evolution of the understanding of response distortion is one of the themes that we explore.

We will document early practical concerns about response distortion in the initial days of personality inventories. This phase of research focused on documenting the problem of faking and identifying people who were thought to be faking. Then we discuss the emergence of theoretical research that investigated causes and individual differences related to faking and related constructs. It is at this stage that researchers began to distinguish between socially desirable responding and faking. Next, we summarize response distortion research up to contemporary times. Finally, we provide some implications about directions that research should take in the future. We emphasize the early stages of response distortion research given that there have been several reviews of contemporary research on research distortion (e.g., Arthur, Woehr, & Graziano, 2001; Hough, 1998; Levin & Zickar, 2002; Smith & Robie, 2004) and there has been relatively little analysis of research conducted before 1970.

We identified several themes that have persisted since psychologists first became concerned about response distortion. First, despite attempts by psychologists to explain or minimize the effects of distortion, managers persist in doubting the accuracy of personality tests due to faking (see Rees & Metcalfe, 2003). Second, despite decades of research, psychologists still struggle with the meaning of response distortion and its consequences. It is our hope that a consideration of the historical record will help broaden contemporary psychologists' understanding of this important topic. Before getting into the specific historical record of response distortion research, it is important to briefly review the introduction of personality testing into industry.

Invention of Personality Testing in Industry

Personality testing is thought to have begun with Robert S. Woodward's Personal Data Sheet (WPDS; Woodworth, 1917), which was designed to identify World War I military recruits who would be susceptible to emotional breakdowns in the presence of military combat (commonly referred to as "shell-shock"). After the war, the WPDS and other similar inventories were used in industry to identify applicants who would likely have adjustment and emotional problems in the workforce. In the 1920s and 1930s, business organizations were concerned about hiring "bad apples" and malcontents who might lower morale and more importantly might agitate for labor unions (see Zickar, 2001). The WPDS and other personality inventories provided needed tools to help screen out employees who suffered from anger and emotional unease. In the early 1930s, personality inventories, such as the Bernreuter Personality Inventory (BPI) and the Humm-Wadsworth Temperament Scale (HWTS), were created to measure a wider range of personality traits that would be explicitly linked to work-place performance. The Bernreuter and Humm-Wadsworth inventories were the first personality inventories to be marketed directly to personnel managers for use in hiring. These tests became popular in industry, with articles in scientific and trade journals touting the benefits that accrued from using personality tests. For more detail about the early days of personality testing, we refer you to Gibby and Zickar (2005), Goldberg (1971), and Parker (1991).

Early Awareness of Fakability

Although personality inventories became popular, researchers and industrial clients expressed concerns about the fakability of personality tests. For example, noted I-O psychologist Morris Viteles commented on "the great flexibility or modifiability of many personalities" (Viteles, 1932, p. 245). Vernon (1934) addressed issues with many of the early personality inventories, and raised the point that test-takers may be motivated to fake their responses. Early experimental studies were conducted on various personality inventories to demonstrate that respondents directed to fake attained higher scores than those not instructed to fake. For example, Hendrickson (1934) reported that teachers who were told to take the BPI as if they were applying for a job were reported to be significantly higher on stability, dominance, extroversion, and self-sufficiency than when completing the inventory under more neutral conditions. Bernreuter (1933), Kelly, Miles, and Terman (1936), and Ruch (1942) found similar results with student populations and different personality tests.

Steinmetz (1932) found that scores on the Strong Vocational Inventory Blank could be manipulated to fit an occupation selected at random. This research was important in that it suggested that respondents, when instructed, could tailor their responses to best fit different occupations. Respondents were not just responding in a way that made them seem the best of all possible humans; they were responding in ways that they believed test-administrators wanted them to answer. Wesman (1952) found similar results on a personality test, in that students changed their responses when applying for a sales job compared to a librarian job. Many years later, Kroger and Turnbull (1975) replicated Steinmetz's results with the MMPI, finding that respondents answered differently if they were told to answer as a creative artist or an Air Force officer. I-O psychologists in the 1990s would later verify that applicants fake according to the desirability that an item has for a particular job, not social desirability in general (see Kluger & Colella, 1993; Miller, 2001).

Although the consequences of faking were yet to be understood, some psychologists speculated on the effects faking would have on a test's usefulness. Bennett and Gordon (1944) were concerned that faking on the BPI and other personality inventories may have been one of the causes of the generally poor validity of these inventories. Reiterating this idea of respondents' possible motivation to distort responses, Ghiselli and Brown (1948) asserted that "a person's performance on a [personality] test or inventory cannot be expected to be divorced from the purposes and motives that actuate him to take it" (p. 169). Early researchers did not have an assessment of the consequences of response distortion on the validity of personnel decisions; they did, however, have a suspicion that faking mattered. The logical next step in this line of research was to develop techniques that would identify those who were faking.

Identification of Fakers

In what appears to be the first attempt to develop a faker identification technique, Humm and Wadsworth (Humm, 1944; Humm & Wadsworth, 1934) created a No-count index of the HWTS to determine bias in frankness of response among respondents. With this index, respondents who answered "No" more often than most respondents were flagged as potential fakers. This technique anticipated other approaches that determine faking based on normative data; according to Humm and Humm's approach, responses that fell outside the pattern of typical response patterns would be suspect.

Similarly, Ruch (1942) examined an experimental data set in which respondents were instructed to answer honestly and then were instructed to respond another time as if they were applying for a sales job. He created an honesty scale based on responses to items that introverts (as identified when responding honestly) were successful in manipulating (when responding for a sales job). The author himself identified one of the problems with the index. The technique was useful in identifying introverts who were faking to get the sales job. However, one limitation of the honesty scale was that it was not able to identify true extroverts who were not faking. This would be a problem that would plague many subsequent faking detection techniques. It is extremely difficult to distinguish between somebody who is naturally high on a desirable trait versus somebody who achieves that high score through response distortion.

There were varying degrees of understanding of what to do with the information that was collected from the detection scales. Rosenzweig (1934) proposed that scores of personality inventories could be corrected based on an understanding of the factors that influence scores; unfortunately he never investigated this idea to determine the utility of these hypothetical corrections. Meehl and Hathaway (1946) created a scale, the K scale, on the Minnesota Multiphasic Personality Inventory (MMPI) that was designed to correct faking responses. Although the K scale was the first scale explicitly designed to correct for response distortion, evaluations of its utility provided mixed results. For example, Ruch and Ruch (1967) found that using the K scale correction reduced the validity of the MMPI. Later research would show that personality scores corrected for one form of distortion, social desirability, did not approximate personality scores when responding honestly (Ellingson, Sackett, & Hough, 1999).

When personality test developers admitted that their tests could be faked, they often claimed that detection scales could be used to identify those who were dissimulating (e.g., Humm, 1944). Other researchers were not as sure of detection scales. In fact, Ghiselli and Brown (1948) argued that such claims were inflated and that if such identification were possible, it would take a prohibitive amount of time and require trained professionals to verify the claims. Given the difficulty of detecting faking, Ghiselli and Brown (1948) believed that personality tests had more use for placement decisions than for selection decisions, because placement decisions often have less stakes compared to selection decisions that typically have higher stakes. As will be discussed later, research on more sophisticated detection techniques has continued, though the results often lead to as pessimistic conclusions as Ghiselli and Brown (1948).

Preempting Faking

Another strategy of addressing faking was to attempt to preempt the occurrence of faking. Researchers attempted to preempt faking by encouraging test-takers to be frank and objective when responding. Research suggested that when respondents felt that their information would be held in confidence that their responses would be most honest and accurate. For example, Spencer (1938) found that administration of a personality test without requiring identification information on the test or its answer sheet resulted in frank responses. He advocated that test administrators not require respondents to sign their name on answer sheets and that administrators stress the confidentiality of individual responses. This advice would help researchers who did not need to identify individual respondents; however, in personnel selection, this advice was not helpful as identification of individual responses was necessary. As Meehl and Hathaway (1946) pointed out, asking respondents to be honest and frank with their responses was not effective. Respondents who were bent on faking would continue to do so. Others for whom faking was more unconscious would also probably be little influenced by explicit instructions to be honest (Meehl & Hathaway, 1946, p. 527). Some researchers, such as Moore (1942), blamed the way the tests were constructed for faking. Specifically, Moore noted that "their form of construction, whether of the yes-no or of the multiple-choice type encourages deceit" (Moore, 1942, p. 165). Given this criticism, the hope was that alternative wordings and formats of items might be less susceptible and less encouraging to fake. As early as the late 1930s, research was conducted to come up with alternative formats. The Kuder Preference Record, which measured vocational interests, phrased questions so that respondents had to choose between two options that were roughly equal in desirability (Kuder, 1939). For example, respondents might be asked to choose whether they would rather:

Take a photograph of a champion swimmer.

Or

Take a photograph of a table you would like to make.

Although the Kuder Preference Record measured vocational interests, researchers working on personality tests soon followed the direction of Kuder and worked at developing forced-choice personality tests.

More specifically, the Annual Report to the National Research Council's Committee on Selection and Training of Aircraft Pilots, 1939-1940 described work by Rethlingshafer on

constructing a test which eliminates the possibility of checking "good" answers. When this research is completed, it should be possible to lay before a candidate a test in which choices are to be made between items of equal "face validity." (National Research Council, 1940, p. 3)

Such a forced choice test could be considered one of the first attempts to preempt respondent distortion of personality assessment, and could be viewed as the forerunner of the most well-known forced choice personality assessments, such as the Edwards Personal Preference Schedule (PPS) (Edwards, 1957).

The PPS had 210 different pairs of statements in which the competing statements were roughly matched in social desirability. For example, as cited in Anastasi (1982, p. 516), a demonstration item was:

(a) I like to talk about myself to others.
(b) I like to work toward some goal that I have set for myself.

There were 15 separate personality scales that were measured within the PPS. The PPS operationalized personality based on needs; example needs measured by the PPS included Need for Achievement, Need for Change, and Need for Heterosexuality.

Although forced-choice personality inventories were thought to be able to preempt faking, they had their own problems. First, respondents often became frustrated with what they viewed as difficult choices. Edwards reported, "I have on occasion been maligned by a few such subjects. It is tempting to interpret these expressions of hostility as testimonials to the success with which the factor of social desirability has been controlled in the PPS" (Edwards, 1957, p. 60). In addition, construction of forced-choice inventories are extremely time-consuming given that the test constructor needs to match roughly pairs on social desirability. Also, research showed that forced-choice personality inventories were not immune to faking (Waters, 1965). Finally, forced choice inventories have a degree of ipsativity in them that makes traditional psychometrics difficult to implement; with ipsative measures, it is possible to compare scores on different personality traits within an individual but not across individuals (see Hicks, 1970; Meade, 2004).

One additional attempt to preempt faking was to rely on subtle items. With subtle items, the intent of the item-writer was disguised; test-takers would have difficulty determining what was the most socially desirable option. Most of the work on developing personality scales based on subtle items was done on the MMPI (Hanley, 1956; Wiener, 1948). Recent researchers seem to have soured on subtle personality items; Weed, Ben-Porath, and Butcher (1990) found that subtle items were less valid indicators than obvious items. They recommended that scales based on subtle items should not be used as indicators of personality or psychopathology. Subtle items have not been tried extensively in personality tests used for personnel selection, although they have been tried with success in biodata tests (Becker & Colquitt, 1992). One of the reasons is that managers are reluctant to ask candidates questions that are not job relevant, which is often the case with subtle personality items. Although it is tempting to suggest that subtle items might provide the answer to faking woes, based on research in clinical psychology applications, the use of subtle personality items seems to be just another in the line of failures in preempting faking.

Distinguishing Constructs: Response Sets, Social Desirability, and Faking

Early research on faking was aimed primarily at figuring out the extent of the problem and documenting that personality (both industrial and clinical) and vocational interest inventories were influenced by faking; follow-up research was aimed at techniques that might help identify fakers or preempt faking. The next wave (1940s through 1960s) of research was conducted to better understand the psychological nature of faking itself. Much of the resurgence of faking research in the 1980s and 1990s attempted to address many of the same theoretical issues that psychologists were grappling with in the 1940s and 1950s.

Cronbach might be the first person to address the concept of a *response set* (see Cronbach, 1946, 1950). Cronbach conceptualized response sets as habitualized tendencies to respond to items in a particular manner, *regardless of the content of the item*. Although there were many different response sets identified by Cronbach and others, the acquiescence response set is the one most relevant to this chapter. Cronbach hypothesized that there would be individuals who would be likely to acquiesce to items, regardless of whether the item reflected their true personality state or attitude. Much follow-up research was conducted to attempt to both correct for acquiescence and other response sets (see Block, 1965 for an excellent summary). In addition, researchers attempted to understand the nature of acquiescence by researching its correlates. In particular, some researchers noted that individuals who scored high on authoritarianism tended to display acquiescence response sets (Bass, 1955; Chapman & Campbell, 1957). In general, however, the response set literature was atheoretical; researchers focused on identifying people engaging particular response sets and statistical methods for controlling those sets. Although research on acquiescence was prevalent in the 1950s and 1960s, research on another response set, social desirability, soon became more prevalent.

Perhaps the first important article relevant to the study of social desirability was Meehl and Hathaway's 1946 article in *Journal of Applied Psychology* titled "The K Factor as a Suppressor Variable in the Minnesota Multiphasic Personality Inventory." Although the K factor scale was discussed earlier in this chapter as a technique to correct for faking, this article had an even longer lasting contribution to research on response distortion. In this article, Meehl and Hathaway distinguished between conscious motivated faking ("faking" or "lying" in their terms) versus "unconscious self-deception" (1946, p. 525). They described the latter as deception by individuals who "may be consciously quite honest and sincere in their responses" (p. 525). This latter response style is what later became known as social desirability.

This distinction was important in that it identified two motives for response distortion. The conscious faking motive was designated for respondents who knowingly responded falsely to a particular personality inventory because they wanted to avoid being diagnosed with a psychological illness, wanted to get hired for a particular job, or for other reasons. This motive, outright faking, revealed moral dysfunctions and was the type of distortion that was really thought to be the biggest problem, at least in industrial settings. According to Meehl and Hathaway (1946), other types of respondents who responded falsely, did so unconsciously. This unconscious motive was seemingly pervasive and presumably part of being a normal human being. People responded positively on items related to self-evaluation because it made them feel better. Understanding the nature of social desirability and self-deception was something that could be linked back to work by Freud, who posited a complex architecture of the self that was designed to keep negative and sinful information from burdening the consciousness (see Freud, 1938/1941; see also Frenkel-Brunswik, 1939 for further empirical work).

Edwards (1957) built on the work of Meehl and Hathaway's notion of unconscious faking and was the first to use the phrase *social desirability*. In Edwards' conceptualization, social desirability was a property of items; some items were high in desirability and thus evoked responses in the favorable direction (see also, Edwards, 1953). Initial research focused on having raters evaluate items on social desirability and attempting to understand what caused some items to be socially desirable. Follow-up research investigated whether respondents from different populations judged items similarly in terms of social desirability (Edwards, 1957). Edwards found that there were little differences between male and female respondents, between psychiatric patients and alcoholics compared to nondiagnosed, "normal" respondents, and between Norwegians and Americans. Item ratings of social desirability seemed invariant across different populations.

In addition to conceptualizing social desirability as a property of items, Edwards also created a social-desirability scale to measure the extent that individual respondents were responding in socially desirable ways (Edwards, 1957). The Edwards Social-Desirability scale was an amalgamation of items from MMPI scales and the Manifest Anxiety scale (Taylor, 1953). Edwards's social desirability scale was the first of many such scales that would follow.

One of the most important developments in measurement of social desirability was created to address limitations of the Edwards scale. Crowne and Marlowe created their own social desirability scale, the Marlowe-Crowne Social Desirability Scale and conducted a program of studies to validate their scale (see Crowne & Marlowe, 1960, 1964). They identified items that had nearly universally high social desirability even though the items were untrue of most individuals. In addition, the items had to be unrelated to psychopathology, which was not true of the Edwards items given that they were taken from clinical personality scales. The original 33 items from the Marlowe-Crowne inventory are well-known among psychologists given that the scale has been used frequently since its original publication. Items such as "Before voting I thoroughly investigate the qualifications of all the candidates" and "I like to gossip at times [keyed false]" may be some of the most recognizable personality items of all time.

Although Marlowe and Crowne designed their scale to remedy limitations in the Edwards social desirability scale, the authors of the scale came to believe, based on experimental research, that the scale measured need for approval, not social desirability. At the end of their 1964 book, they conclude "the association of defensiveness and protection of self-esteem with dependence on the approval of others appears sufficiently well established now" (Crowne & Marlowe, 1964, p. 206). This awareness that a scale that was designed to measure a response set was actually measuring a substantive personality trait was one that would engage researchers for years.

Over the years, authors have revisited the Marlowe-Crowne scale, developing shorter forms of the original instrument (Barger, 2002; Reynolds, 1982) as well as translating the scale into other languages (e.g., Rudmin, 1999). Marlowe and Crowne dodged the question of whether high scores on social desirability indicated whether they were responding in a way they thought was accurate or whether "their responses represent witting and deliberate deceit" (Crowne & Marlowe, 1964, p. 21). Other researchers, however, posited that there were two components to the Marlowe-Crowne scale, a self-deception component and a conscious faking component (see Millham, 1974; Millham & Kellogg, 1980). This empirically-derived dichotomy was anticipated by Meehl and Hathaway's (1946) research on the MMPI.

Paulhus (1988, 1991) created a scale, the Balanced Inventory of Desirable Responding (BIDR) that measured separately the self-deception and impression management dimensions (see Paulhus, 2002 for an excellent history of the development of his scale). The BIDR scale broke out self-deception into self-deceptive denial and self-deceptive enhancement. Although the Marlow-Crowne scale had been factor-analyzed and mapped on to these two dimensions, the BIDR provided a contribution in that each of the two scales filled out those dimensions with a full array of items and provided relatively uncorrelated measures of these distinct constructs. Research has shown that the Impression Management scale was extremely sensitive to instructions to fake in a simulated job application scenario; the Self Deception scales were not as sensitive (Paulhus, Bruce, & Trapnell, 1995). The BIDR has gained prominence in response distortion research, with over 400 citations of the book chapter that presents much of the original research (Paulhus, 1991). Although the BIDR has introduced a multifaceted measure of the construct formerly known as social desirability, questions still remained in terms of how the social desirability constructs should be treated. Specifically, researchers have grappled with the question of whether social desirability and its related constructs is a nuisance variable that should be controlled or whether it represents an important construct and personality trait of its own.

Is Social Desirability Such a Bad Thing?

The question of whether social desirability is a good thing (or not) has been pondered by psychologists for a long time. Ruch (1942) found that students who were instructed to fill out the Bernreuter personality inventory as if they were applying for a sales job had Extroversion scores that were at the 98% level in the test manual's norms. Ruch speculated that "this type of dishonesty is just as valuable as real extroversion in the psychological make-up of a salesman" (p. 232). Some researchers have posited that social desirability and faking would be a positive characteristic that might be desirable for certain jobs. Similarly to Ruch's logic, the thought is that some jobs require that people adapt to specific situations and demands. People who are able to fake or who are high in social desirability might be better able to adapt to those situations.

To test this hypothesis, Ones, Viswevaran, and Reiss (1996) derived, meta-analytically, the correlation between social desirability and various job performance criteria. Social desirability was related to training performance ($\rho = .22$) but was unrelated to task performance ($\rho = .00$), measures of counterproductive behaviors ($\rho = -.03$) and job performance ($\rho = .01$). Their meta-analytic results suggest that social desirability is more of a nuisance variable than a substantive variable. Although there may be some jobs where social desirability scores would be related to job performance, it appears, based on their results, that there are no general trends.

The previous section summarized the some of the earliest research on impression management. Much of this research focused on determining whether personality tests could be faked (they could) and whether people who were faking could be identified. Finally, researchers became interested in understanding the nature of the construct of faking and identifying its various components. These threads of research have continued throughout the recent past. Although we cannot summarize recent research in as much detail as the earlier research, we present an overview of recent research.

THE LAST 15 YEARS OF FAKING AND
SOCIAL DESIRABILITY RESEARCH

Although this chapter focuses mainly on research conducted prior to 1970, we felt it important to review briefly some of the research conducted on response distortion in the more recent past. We identify two primary lines of research that have consumed researchers' efforts and attention: further attempts to identify fakers and more research to investigate the effects of faking and response distortion on the value of personality testing. Much of the recent interest in faking has centered on the last theme and so we focus on that topic first.

Does Faking Matter?

Leatta Hough and colleagues published a multisample, military-based investigation of response distortion and faking that attempted to address the question of whether faking made a difference in personality test validities (Hough, Eaton, Dunnette, Kamp, & McCloy, 1990). They analyzed an experimental study in which some participants were instructed to fake and others were instructed to be honest, as well as a comparison of applicant respondents with incumbent respondents. Throughout their research, they found evidence that people did engage in response distortion when instructed to do so, but that there was little difference in mean scores between applicants, incumbents, and individuals instructed to respond honestly. Most importantly, they found that criterion-related validities were similar for individuals high in social desirability versus those low in social desirability. Their results were intriguing: people can fake personality inventories if instructed to do so but that they do not do so in large numbers if in an applicant-like setting (i.e., the mean of applicants was not much different from the mean of incumbents). And, perhaps most importantly, even if faking (as operationalized by a social desirability scale in their study) occurs, it does not reduce predictive validity substantially.

Their study prompted much additional research to investigate whether faking and social desirability really made a difference. One promising line of research has investigated whether the correlation coefficient is an insensitive way to determine the effects of faking. Certain researchers have speculated that faking may impact rank-ordering of candidates at the top end of the distribution, though the correlation coefficient may be relatively immune to these changes in rank-ordering given that they occur only in a narrow rank of the overall distribution of test-scorers. Zickar and colleagues (Zickar, 2000; Zickar, Rosse, & Levin, 1996) used a Monte-Carlo computer simulation in which they varied the number of fakers, the amount of faking, and the overall test validity. They found that the observed validity correlation was relatively insensitive to the presence of faking even though the rank-ordering of candidates changed in the presence of faking. When faking occurred, those who were faking were disproportionately ranked among the top applicants. Similar results were reported by Douglas, McDaniel, and Snell (1996). These results were paralleled by other studies that found changes in ordering of test-scores in the presence of faking. For example, Rosse, Stecher, Miller, and Levin (1998) found that applicants who had the highest conscientiousness scores were much more likely to have high response distortion scores than applicants with modest conscientiousness scores. Christiansen, Goffin, Johnston, and Rothstein (1994) found that personality test-scores corrected for scores on a social desirability scale had a different rank ordering than when the scores were uncorrected. These results suggest that faking does matter, although its effects may hard to determine given traditional correlation-based statistics. As will be discussed in the final section, however, determining the precise effects of faking is difficult to do given inherent limitations of most faking methodologies.

Faking Detection Revisited

The search for a method to detect and, hence, control faking has continued unabated since Rosenszweig's (1934) initial suggestion that measures could be developed to control for response distortion. Faking scales, similar in concept to Meehl and Hathaway's (1946) MMPI-based scales as well as later social desirability scales, have proliferated. These scales go by different names: Unlikely Virtues (Hurtz & Alliger, 2002), Social Desirability (Stober, 2001), Lie Scales (Elliott, 1981), and Impression Management (Seisdedos, 1996). Each of these scales, however, has been proposed to be able to detect who is faking and who is not. In addition, more sophisticated techniques have been studied, such as using item response theory based appropriateness measurement (Zickar & Drasgow, 1996) and response latencies (Vasilopoulos, Reilly, & Leaman, 2000). Many of these techniques show promise because people instructed to fake generally score higher on these detection devices than those instructed to respond honestly; in addition, the mean differences on these detection devices across fakers and honest respondents is generally larger than the mean difference on substantive personality scales.

Despite these promises, there are substantial barriers that limit the effectiveness of these detection devices. First off, to judge the effectiveness of a faking detection device, one needs to look beyond mean differences. Zickar and Drasgow (1996) discuss the need to investigate the effectiveness of evaluating the number of fakers correctly identified with a technique (i.e., *correct rejections* in signal detection theory [SDT] terminology) with the number of honest respondents incorrectly classified as fakers (i.e., *false positives* in SDT terminology). The rate of false positives is extremely important for testing agencies because a false positive results in falsely accusing an honest individual of misrepresenting themselves. Most evaluations of personality detection techniques do not consider false positives. One exception was Zickar and Drasgow's (1996) evaluation of a social desirability scale and an appropriateness measurement technique, which suggested that false positive rates were too high for either of these techniques to be used operationally.

In addition to prohibitively high false positive rates, other research demonstrates that the faking detection scales can themselves be faked. If respondents are coached on the type of scale items that comprise a personality inventory, they can successively avoid endorsing items on an Unlikely Virtues scale (Hurtz & Alliger, 2002). In addition, Robie, Curtin, Foster, Phillips, Zbylut, and Tetrick, (2000) demonstrated that individuals coached on how response latencies could be used to detect faking were able to avoid detection. If the faking detection techniques can be themselves faked by sophisticated test-takers, the utility of these techniques is limited.

The odds that respondents are sophisticated enough to know how to avoid faking detection have increased over the years. Wide-scale awareness of the fakability of personality inventories began in 1956 with the publication of William Whyte Jr.'s immensely popular book *The Organization Man*. This book assailed personality testing as a way for companies to weed out individuals who did not conform to their strict rules of behavior:

> the [personality] tests, essentially, are loyalty tests, or rather, tests of potential loyalty. Neither in the questions nor in the evaluation of them are the tests neutral; they are loaded with values, organization values, and the result is a set of yardsticks that reward the conformist, the pedestrian, the unimaginative—at the expense of the exceptional individual without whom no society, organization or otherwise, can flourish. (p. 182)

As part of the book's assault on personality testing, there was an appendix titled "How to cheat on personality tests" that instructed potential applicants on how to respond to typical items on personality scales. For example, Whyte instructed the potential applicant that "when asked for word associations or comments about the world, give the most conventional, run-of-the-mill, pedestrian answer possible" (1956, p. 405). Whyte's book inspired other criticisms of personality testing. For example, in 1965 Charles Alex published a book *How to Beat Personality Tests* that was popular enough to justify two printings of a cheap paperback version.

The "how to" guides to faking personality tests appear to be a mainstay now. Books exist such as *Ace the Corporate Personality Test* (Hoffman, 2000) as do newspaper articles titled "Faking your type to 'pass' a personality test" (Song, 2005). Employers will no longer be able to count on unsophisticated test-takers who have little experience with pre-employment personality tests. Some applicants will be quite sophisticated in terms of knowing how to answer items in specific directions to get hired. For example, Hoffman (2000) provides specific guidance on how to respond to the "softball questions" on lie scales and discusses the best way to respond to items that measure specific personality traits such as conscientiousness and stress tolerance. Although we predict researchers will continue to search for a detection technique that will be able to be used operationally, review of early and recent research does not give much hope.

FUTURE DIRECTIONS FOR RESEARCH SUGGESTED BY PAST RESEARCH

There are several directions for future research that we believe will pay dividends for personality researchers. Like much research in I-O psychology, these ideas have been, for the most part, suggested and anticipated by early researchers. Development of more sophisticated (and precise) data analytic techniques as well as the slow but persistent accumulation of knowledge and theory means, however, that there is a high probability that additional research in these areas will pay more dividends.

Antecedents of faking. Vernon (1934) laid out an agenda for understanding why some respondents engage in large amounts of faking and others seem to respond honestly in nearly all situations. He suggested looking at competition, economic reasons, personal interest, interest in the material of the test, among other factors. McFarland and Ryan (2000) examined how certain personality characteristics influenced respondent score inflation and found that integrity, neuroticism, and conscientiousness were related to faking. Griffith, Malm, English, Yoshita, Gujar, and

Monnot (2004) provide additional support for these variables as well as locus of control, temporal cognitive bias, and situational influence in chapter 6 of this text. An integrated model of faking, however, should include these and additional variables.

Preempting faking.   Minimizing the effects of faking by minimizing the amount of faking is a research strategy that has engaged people since the early days of faking research. Dwight and Donovan (2003) carry on this research by investigating the effects of warning applicants that their responses can be identified if they fake. Their results suggest that with such warnings, respondents do fake less. We suspect that this approach will not work in the long run because there is no reliable method for detecting fakers. Respondents will realize that warning applicants not to fake are generally empty threats. In the long run new warning designs that are less threatening may prove to be useful (e.g., Pace and Borman's work in chapter 12).

Forced-choice methodology has some promise. Christiansen, Burns, and Montgomery (2005) present some research that shows that forced-choice inventories are better able to prevent faking than traditional personality tests. Although forced choice inventories present their own problems (ipsativity and negative applicant reactions), recent sophisticated psychometric theory might hold some potential for solving some of these problems (see McCloy, Heggestad, & Reeve, 2005). Meade (2004) presents an excellent analysis of a forced-choice personality instrument and discusses some solutions to the difficult measurement issues.

Corroborating personality tests.   One of the strategies for dealing with faking is to corroborate suspect scores with ancillary scores. For example, people who score high on a faking detection scale could have additional interview questions that probe the suspect areas. For example, an individual who scores very high on Conscientiousness might be asked additional questions to get at that dimension. To do so, researchers should determine how best to ascertain personality information from interviews. In addition, they should study the role of impression management in interviews and how that relates to impression management on personality test items. This advice is consistent with Ghiselli and Brown's (1948) advice that to detect faking would require a skilled professional to verify and evaluate personality test-scores; to do so competently, skilled professionals would need access to additional information beyond that which is present in the personality test-scores.

Further theoretical refinement.   The final future research direction that needs to be addressed is the most important. As can be seen throughout this chapter, there has been long-standing confusion over the nature of response distortion on personality tests. We have learned much about the nature of faking over the past 70 years of research. Distinctions between conscious, motivated faking and unconscious socially desirable responding were first discussed by Meehl and Hathaway (1948) and empirically measured by Paulhus and colleagues (e.g., Paulhus, 1991). These distinctions have been an important advance. It is unclear, however, how this dichotomy maps on to the different operationalizations of faking that have occurred. For example, participants who are instructed to fake in an experimental study respond differently than participants who are instructed to respond honestly. Unfortunately it is difficult to map the faking processes used by instructed fakers on to the dichotomy proposed by Paulhus (1991). Important questions remain about differences in faking processes used by applicants versus those used by instructed experimental fakers versus those who are responding without any incentive to fake. In addition, much of our research has assumed that everyone who fakes in a particular study uses the same faking process; one study using a latent class version of item response theory suggests that this assumption is not true (Zickar, Gibby, & Robie, 2004).

To advance faking research, we need an accepted taxonomy of faking similar to the Big Five taxonomy for personality traits. This taxonomy could be used to help structure the accumulation of knowledge in faking research. Currently, researchers often use the concepts of faking, social desirability, response distortion, and impression management interchangeably. This construct confusion makes it difficult to assess the generalizability and value of any particular research study, given that two researchers may use the same term to refer to completely different response processes. Coming up with an accepted taxonomy will not resolve all future issues in faking research, but it will be an important next step to help advance the area of faking research far beyond the current research base.

CONCLUSIONS

As demonstrated in this chapter, early researchers and users of personality inventories knew much about faking on personality inventories. In fact, it is sort of disheartening to see how studies conducted in the 1930s and 1940s are very similar to many of the faking studies conducted in the recent past. Current topics related to applicant warnings, faking detection, and distinguishing between motivated and unconscious response distortion were issues grappled by early researchers. Although early researchers used much less sophisticated statistical methods and their accumulated theory and knowledge base were much less than what is available today, there is still much to learn from digging through their work. Failure to consider their early work results in unnecessary reinvention. Although it is trite to repeat the common dictum that those who are not aware of history are doomed to repeat it, we believe that reviewing the historical record on faking research helps provide a context that will better advance current and future research.

NOTE

1.   Nine hundred and seven articles were identified on PsychInfo using the keywords "social desirability," "faking," "impression management," and "response distortion."

REFERENCES

Alex, C. (1965). *How to beat personality tests*. New York: ARC Books.

Anastasi, A. (1982). *Psychological testing* (5th ed.). New York: McMillan.

Arthur, W., Woehr, D. J., & Graziano, W. G. (2001). Personality testing in employment settings: Problems and issues in the application of typical selection practices. *Personnel Review, 30,* 657-676.

Barger, S. D. (2002). The Marlowe-Crowne affair: Short forms, psychometric structure and social desirability. *Journal of Personality Assessment, 79,* 286-305.

Bass, B. M. (1955). Authoritarianism or acquiescence? *Journal of Abnormal and Social Psychology, 51,* 616-623.

Becker, T. E., & Colquitt, A. L. (1992). Potential versus actual faking of a biodata form: An analysis along several dimensions of item type. *Personnel Psychology, 45,* 389-406.

Bennett, G. K., & Gordon, H. P. (1944). Personality test scores and success in the field of nursing. *Journal of Applied Psychology, 28,* 267-278.

Bernreuter, R. G. (1933). Validity of the personality inventory. *Personnel Journal, 11,* 383-386.

Block, J. (1965). *The challenge of response sets: Unconfounding meaning, acquiescence, and social desirability in the MMPI.* New York: Irvington.

Chapman, L. J., & Campbell, D. T. (1957). Response set in the F scale. *Journal of Abnormal and Social Psychology, 54,* 129-132.

Christiansen, N. D., Burns, G., & Montgomery. G. E. (2005). Reconsidering the use of forced-choice formats for applicant personality assessment. *Human Performance, 18,* 267-307.

Christiansen, N. D., Goffin, R. D., Johnston, N. G., & Rothstein, M. G. (1994). Correction the 16PF for faking: Effects on criterion-related validity and individual hiring decisions. *Personnel Psychology, 47,* 847-860.

Cronbach, L. J. (1946). Response sets and test validity. *Educational and Psychological Measurement, 6,* 475-494.

Cronbach, L. J. (1950). Further evidence on response sets and test design. *Educational and Psychological Measurement, 10,* 3-31.

Crowne, D. P., & Marlowe, D. (1960). A new scale of social desirability independent of psychopathology. *Journal of Consulting Psychology, 24,* 349-354.

Crowne, D. P., & Marlowe, D. (1964). *The approval motive: Studies in evaluative dependence.* New York: Wiley.

Dwight, S. A., & Donovan, J. J. (2003). Do warnings not to fake reduce faking? *Human Performance, 16,* 1-23.

Douglas, E. F., McDaniel, M. A., & Snell, A. F. (1996, August). *The validity of non-cognitive measures decays when applicants fake.* Paper presented at the meeting of the Academy of Management, Nashville, TN.

Edwards, A. L. (1953). The relationship between the judged desirability of a trait and the probability that the trait will be endorsed. *Journal of Applied Psychology, 37,* 90-93.

Edwards, A. L. (1957). *The social desirability variable in personality assessment and research.* New York: Dryden.

Ellingson, J. E., Sackett, P. R., & Hough, L. M. (1999). Social desirability corrections in personality measurement: Issues of applicant comparison and construct validity." *Journal of Applied Psychology, 84,* 155-166.

Elliott, A. G. (1981). Some implications of lie scale scores in real-life selection. *Journal of Occupational Psychology, 54,* 9-16.

Frenkel-Brunswik, E. (1939). Mechanisms of self-deception. *Journal of Social Psychology, 10,* 409-420.

Freud, S. (1941). Splitting of the ego in the defensive process. *International Journal of Psychoanalysis, 22,* 65-68. (Original work published 1938)

Ghiselli, E. E., & Brown, C. W. (1948). *Personnel and industrial psychology.* New York: McGraw-Hill.

Gibby, R. E., & Zickar, M. J. (2005). *A history of the early days of personality testing in american industry: An obsession with adjustment.* Unpublished manuscript.

Goldberg, L. R. (1971). A historical survey of personality scales and inventories. In P. McReynolds (Ed.), *Advances in psychological assessment* (Vol. 2, pp. 293-336). Palo Alto, CA: Science and Behavior Books.

Griffith, R. L., Malm, T., English, A., Yoshita, Y., Gujar, A., & Monnot, M. (2004, April). Individual differences and applicant faking behavior: One of these applicants is not like the others. In N. D. Christiansen (Chair), *Beyond Social Desirability in Research on Applicant Response Distortion.* Symposium conducted at the 19th annual conference for the Society of Industrial and Organizational Psychology, Chicago, IL.

Hanley, C. (1956). Social desirability and responses to items from three MMPI scales: D, Sc, and K. *Journal of Applied Psychology, 40,* 324-328.

Hendrickson, G. (1934). Some assumptions involved in personality measurement. *Journal of Experimental Education, 2,* 243-249.

Hicks, L. E. (1970). Some properties of ipsative, normative, and forced normative measures. *Psychological Bulletin, 74,* 167-184.

Hoffman, E. (2000). *Ace the corporate personality test.* New York: McGraw-Hill.

Hough, L. M. (1998). Effects of intentional distortion in personality measurement and evaluation of suggested palliatives. *Human Performance, 11,* 209-244.

Hough, L., Eaton, N., Dunnette, M., Kamp, J., & McCloy, R. (1990). Criterion-related validities of personality constructs and the effect of response distortion of those validities. *Journal of Applied Psychology, 75*, 581-595.

Humm, D. G. (1944). Discussion of Dorcus' study of the Humm-Wadsworth Temperament Scale. *Journal of Applied Psychology, 28*, 527-529.

Humm, D. G., & Wadsworth, G. W., Jr. (1934). *The Humm-Wadsworth temperament scale.* Los Angeles: Humm and Associates.

Hurtz, G. M., & Alliger, G. M. (2002). Influence of coaching on integrity test performance and unlikely virtues scale scores. *Human Performance, 15*, 255-273.

Kelly, E. L., Miles, C. C., & Terman, L. M. (1936). Ability to influence one's score on a pencil-and-paper test of personality. *Character and Personality, 4*, 206-215.

Kluger, A. N., & Colella, A. (1993). Beyond the mean bias: The effect of warning against faking on biodata item variances. *Personnel Psychology, 46*(4), 763-780.

Kroger, R. O., & Turnbull, W. (1975). The invalidity of validity scales: The case of the MMPI. *Journal of Consulting and Clinical Psychology, 43*, 48-55.

Kuder, G. F. (1939). *Preference record.* Chicago: Science Research Associates.

Levin, R. A., & Zickar, M. J. (2002). Investigating self-presentation, lies, and bullshit: Understanding faking and its effects on hiring decisions using theory, field research, and simulation. In F. Drasgow & J. Brett (Eds.), *The New Millennium of Work* (pp. 253-276). San Francisco: Jossey-Bass.

McCloy, R. A., Heggestad, E. D., & Reeve, C. L. (2005). A silk purse from the sow's ear: Retrieving normative information from multidimensional forced-choice items. *Organizational Research Methods, 8*, 222-248.

McFarland, L. A., & Ryan, A. M. (2000). Variance in faking across noncognitive measures. *Journal of Applied Psychology, 85*, 812-821.

Meade, A. W. (2004). Psychometric problems and issues involved with creating and using ipsative measures for selection. *Journal of Occupational and Organizational Psychology, 77*, 531-552.

Meehl, P. E., & Hathaway, S. R. (1946). The K Factor as a Suppressor Variable in the Minnesota Multiphasic Personality Inventory. *Journal of Applied Psychology, 30*, 525-564.

Miller, C. E. (2001). *The susceptibility of personality selection tests to coaching and faking.* Unpublished doctoral dissertation, University of Akron, OH.

Millham, J. (1974). Two components of need for approval score and their relationship to cheating following success and failure. *Journal of Research in Personality, 8*, 378-392.

Millham, J., & Kellogg, R. W. (1980). Need for social approval: Impression management or self-deception? *Journal of Research in Personality, 14*, 445-457.

Moore, H. (1942). *Psychology for business and industry* (2nd ed.). New York: McGraw-Hill.

National Research Council (1940). *Annual Report to the National Research Council's Committee on Selection and Training of Aircraft Pilots.* Available at the University of Akron's Archives of the History of American Psychology.

Ones, D. S., Viswesvaran, C., & Reiss, A. D. (1996). Role of social desirability in personality testing for personnel selection: The red herring. *Journal of Applied Psychology, 81*, 660-679.

Parker, J. D. A. (1991). *In search of the person: The historical development of American personality psychology.* Unpublished doctoral dissertation, Toronto, Ontario, Canada, York University.

Paulhus, D. L. (1988). *Manual for the Balanced Inventory of Desirable Responding* (BIDR-6). Unpublished manual, University of British Columbia.

Paulhus, D. L. (1991). Measurement and control of response bias. In J. P. Robinson, P. R. Shaver, & L. S. Wrightsman (Eds.), *Measures of personality and social psychological attitudes* (pp. 17-59). New York: Academic Press.

Paulhus, D. L. (2002). Socially desirable responding: The evolution of a construct. In H. I. Braun, D. N. Jackson, & D. E. Wiley (Eds.), *The role of constructs in psychological and educational measurement* (pp. 49-69). Mahwah, NJ: Erlbaum.

Paulhus, D. L., Bruce, M. N., & Trapnell, P. D. (1995). Effects of self-presentation strategies on personality profiles and structure. *Personality and Social Psychology Bulletin, 21,* 100-108.

Rees, C. J., & Metcalfe, B. (2003). The faking of personality questionnaire results: Who's kidding whom? *Journal of Managerial Psychology, 18,* 156-165.

Reynolds, W. M. (1982). Development of reliable and valid short forms of the Marlowe-Crowne Social Desirability Scale. *Journal of Clinical Psychology, 38,* 119-125.

Robie, C., Curtin, P. J., Foster, T. C., Phillips, H. L., Zbylut, M., & Tetrick, L. E. (2000). The effects of coaching on the utility of response latencies in detecting fakers on a personality measure. *Canadian Journal of Behavioural Science, 32,* 226-233.

Rosenzweig, S. (1934). A suggestion for making verbal personality tests more valid. Psychological Review, *41,* 400-401.

Rosse, J. G., Stecher, M. D., Miller, J. L., & Levin, R. A. (1998). The impact of response distortion on preemployment personality testing and hiring decisions. *Journal of Applied Psychology, 83,* 634-644.

Ruch, F. L. (1942). A technique for detecting attempts to fake performance on the self-inventory type of personality test. In Q. McNemar & M. A. Merrill (Eds.), *Studies in personality* (pp. 229-234). New York: McGraw-Hill.

Ruch, F. L., & Ruch, W. W. (1967). The K Factor as a suppressor variable in predicting success in selling. *Journal of Applied Psychology, 51,* 201-204.

Rudmin, F. W. (1999). Norwegian short-form of the Marlowe-Crowne Social Desirability Scale. *Scandinavian Journal of Psychology, 40,* 229-233.

Seisdedos, N. (1996). The "IM" (Impression Management) Scale. *European Review of Applied Psychology/Revue Europeenne de Psychologie Appliquee, 46,* 45-55.

Smith, D. B., & Robie, C. (2004). The implications of impression management for personality research in organizations. In B. Schneider & D. B. Smith (Eds.), *Personality and organizations* (pp. 111-138). Mahway, NJ: Erlbaum.

Song, K. M. (2005, June 22). Faking your type to "pass" a personality test. *Seattle Times,* p. F1.

Spencer, D. (1938). Frankness of subjects on personality measures. *Journal of Educational Psychology, 29,* 26-35.

Steinmetz, H. L. (1932). Measuring ability to fake occupational interest. *Journal of Applied Psychology, 16,* 123-130.

Stober, J. (2001). The Social Desirability Scale-17 (SDS-17): Convergent validity, discriminant validity, and relationship with age. *European Journal of Psychological Assessment, 17,* 222-232.

Taylor, J. A. (1953). A personality scale of manifest anxiety. *Journal of Abnormal and Social Psychology, 48,* 285-290.

Vasilopoulos, N. L., Reilly, R. R., & Leaman, J. A. (2000). The influence of job familiarity and impression management on self-report measure scale scores and response latencies. *Journal of Applied Psychology, 85,* 50-64.

Vernon, P. E. (1934). The attitude of the subject in personality testing. *Journal of Applied Psychology, 18,* 165-177.

Viteles, M. S. (1932). *Industrial psychology.* New York: Norton.

Waters, L. K. (1965). A note on the "fakability" of forced-choice scales. *Personnel Psychology, 18,* 187-191.

Weed, N. C., Ben-Porath, Y. S., & Butcher, J. N. (1990). Failure of Wiener and Harmon Minnesota Multiphasic Personality Inventory (MMPI) subtle scales as personality descriptors and as validity indicators. *Psychological Assessment, 2,* 281-285.

Wesman, A. G. (1952). Faking personality test scores in a simulated employment situation. *Journal of Applied Psychology, 36,* 112-113.

Whyte, W. H., Jr. (1956). *The organization man.* Garden City, NY: Doubleday Anchor Books.

Wiener, D. R. (1948). Subtle and obvious keys for the Minnesota Multiphasic Personality Inventory. *Journal of Consulting Psychology, 12,* 164-170.

Woodworth, R. S. (1917). *Personal data sheet.* Chicago: C. H. Stoelting.

Zickar, M. J. (2000). Modeling faking on personality tests. In D. Ilgen & C. L Hulin (Eds.), *Computational modeling of behavior in organizations* (pp. 95-108). Washington, DC: American Psychological Association.

Zickar, M. J. (2001). Using personality inventories to identify thugs and agitators: Applied psychology's contribution to the war against labor. *Journal of Vocational Behavior, 59,* 149-164.

Zickar, M. J., & Drasgow, F. (1996). Detecting faking on a personality instrument using appropriateness measurement. *Applied Psychological Measurement, 20,* 71-87.

Zickar, M. J., Gibby, R. E., & Robie, C. (2004). Uncovering faking samples in applicant, incumbent, and experimental data sets: An application of mixed-model item response theory. *Organizational Research Methods, 7,* 168-190.

Zickar, M. J., Rosse, J., & Levin, R. (1996, May). Modeling of faking in a selection context. In C. L. Hulin (Chair), *The third scientific discipline: Computational modeling in organizational research.* The meeting of the Society of Industrial Organizational Psychology, San Diego, CA.

CHAPTER 3

SEVEN NESTED QUESTIONS ABOUT FAKING ON PERSONALITY TESTS

An Overview and Interactionist Model of Item-Level Response Distortion

Robert P. Tett, Michael G. Anderson, Chia-Lin Ho, Tae Seok Yang, Lei Huang, and Apivat Hanvongse

Calling someone a liar, a cheat, or a fake is typically not undertaken lightly; being called one can be cause for assault. The severity of such acts and accusations derives from a breakdown of trust: interdependence between friends, family, and cohorts improves one's chances to survive and flourish, and insincerity, when detected, can destroy beneficial alliances. Accordingly, lying, cheating, and faking fall, for the most part, on the dark side of human nature, and efforts to prevent, identify, and predict such behaviors are eagerly if not passionately pursued.

Faking in psychological assessment has been a recurring theme since the advent of self-report measures (Helmes, 2000; Paulhus, 2002). Psychological constructs targeted for assessment reduce to two general types: char-

A Closer Examination of Applicant Faking Behavior, 43–83
Copyright © 2006 by Information Age Publishing
All rights of reproduction in any form reserved.

acteristics permitting measurement involving objectively right and wrong answers (e.g., ability, skill, knowledge) and traits requiring self-description (e.g., personality, attitudes, values). The former carry the advantage of verifiability; the latter are inherently more subjective. A critical implication is that people tend not to fake on ability, skill, or knowledge tests (they may deliberately underperform in rare cases where high scores yield undesired outcomes; e.g., in military drafts). Asking people to describe themselves, on the other hand, is an open invitation to fake good or bad, especially when responses could be used to benefit or harm the respondent.

Employment settings are one place where test-takers are likely to see value in making a favorable impression. Faking can be expected to undermine both validity and the sense of fairness regarding selection decisions based on test scores. Despite the plausibility of these expectations, research on faking has yielded equivocal conclusions, ranging from "faking is a serious problem in need of control" to "faking doesn't matter." This lack of agreement leaves selection specialists with little more than guesses, biases, and post-hoc rationales, all with potential for serious errors from test use: in few jobs is it advantageous to hire a liar or a cheat, especially if more qualified applicants were edged out by less qualified applicants who faked their way to the top of the list.

Our goals in this chapter are twofold. First, we present an overview of the faking literature targeting relevant research and discussion mostly spanning the last 20 years. Our analysis is uniquely organized by seven nested yes/no questions to which an answer of "no" renders moot all subsequent questions and to which an answer of "yes" begs further inquiry. Second, in light of that review, we present a trait-situation interactionist model of item-level responding as a guide for further research. Our presenting a model in the latter part of the chapter is a clue to our overall conclusion from the review in that models of faking are of little value unless faking itself is judged a tractable problem. Thus, our answer to each of the seven questions is "yes," albeit qualified in some cases, conditional on certain key factors.

A Hierarchical Overview of Personality Response Distortion Research in Personnel Selection Settings

Our aim in this section is to walk the reader through the faking literature, using seven questions intended to clarify the conditions under which faking research has yielded, and is likely to continue to yield, fruitful and important findings. As depicted in Figure 3.1, later questions assume "yes" answers to earlier questions. It could be argued that evidence in support of subsequent "yes" answers renders prior questions moot. For

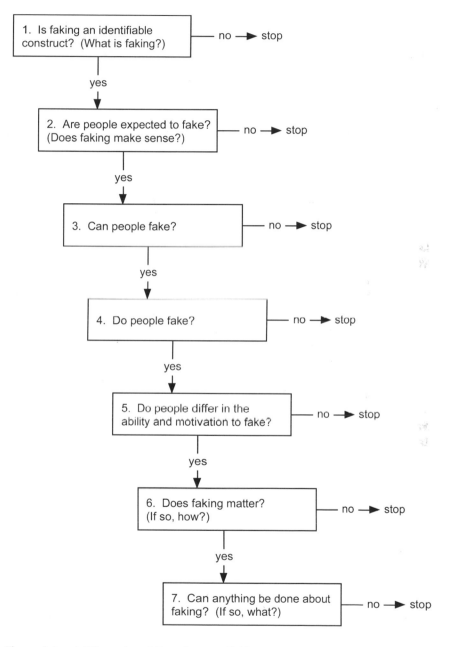

Figure 3.1.   A Hierarchy of Questions on Faking.

example, if some people *do* fake (question 4) it presupposes that they *can* fake (question 3). Posing all questions has merit nonetheless. For one thing, the yes/no format can mask a truer complexity offering richer, more complete answers to later questions. Whether or not people *do* fake (question 4), for example, depends on what one means by faking (question 1). Similarly, how one might solve a problem with faking (question 7) might depend on what the problem is (question 6). Secondly, the question sequence, especially in the heart of the order, has historical significance: whether or not people can fake has more recently given way to the more practical question of whether or not they actually do. Thus, our question sequence offers a convenient heuristic for organizing the faking literature, with value beyond that stemming from simple logic.

Question 1: Is Faking an Identifiable Construct?

Reasoned discussion on any topic requires some agreement on the nature of core constructs. Clearly, if we cannot agree as to what faking is, there is little hope of identifying its causes, its implications for assessment and personnel selection, and means of dealing with any unwanted consequences. Our answer to the related question, "What is faking?" can frame our answers to later questions.

Colloquial usage of faking invokes themes of "lying," "forgery," and "hoax." In personality assessment, terms most commonly associated with faking are response distortion, impression management, displaying unlikely virtues, dissimulation, self-enhancement, and socially desirable responding (Hough, Eaton, Dunnette, Kamp, & McCloy, 1990; Rosse, Stecher, Miller, & Levin, 1998; Viswesvaran & Ones, 1999). The last of these terms has emerged as a general construct, defined by Paulhus (2002) as "the tendency to give overly positive self-descriptions" (p. 50). The "overly" part explicitly captures departure from the individual's true nature. Thus, high scores on measures of generally desirable qualities due to true high standing on those qualities must be distinguished from high scores due to the respondent's overestimation. Desirability bias denotes the latter case, valid assessment denotes the former, and separating the two in measurement is a critical research challenge.

The meaning of social desirability as error has not been universally endorsed. There has been protracted debate on this issue spanning roughly 50 years. For some (e.g., Block, 1965; McCrae & Costa, 1983), social desirability is a meaningful trait construct, control of which in the measurement of other traits undermines the validity of that measurement. This "valid variance" argument draws from Crowne and Marlowe's (1964) advancement of social desirability as need for approval. Similarly, Hogan and Hogan (1995) argue that personality tests, regardless of the construct overtly targeted, also measure social skill, which can add to validity as social skill contributes to job performance. Jackson (1984) more generally described a high scorer on the Desirability scale of the *Personality Research Form* as someone who "consciously or unconsciously, accurately or inaccurately, presents favorable picture of self in response to personality statements" (p. 7). The arguments regarding social desirability as a response bias versus a substantive trait are complex (Helmes, 2000) and continue to evolve (e.g., Paulhus, 2002). For present purposes, we adopt Paulhus' (1986, 2002) view of social desirability as bias (i.e., systematic error), as this perspective has driven much of the faking literature and warrants close attention in the use of personality tests in applied settings.

A second and widely adopted distinction championed by Paulhus (1986) and reflected in others' work (e.g., Sackheim & Gur, 1978) breaks social desirability into two distinct facets of self deception and impression management. Self-deception is "any positively biased response that respondents actually believe to be true" (p. 146). It is honest and unconsciously motivated (Paulhus, 1986), consistent with Damarin and Messick's (1965) "autistic bias," the unconscious tendency to distort responses to be consistent with self-attitudes. Impression management, by comparison, is a "conscious attempt to create a favorable impression in some audience" (Paulhus, 1986, p. 146). It is deliberate and insincere, what most people identify as "faking." The distinction between self-deception and impression management offers a rudimentary conceptual framework for further inquiry. A neglected point in faking research is that both facets denote error in estimating a person's true standing on a trait. Whether unconscious and honest or deliberate and insincere, both deserve our attention in attempting to improve predictions of future work behavior based on self-report.

Table 3.1 presents a summary of dominant labels and themes falling under Paulhus' two-factor conceptualization of social desirability. Three points are worth noting. First, similar labels have been used in different ways and similar constructs have been given different labels. This is expected in any active research domain. Second, the conceptualization of social desirability, especially that offered by Paulhus (e.g., 2002), continues to evolve. Most recent developments include bifurcation within each of the two noted facets into conscious and unconscious components. Such developments are largely ignored in the current undertaking as they are too recent to have had appreciable impact on the literature and are otherwise largely untested. Third, and most importantly, there is general consensus among researchers that social desirability bias comes in two forms that may be distinguished on the basis of honesty and deliberateness. Our answer to question 1, then, is a qualified "yes."

Table 3.1. Conceptualizations of Two Types of Response Distortion

Authors	Self-Deception	Impression Management
Damarin and Messick (1965)	"Autistic Bias": tendency to distort response to be consistent with self-attitude	"Propagandistic Bias": purposive, systematic distortion aimed at a specific audience
Sackeim and Gur (1978)	(a) Individual holds 2 contradictory beliefs, (b) 2 contradictory belief held simultaneously, (c) individual is unaware of holding one of the belief, (d) unawareness of one belief is motivated (p. 150)	
Paulhus (1984)	Self-deception (unconscious)	Other-deception (conscious)
	Self-deception	Impression management
Paulhus (1986)	any positively biased response that the respondent actually believes to be true (pp. 145-147)	conscious attempt to create a favorable impression in others (pp. 145-147)
	4 guises: (a) distortion of negative traits into positive, (b) omission of major traits in free description, (c) justification of defects, (d) minimizing the importance of defects (p. 146)	3 usages: (a) strategic simulation (Edwards, 1970; Jones & Pittman, 1982); (b) motive (Crowne & Marlowe, 1964); (c) skill (Collins, Paulhus, & Graziano, 1983)
	less conscious' attempt to look good to oneself (pp. 145-147)	lying; conscious, purposeful, deception to others (pp. 145-147)
	"autistic bias": tendency to distort responses to be consistent with self-attitude (pp. 145-147)	"propagandistic bias": "purposive, systematic distortion aimed at a specific audience" (pp. 145-147)

Author	Self-deception	Impression management
Furnham (1986)	"motivated unawareness" of one of two conflicting cognitions (p. 150) unconscious defensive mechanism; Sackheim and Gur (1978) sincerity	insincerity Faking; lying; dissimulation, social desirability; more specific; deliberate other-deception (not self-deception) (p. 385)
Paulhus (1988)	Self-deceptive enhancement and self-deceptive denial	Impression management
Furnham (1990)		motivational distortion or faking to achieve a particular profile (p. 2)
Paulhus and Reid (1991)	biased, but honest self-description (p. 307)	favorable self-descriptions to others; enhancement claiming of positive attributes versus denial or repudiation of negative attitudes (p. 307)
Holden et al. (1992) Seisdedos (1993)		Responsive to audience effects decision about the congruence between the stereotypical (Faking is) an intelligent form of a subject's adaptation (p. 93)
Paulhus and John (1998)	unconscious "defense in personality"; (Sackeim & Gur, 1978, p. 1029) Egoistic bias and moralistic bias; both self-deceptive tendencies operating unconsciously (p. 1041)	conscious "defense in personality"; (Sackeim & Gur, 1978, p. 1029)

Study	Definition
Stark et al. (2001)	Faking is intentional distortion of responses to personality items (p. 943)
Paulhus (2002)	Alpha Self-deceptive enhancement Induces error; conscious, and unconscious process
	Gamma Self-deceptive denial/impression management (p. 62) Induces error; conscious, and unconscious process
Helmes and Holden (2003)	related to "the sensitivity of social signals, along with social recognition construct" (p. 1017)
Griffith et al. (2005)	closely related to the construct of faking (p. 11) honest, but overtly positive self presentation of oneself; associated with healthy adjustment and confidence and related to job performance (Barrick & Mount, 1996, p. 11)
	an individual's propensity to intentionally distort responses to alter the perceptions of others (p. 11)
	faking referred to in the literature as response distortion, impression management, social desirability, displaying unlikely virtues, and self enhancement (p. 2)

Question 2: Are People Expected to Fake?

Research on social desirability is driven in large part by the idea that presenting an overblown impression of oneself to others is as natural as self-defense. One does not need an elaborate rendering of conscious and unconscious motives interacting with self-perceptions and situational forces to propose that, given the opportunity, the average person will tend to put his or her best foot forward and hide imperfections, and that some will lie if the probability of getting caught is low. This is especially obvious when the stakes are high, as occurs in selection settings where everyone knows that getting the job hinges on how appropriate one is perceived. Presenting a good image when it counts is simply human nature.

The strength of such beliefs is clearly evident in the faking literature (Seisdedos, 1993). Krahe (1989), for example, asks "why should respondents complete, in an honest, undistorted and ostensibly 'naïve' way, a test whose aims are fully transparent to them and in whose outcome they often have a vested interest?" (p. 442). Likewise, Mahar, Cologon, and Duck (1995) suggest "there is clear incentive for the respondent to present a false impression of themselves via their test profiles" (p. 605). Hough et al. (1990) note that the possibility of response distortion is often cited as a main argument against using personality tests in hiring, owing to expectations that validity will be compromised. Similar observations are offered by Rosse et al. (1998), Smith and Ellingson (2002), Costa (1996), and Hogan and Hogan (1995). Using a detailed survey, Rees and Metcalf (2003) found that 36% of 190 U.K. human resources practitioners believe that personality tests are easy to fake, and that, averaging across those surveyed, more than half of test-takers are expected to fake.

All such convictions, of course, fall short of proving that faking can and does occur, that it matters and warrants control. They offer, nonetheless, a compelling impetus for investigation into faking. If we did not expect faking in selection contexts based on natural inclinations and surmise that it undermines accuracy in predicting behavior from self-descriptions, it is doubtful there would be an identifiable literature on faking. Expectations can be fooled but the burden of showing that faking does not occur or that it does not matter is not readily cast aside. It falls to researchers contradicting commonly held beliefs about faking to explain further why our expectations are wrong. Thus, our answer of "yes" to question 2 drives further inquiry but just as importantly, it presents a standard of rational shared belief against which the evidence bears scrutiny.

Question 3: Can People Fake?

Whether or not people can fake their responses to personality tests has been a frequent focus of research, much of it conducted over the last 15 years (Arthur, Woehr, & Graziano, 2000; Furnham, 1990a, 1990b; Griffith, Snell, McDaniel, & Frei, 1998; Holden, Kroner, Fekken, & Popham, 1992; Hough et al., 1990; Mahar et al., 1995; Ones & Viswesvaran, 1998; Rosse et al., 1998; Vasilopoulos, Reilly, & Leaman, 2000; Windle, 1954). Notably, this ability-oriented question does not readily apply to self-deception. To the degree that ability plays a role in self-deception, it is a *lack* of ability to see the true self that comes into play. Thus, when we ask if people can distort responses to personality tests, impression management is our target, not self-deception.

There is broad consensus that personality tests can be faked (e.g., Dunnette, Koun, & Barber, 1981; Furnham & Craig, 1987; Gillis, Rogers, & Dickes, 1990; Hinrichsen, Gryll, Bradley, & Katahn, 1975; Hough & Furnham, 2003; Krahe, 1989; Ryan & Sackett, 1987; Schwab, 1971; Stark, Chernyshenko, Chan, Lee, & Drasgow, 2001; Thornton & Gierasch, 1980; Walker, 1985; Zickar & Drasgow, 1996). The dominant method in "can-fake" studies is the controlled experiment comparing scale means derived under straight-take (i.e., honest) instructions and those derived under fake-good or fake-bad instructions. Viswesvaran and Ones (1999) offer meta-analytic mean effect sizes for both within- and between-subjects designs involving the Big Five personality dimensions, based on a total of 51 studies. Their results, summarized in Table 3.2, show stronger effect sizes for fake-bad instructions and for within-subjects designs. Mean effect sizes are also stronger for Emotional Stability and a little weaker for Agreeableness.

As noted by Viswesvaran and Ones (1999), within-subjects designs offer more powerful and credible results (for a full discussion of this issue see Chapter 4). Averaging those results across dimensions yields mean effect sizes of .78 for fake-good studies and –2.79 for fake-bad studies. Faking bad is unlikely to occur in natural assessment settings (perhaps when test scores will be used to decide who to place in an unwanted job). Faking-good is far more pertinent to discussions of test use in employment settings. A .78 effect size suggests that test-takers who fake during screening as much as they can will substantially increase their chances of being hired with top-down selection relative to those who do not fake.

A further related question is how much people can fake personality test scores to match particular jobs or roles. Furnham (1990a) argued that studies on global faking have limited ecological validity, as real job applicants have a specific job in mind when they complete screening measures. Krahe (1989) found that, when instructed, respondents could score high

Table 3.2.   Summary of Viswesvaran and Ones' (1999) Meta-Analytic Mean Effect Sizes for the Big Five Personality Dimensions Under Fake-Good and Fake-Bad Instructions

| | Faking Instructions | | | | | |
| | Fake Good | | | Fake Bad | | |
Big Five Dimension	Within	Between	Weighted Average[b]	Within	Between	Weighted Average[b]
Emotional stability	.93	.64	.76	-3.34	-1.00	-2.29
	(921)	(1,357)		(735)	(603)	
Extraversion	.54	.63	.61	.91	-1.95	-1.80
	(391)	(1,122)		(23)	(419)	
Openness to Experience	.76	.65	.68	-1.25		-1.25
	(259)	(614)		(23)		
Agreeableness	.47	.48	.48	-1.14	-1.83	-1.71
	(408)	(1,009)		(116)	(552)	
Conscientiousness	.89	.60	.66	-2.54	-1.90	-2.10
	(723)	(2,650)		(383)	(848)	
Weighted average	.78	.60	.65	-2.79	-1.67	-2.06

From Viswesvaran and Ones (1999).
[a]Total N per mean effect size is shown in parentheses.
[b]Weighted averages are average mean effect sizes weighted by total Ns.

or low on social orientation and achievement orientation to fit or not fit a defined role. Hough et al. (1990) reported similar findings involving army personnel, and, using a within-subjects design, Mahar et al. (1995) found that MBTI profiles differed significantly under general instructions to fake-good versus instructions to match the profile for psychiatric nurse.

In sum, the literature strongly supports the idea that test-takers can, on average, readily manipulate their scores on self-report personality measures when instructed to do so, and are further able to distort their responses with respect to specific jobs.

Question 4: Do People Fake?

Research on whether or not people can fake on personality tests is driven by the assumption that, if they can, they will. A number of researchers have challenged that assumption (Arthur et al., 2000; Hogan & Hogan, 1995; Hough & Schneider, 1996; Ones & Viswesvaran, 1998;

Schwab, 1974; Viswesvaran & Ones, 1999). Mahar et al. (1995) noted that asking respondents to match stereotypical job-related profiles fails to capture true applicant mindsets, and Kirton (1991) argued that most respondents are hard pressed to fake because the traits likely to be required in a particular company are rarely obvious. Deliberate faking may be less common than many have presumed.

In support of the "no-faking" position, Dunnette, McCartney, Carlson, and Kirchner (1962) reported that *Adjective Checklist* scores varied little between sales job applicants and incumbents, with only one out of seven tending to distort their self-descriptions. Similarly, Abrahams, Neuman, and Githens (1971) found that mean scores on the *Strong Vocational Interest Bank* (SVIB) for scholarship applicants were similar to those of nonapplicants when the SVIB was to be used for selecting scholarship winners. In a large military sample, Hough et al. (1990) reported that test score differences between applicants and incumbents were smaller than those reported in instructed faking studies (i.e., honest vs. fake-good). Whether enlisted personnel are as motivated to fake as are nonmilitary applicants, however, is unclear (Griffith, Chmielowski, & Yoshita, in press).

A growing number of studies suggest that faking does in fact occur during employment screening. Hough (1998) reported mean effect sizes from .04 to .56 between applicants and incumbents on a number of personality measures in three studies. Similarly, Rosse et al. (1998) found that personality scale scores were higher among job applicants than among job incumbents, such that 18% of applicants' scores completely exceeded the range of incumbents' scores. Stark et al. (2001) reported similar findings. Griffith et al. (in press) asked 141 temp workers who had completed a conscientiousness measure in the screening process to complete the same measure a second time honestly, with assurances of confidentiality, yielding an effect size of .60.

To permit comparisons with mean effect sizes from directed faking studies, reported in Table 3.2 (from Viswesvaran & Ones, 1999), we calculated weighted mean *d* statistics from data reported in Hough (1998), Rosse et al. (1998), Stark et al. (2001), and Griffith et al. (in press) using bare bones meta-analysis (Hunter & Schmidt, 2004). Results organized by the Big Five are summarized in Table 3.3. The small number of "do-fake" samples limits the reliability of the comparisons. We offer the following tentative observations.

First, applicants tend to score higher, on average, than incumbents on all personality dimensions except Openness. Why the mean effect size for Openness is so much weaker compared to that for fake-good conditions is not clear. Notwithstanding second-order sampling error, it may be that Openness is desirable in general but perceived as less so by applicants. The same may hold for Extraversion, which also yielded a markedly

**Table 3.3.   Bare-Bones Meta-Analytic Mean Effect Sizes for
the Big Five Personality Dimensions in
Applicant Versus Incumbent Samples**

Big Five Dimension	K	Total N	Weighted Mean d
Emotional stability	3	3,353	.61
Extraversion	3	28,337	.11
Openness to experience	4	29,292	-.01
Agreeableness	2	2,408	1.33
Conscientiousness	5	43,889	.70
Weighted mean d for all 5			.35
Weighted mean d without openness			.52

weaker effect size in the do-fake studies than under fake-good instructions. Second, averaging across the Big Five dimensions yields an overall effect size smaller than that reported on average for instructed fake-good conditions. It appears that applicants fake good on personality tests about half as much as they are capable of faking. One reason for this may be that those attracted to the job are above average on job-related traits, limiting opportunity to fake good. (Another possibility is offered below, in light of the proposed interactionist model.) Third, the degree of actual faking, nonetheless, appears high enough to warrant concerns about validity and selection decisions. Finally, applicants tend to inflate their scores on measures of traits most would expect to be related to job performance: conscientiousness, emotional stability, and agreeableness.

Question 5: Do People Differ in the Ability and Motivation to Fake?

Comparing means derived under straight-take versus fake-good or fake-bad instructions shows how much test takers *can* fake; comparing means of applicants and incumbents shows how much they actually *do* fake. Concerns about faking rest on the further expectation that some people fake more than others do (Furnham, 1986; McCrae & Costa, 1983). If all test-takers distort to the same degree, their relative standing on the measure, and hence the validity of that measure, is unaffected. The short answer to the question of whether people differ in the ability and motivation to fake is "yes." Furnham (1986) argued that such individual differences are inevitable. Similar points have been made by Hough and Ones (2001), McFarland and Ryan (2000), Mueller-Hanson, Heggestad, and Thorton (2003), Rees and Metcalfe (2003), Rosse et al. (1998), and

Viswesvaran and Ones (1999). Deeper issues include the sources of those differences and the implications for the meaning of response distortion. Complexities arise, in particular, in distinguishing between the ability and motivation to fake, between self-deception and impression management as types of distortion, and the effects of situations that interact with abilities and motives to distort.

With respect to the ability to fake, Vasilopoulos et al. (2000) suggested that fakers vary in the abilities to create good-worker schemas and to identify traits assessed by a given scale. Furnham (1984) reported that test-takers identified between 10% and over 90% (mean = 54%) of 23 neuroticism items on the Eysenck Personality Inventory. As noted earlier, self-deception can be understood as inaccuracy in self-perception. It appears then that multiple abilities are involved in response distortion and test-takers are likely to vary in all of them.

Motivation to distort has been examined in terms of substantive traits (e.g., integrity; McFarland & Ryan, 2000), interests (Krahe, 1989), values (Paulhus & John, 1998), and sex differences (Hartshorne & May, 1928; Mahar et al. 1995; Ones & Viswesvaran, 1998). Regarding interests, for example, Krahe (1989) suggested that people seeking feedback based on test scores are less likely to fake.

Using meta-analysis, Ones, Viswesvaran, and Reiss (1996) found that social desirability, undifferentiated with respect to facets, correlates between .00, with Openness, and .27, with Emotional Stability. Stronger results have been reported more recently for desirability facets. Barrick and Mount (1996) found that the Big Five correlated from .01 to .44 with impression management in two samples of truck driver applicants. Averaging across samples yielded values exceeding .30 for all traits except Extraversion. Rosse et al. (1998) reported correlations between impression management and Big Five facets, averaging from .08 (Extraversion, Openness) to .54 (Emotional Stability) in applicants and from -.20 (Openness) to .33 (Emotional Stability) in incumbents. Impression management has also been linked positively to social recognition and social adroitness (Helmes & Holden, 2003), as well as integrity, conscientiousness, and emotional stability (McFarland & Ryan, 2000).

Self-deception has been suggested as more likely to show relations with substantive traits (Furnham, 1986). Barrick and Mount (1996) reported correlations ranging from .17 to .54 for the Big Five in truck driver applicants, with Emotional Stability the strongest correlate. Helmes and Holden (2003) found that self-deception correlates with both self-esteem and value orthodoxy (i.e., conservative values). That social desirability is linked to multiple individual difference variables shows that it is itself an individual difference variable.

Person-situation interactionism holds that situations can have main effects on behavior as well as interactions with personological traits. Support for trait-situation interactions is substantial (e.g., Bowers, 1973; Ekehammar, 1974; Endler & Magnusson, 1976; Epstein & O'Brien, 1985; Pervin, 1985; Snyder & Ickes, 1985; Tett & Burnett, 2003; Weiss & Adler, 1984) and there is no reason to expect that traits underlying response distortion should not operate by similar forces. In support, Stark et al. (2001) found that impression management scales appear to measure different constructs in applicant and nonapplicant settings. Rosse et al. (1998) showed that individual differences in response distortion are more pronounced under applicant conditions (Leary & Kowalski, 1990; Paulhus, 1991). Hough (1998) reported similar findings in three samples. Viswesvaran and Ones (1999) suggested that people more field dependent or with higher need for approval may be more likely to fake when instructed.

The noted complexities make it difficult to draw firm conclusions about the meaning and importance of socially desirable responding and, in particular, whether it warrants control as a source of error or can be safely ignored in making decisions based on test scores. In the second part of this chapter, we propose an interactionist model of test item responding intended to frame the complexities articulated above. For now, in the context of our hierarchical overview, we conclude that people indeed vary in abilities and motives directly relating to response distortion and that expression of such individual differences depends in part on the situation.

Question 6: Does Faking Matter?

As others have noted (Hogan & Nicholson, 1988; Zerbe & Paulhus, 1987), just because people can and do fake, it does not necessarily follow that faking affects relevant selection outcomes to any appreciable degree. Two important outcomes involving personality tests are (a) test validity, and (b) the accuracy of selection decisions. Although these outcomes are conceptually linked, research cited below offers different answers to the question of faking effects. Our review is structured in terms of the classic triarchic approach to validity and of selection decisions.

Content-Related Evidence

The content of a personality test derives from the meaning of its items. Content validity may be inferred directly by the writing of items targeting an identifiable construct (Anastasi & Urbina, 1997). Documenting content validity, however, requires assessment by independent expert judges. Personality researchers have long been aware of the fact that personality traits vary in desirability and that personality items, accordingly, also vary in desirability (Bentler, Jackson, & Messick, 1971; Edwards, 1953, 1957; Jackson & Helmes, 1979; Krueger, 1998; Messick & Jackson, 1961; Morf & Jackson, 1972; Taylor & Brown, 1988). Assessing the desirability value of personality items (e.g., on a 9-point scale) would be helpful in content validation of personality tests used in selection. Items with extreme desirability values evoke ambiguous responses (substance vs. style). Few personality measures used in personnel selection have been subjected to formal content validation and fewer have been subjected to such validation targeting social desirability per se. Future research involving this approach to validation may further understanding of the role of response distortion in employment settings.

Construct-Related Evidence

Inferences based on test scores can be valid only to the degree that the corresponding measure captures the intended construct (Guion, 1998). Faking on personality tests is expected to degrade validity by introducing construct-irrelevant variance (Griffith et al., 1998; Rosse et al., 1998). Evidence reviewed earlier shows that people can enhance their scores on personality tests if asked and actually do if motivated (as in applicant settings). In addition, people vary in their ability and propensity to distort. It follows that inferences based on personality test scores are compromised because at least some test score variance is attributable to reliable nontargeted sources. The severity of the problem, however, is unclear. In addition to mean differences, reviewed above, researchers have assessed the impact of faking on the construct validity of personality measures in terms of convergent and discriminant validity, factor structure, and item-level analysis.

Assessing convergent and discriminant validity entails determining whether a given scale correlates stronger with other scales designed to measure the same or similar constructs than it does with scales designed to measure different constructs (Guion, 1998). Research on the effects of response distortion on convergent and discriminant validity regarding personality tests is scant. One notable exception is Ones et al. (1996), who partialed social desirability from linkages among measures of the Big Five based on meta-analytic mean correlations. The original linkages remained largely intact, suggesting minimal effect due to overlapping susceptibility to response distortion. Several methodological limitations, however, weaken that conclusion. First, Ones et al. (1996) failed to distinguish between self-deception and impression management, which may have unique causes and effects (Rosse et al., 1998). Second, they also failed to distinguish between settings where there is high versus low motivation to distort (i.e., applicant vs. incumbent). Combining effects across distortion types and across settings may have clouded the effects of response distortion on relations among the Big Five. Third, response distortion may not affect correlations for an entire sample, but may influence rank order differences at the extremes of the distribution, such that correlation is insensitive to distortion effects (Christiansen, Goffin, Johnston, & Rothstein, 1994; Rosse et al., 1998; Zickar & Drasgow, 1996). More research is needed on the effect of response distortion on the convergent and discriminant validity of personality measures.

Some researchers have used factor analysis to address the construct validity issue regarding response distortion (Smith & Ellingson, 2002; Schmit & Ryan, 1993; Zickar & Robie, 1999). Smith and Ellingson (2002) compared trait loadings in applicant and student groups. Social desirability showed a uniform pattern of relationships with personality across the two groups, suggesting that it represents meaningful content overlap with personality traits and not response distortion. Schmit and Ryan (1993), on the other hand, found using confirmatory factor analysis that the five-factor model fit NEO data in students much better than in applicants. The latter yielded an "ideal-employee" factor, heterogeneous with respect to substantive traits, consistent with a response style interpretation. Similarly, Zickar and Robie (1999) reported differences between applicant and nonapplicant samples in the number of latent factors, error variance, and relationships between latent factors, concluding that faking good distorts the measurement properties of personality instruments. Although mixed, results taken together suggest that response distortion can indeed affect the factor structure of personality tests, challenging inferences of construct validity.

The effects of response distortion on personality test construct validity have also been assessed using item-level analysis (Stark et al., 2001; Zickar & Drasgow, 1996; Zickar & Robie, 1999). Stark et al. (2001) used item response theory (IRT) to compare item functioning of job applicants versus nonapplicants and subgroups created by impression management scores. They found that faking distorts construct validity by producing differences in item functioning across groups. Similarly, Zickar and Robie (1999) found that mean IRT thetas were lower for the honest responding condition compared to the faking condition, suggesting that faking increases scores on personality scales. Thus, IRT results reveal damaging effects of faking on the construct validity of personality tests.

Criterion-Related Evidence

Of all the types of validity evidence, criterion-related evidence is the most well-documented for most types of test, including personality measures. Although the importance of construct validity is affirmed in the EEOC's Uniform Guidelines, the legal defensibility of selection systems depends almost entirely on criterion-related validity (Guion, 1998). Some studies have found that faking does influence the criterion validity of personality tests (Dunnette et al., 1962; Norman, 1963; Schmit & Ryan, 1993; Schmit, Ryan, Stierwalt, & Powell, 1995; Zickar & Drasgow, 1996). Others, however, contradict such claims (Barrick & Mount, 1996; Becker & Colquitt, 1992; Christiansen et al., 1994; Hough, 1998; Hough et al., 1990; Hough & Ones, 2001; Mueller-Hanson et al., 2003; Ones & Viswesvaran, 1998; Ones et al., 1996; Rosse et al., 1998).

Most criterion-related validity studies adopt one of three research paradigms, each entailing comparison between two groups: (a) test-takers instructed to fake good versus answer honestly, (b) test-takers high versus low on impression management, or (c) applicants versus nonapplicants (Stark et al., 2001). Dunnette et al. (1962) instructed salesmen to complete the *Adjective Checklist* honestly and then to "beat" the test by scoring well. Faking severely distorted the validity of the *Checklist* in predicting sales performance. Salesmen completing the tool in a selection context, however, did not appreciably distort their responses. The authors suggested that the small amount of distortion may have been enough to attenuate predictive validity. Hough et al. (1990) compared the predictive validity of temperament scales among soldiers high versus low on response distortion. Very small differences were observed, prompting the conclusion that socially desirable responding does not influence the criterion-related validity of temperament scales. In a longitudinal study of truck driver applicants, Barrick and Mount (1996) found that, although applicants distorted their responses to personality tests, intentional distortion did not attenuate predictive validity.

Several meta-analytic studies have addressed the influence of response distortion on the criterion-related validity of personality tests. Ones, Viswesvaran, and Schmidt (1993) compared the criterion-related validities of integrity tests in applicant and incumbent samples for counterproductive behaviors and overall job performance. The validities for applicants and incumbents in relations with overall job performance, corrected for artifacts, were .40 and .29, respectively, suggesting that validity under conditions expected to motivate faking is not attenuated (it is, in fact, 38% larger). Ones et al. (1996) assessed the roles of social desirability as a predictor, mediator, and suppressor with respect to criterion validity. Null effects were observed on all three counts, leading the researchers to conclude "the reservation of industrial-organizational psychologists about using personality inventories for personnel selection because of the potential of social desirability is unfounded" (p. 671).

Limitations in Ones et al.'s (1996) study raised earlier in regard to construct validity (i.e., confounding self-deception and impression management, confounding research settings, limits in the use of correlation) may be relevant here as well. Rosse et al. (1996) cite those and other concerns: skewed distributions, low selection rates, and modest predictive validities may adversely affect the power of correlation to assess the impact of distortion; and job type, which could moderate distortion effects, was ignored.

To address the moderating effects of differential motivation to fake, Hough et al. (1990) assessed the influence of response distortion on personality-job performance relations derived from three study paradigms: (a) directed faking, (b) concurrent validation, and (c) predictive validation using job applicants. Results showed that fake-good instructions seriously undermine the criterion-related validity of personality tests. Differences in validity between applicant and incumbent samples, however, were small. All told, research to date suggests that response distortion in realistic settings does not appreciably undermine the criterion-related validity of personality tests.

Selection Decisions

Looking beyond criterion validity per se, some researchers have examined directly how response distortion affects the decision of whom to hire. Rosse et al. (1998) found that applicants scoring highly on impression management were over-represented among applicants ranked highest on targeted personality traits. For example, with a selection ratio of .20, 44% of those who would be hired based on conscientiousness scores had extreme impression management scores (i.e., $z > 3.0$). With a .05 selection ratio, 88% of the prospective hires showed extreme response distortion. Using a within-subjects design, Griffith et al. (in press) administered a conscientiousness measure to job applicants and again after hire under conditions of strict confidentiality. The resulting effect size of .60 translated into notable differences in prospective hiring decisions. For example, with a .50 selection ratio, 31% of applicants hired using the first (biased) test scores would not be hired using the more honest scores. With a .10 selection ratio, decision errors rose to 66%.

Christiansen et al. (1994) corrected personality test scores for faking among 495 supervisors to see what would happen to criterion-related validity and selection decisions using a top-down strategy. Correction had little effect on criterion validity, but did affect selection decisions. For example, with a .50 selection ratio, 6% of those hired based on uncorrected scores would not be hired using corrected scores. With a .10 selection ratio, the "percentage of discrepant hires" was 14%. Christiansen et al. cautioned that controlling for faking may not be legally defensible until links with criterion-related validity are more firmly established. All told, evidence to date suggests that faking influences selection decisions based on personality assessments and that the effect strengthens as the selection ratio decreases.

We offer the following observations bearing on question 6. First, the role of social desirability in content validation of personality tests used for selection has received little empirical attention and warrants more. Second, studies of mean differences, confirmatory factor analyses, and IRT in applicant and incumbent samples suggest that response distortion adversely influences the construct validity of personality tests. Third, the criterion validity of personality tests seems surprisingly robust to response distortion. Fourth, however, such distortion may have a nontrivial impact on personality-based hiring decisions. Fifth, it is not obvious how faking can affect construct validity and hiring decisions without also affecting criterion-related validity. Scientists rely on theory and conceptual integration to guide their efforts and recommendations to practitioners, and the challenge of reconciliation in this case is one most worthy of pursuit in future research.

Question 7: Can Anything Be Done About Faking?

A number of strategies have been proposed to cope with problems engendered by response distortion. Such strategies are either preventive or remedial. Preventive methods attempt to limit respondents' opportunities, abilities, or motives to fake, and include the use of instructions and warnings, subtle items, forced choices, and item selection. Remedial strategies, on the other hand, are reactive, designed to reduce the negative effects of faking that has already taken place. The latter category includes option-keying, response latencies, detection scales, attitudes toward faking, and select-out selection. Each of these particular methods is discussed briefly, below.

Preventive Approaches

Response instructions provided on many personality tests direct respondents to answer as honestly as possible. Eysenck, Eysenck, and Shaw (1974) found that "honesty" instructions reduced response distortion significantly. Few other studies have examined such effects. *Warnings*, a variant of instruction, are intended to discourage distortion by informing respondents that faking can be detected and that those caught faking may suffer negative consequences (e.g., disqualification). Using meta-analysis, Dwight and Donovan (1998) found that warnings reduce distortion by .23

standard deviations. Effects of warnings on criterion validity, however, have not been empirically examined to date (Smith & Robie, 2004).

Subtle items are those for which the targeted trait is not readily apparent to most test-takers. A number of studies suggest that subtle items are not resistant to faking (e.g., Haymaker & Erwin, 1980; Trent, Atwater, & Abrahams, 1986). Such items, moreover, can undermine construct validity (Holden & Jackson, 1979) and may attract legal attention due to lack of face validity (Hogan, Hogan, & Roberts, 1996). Alliger, Lilienfeld, and Mitchell (1996) found using meta-analysis, however, that subtle items are more resistant to coached distortion than are obvious items. *Forced-choice response formats* pair items judged to be similar in social desirability, thereby restricting opportunity to fake. Findings have been mixed. Waters (1965) cited a large number of studies in which the forced-choice strategy fell prey to instructed faking. More recently, however, Jackson, Wroblewski, and Ashton (2000) found that a forced-choice format substantially reduced susceptibility to response distortion in a simulated applicant setting.

Item selection during scale construction can be based on a number of criteria aimed at improving a scale's psychometric properties (cf. Jackson, 1970). One such criterion is a relatively low correlation between the given item and a separate measure of social desirability (Helmes, 2000). Jackson used this approach in constructing the PRF (Jackson, 1984) and *Jackson Personality Inventory – Revised* (JPI-R) (Jackson, 1994). All else being equal, choosing items less susceptible to response distortion is expected to evoke more valid and interpretable responses.

Remedial Approaches

Option-keying assumes a nonlinear relationship between item scores and the criterion. Only those response options that significantly correlate with the criterion are scored and options to the extreme ends may not contribute to a score. For example, on a 5-point agree/disagree scale for a conscientiousness item, the "1" option may be scored -1 and the "4" option may be scored 1, while the remaining options (2, 3, and 5) are scored 0, based on empirical analyses linking each option to the criterion. Kluger, Reilly, and Russell (1991) reported success with this approach applied to biodata items and that option-keying and standard keying capture unique portions of criterion variance.

Response latencies are the target of cognitive approaches to dealing with faking. Using computerized personality tests, Holden et al. (1992) found that response latencies were faster for schema-congruent test answers, and slower for schema-incongruent answers. Vasilopoulos et al. (2000) reported that job familiarity moderated the relationship between faking and response latency: when job familiarity was high, faking was engaged more quickly, and when job familiarity was low, faking took longer. Although response latencies may allow scale adjustments for faking, Robie, Curtin, Foster, Philips, Zbylut, and Tetrick (2000) reported that the utility of response latencies in detecting distortion is nullified by coaching. Further research on the cognitive processes underlying response latencies and coaching effects appears warranted.

Detection scales are included in a number of popular personality tests [e.g., the PRF (Jackson, 1984), the MMPI (Dahlstrom, Welsh, & Dahlstrom, 1972, 1975), the CPI (Gough, 1975), the *Multidimensional Personality Questionnaire* (MPQ) (Tellegen, 1982), and the *Hogan Personality Inventory* (HPI) (Hogan & Hogan, 1995)]. Many studies on response distortion have used Paulhus' (1998) *Balanced Inventory of Desirable Responding* (BIDR) and many others the Marlowe-Crowne (Crowne & Marlowe, 1964). Such scales are the most prevalent methods aimed at detecting response distortion and are used either to eliminate respondents with high scores or to adjust scores on content scales. A major concern with such scales is that truly desirable individuals may be indistinguishable from those overstating their virtues.

Rees and Metcalfe (2003) found that admitting to a high likelihood to fake on personality tests is associated with estimates of the percentage of other respondents expected to fake. This suggests that *attitude toward faking* might be utilized as a proxy for faking intent, offering a new direction for faking research. Mueller-Hanson et al. (2003) proposed a *select-out strategy* to deal with faking in the use of personality tests. Whereas it is difficult to differentiate high-scoring fakers from true high scorers on a personality test, low-scoring respondents are more readily identifiable as unlikely to become high performers, and so may be dropped from consideration. We note that such a strategy implies passive acceptance of the failure to deal with faking, essentially making the best of a bad situation.

Of all the strategies discussed above, warnings may be most preferred because they seem to work while costing little. It is unlikely, however, that they are equally effective for all respondents. For example, warnings may provide a cue for risk-takers to attempt to "get away with" faking good or for autonomous respondents to express rejection of authority. Accordingly, although warnings may be effective by themselves, true faking detection techniques are warranted as a supplement. We note that, by themselves, warnings are an empty threat, akin to faking on the part of the test user. Thus, test users relying solely on warnings face an ironic duplicity. Also, curiously, despite claiming that response distortion is not a problem in the use of personality tests in selection settings, Hough and Ones (2001) simultaneously encourage the use of warnings. In a similar vein, despite downplaying the problem of response distortion, Hogan and Hogan (1995) offer HPI users a "Virtuous" subscale to allow detection of overly desirable responders. How such recommendations might be reconciled on theoretical grounds is not obvious.

Summary

Our hierarchical overview presents a relatively clear path from the conceptual foundations of response distortion in personality testing, through delineation of associated problems in selection settings, to a variety of available solutions. It all boils down to this: it is natural for people to see and present themselves as better than they really are, whether honestly or dishonestly, and personality tests invite accurate as well as distorted self-appraisals. On average, people can fake when told to do so and actually do fake when the stakes are high, but generally not as much as they are capable of faking. Response distortion warrants further analysis with respect to content validity, poses threats to construct validity and to top-down selection decisions, but, surprisingly, appears to have little impact on criterion validity coefficients. Warnings and other preventive strategies, perhaps in combination with selected remedial approaches, may allow better use of personality tests for personnel selection. A reasonable next step is to offer grounds for systemic inquiry into the nature and sources of personality test response distortion. The remainder of this chapter is devoted to that end.

An Interactionist Model of Item-Level Response Distortion

Models of personality item responding are few, and most of those proposed target explanations of response latencies (e.g., Holden et al., 1992; Vasilopoulos et al., 2000). As noted earlier, response latencies offer clues about the faking process and so latency models may play an important role in faking research, particularly from a cognitive perspective. Another approach, taken by Snell, Sydell, and Lueke (1999), posits that response distortion is the product of a complex interplay among abilities, dispositions, and situational factors. We find this approach appealing, as it accounts for the major classes of variables known to be related to response distortion and seems poised to address many of the associated complexities. Snell et al. refer to their model as "interactional" but the precise nature of the interactional process is not specified. We offer below a variation on the interactionist theme that overlaps in many respects with Snell et al.'s pioneering work. Our foundations, however, are somewhat independent, and we focus more closely than they do on the individual personality test item. (Truth be told, we were ashamedly unaware of this work until the latter stages of chapter preparation. We wholeheartedly acknowledge the full magnitude of this earlier contribution, nonetheless.)

In an attempt to clarify problems and issues involving use of personality tests in work settings, Tett and Burnett (2003) proposed a trait-situation interactionist model of job performance. The model holds that workers express their traits in response to trait-relevant situational cues operating at task, social, and organizational levels. Trait expression is intrinsically rewarding and extrinsic rewards follow from the reactions of others. Trait expressive work behavior, valued by how well it meets job demands, is judged as job performance, and performance ratings are influenced by how well the ratee brings out desirable traits in the rater.

Our hierarchical review, especially pertaining to question 5 on individual differences, suggests opportunity to extend Tett and Burnett's (2003) model to consideration of response distortion. The new model posits that any observed response (X) to a given personality item contains three systematic components (i.e., in addition to random errors, which are ignored for the most part in the current discussion). The primary component captures the targeted trait (T), the second component is error due to self-deception (sd, usually positive), and the third is error due to impression management (im, also positive). Extending the nomenclature of classic true score theory,

$$X = T + sd + im + e.$$

Each systematic component (T, sd, and im) is subject to influences from three source categories: abilities (plus knowledge and skills, which are closely tied to abilities), targeted and nontargeted personality traits, and situational factors. Each component has its own relevant sources. For example, abilities/skills relevant to the primary component (T) include reading and writing, or basic computer skills, which are required to respond to the item. Nontargeted personality traits relevant to T include those relating to compliance (e.g., agreeableness), as respondents who do not comply with test instructions will yield no usable data. Situational factors have main effects (i.e., they influence mean scores) as well as moderating effects in interactions with the relevant abilities (knowledge, skills), and nontargeted traits. Situational factors relevant to T include the general testing situation and response instructions. Such factors offer the basic cues for test-takers to utilize the noted abilities/skills (e.g., reading) and express targeted and nontargeted personality traits in completing the test.

Figure 3.2 depicts the functional relationships among the main features of the model. The arrows from situational factors to the paths linking abilities and traits to the item response represent the moderating role

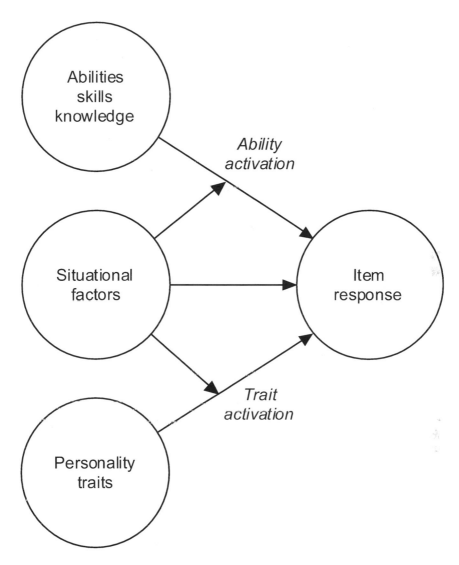

Figure 3.2.   The functional relationships among the main features of the model.

of situations. Trait-situation interactions are labeled as "trait activation," as per Tett and Burnett (2003). Ability-situation interactions, also discussed by Tett and Burnett, are labeled "ability activation." An important implication of the model, one that is implicit in much of the response distortion literature, is that any given item response is multiply determined (Snell et al., 1999). Taken to the population level, item variance can be traced to multiple distinct causes, and efforts to reduce distortion may be directed at any of several interacting mechanisms.

Figure 3.3 offers more complete lists of the various situational factors, abilities, and nontargeted traits that can come into play regarding each of the T, sd, and im components of an individual's response (X) to a given personality item. Variables expected to affect sd and im are in the middle and right columns, respectively. Several noteworthy points about the model are as follows.

1. A given item (e.g., "I always stay calm in stressful situations") can serve as a cue for the targeted trait (Emotional Stability) as well as self-deception and impression management. This "multi-saturation" conceptualization of personality items can account for observed correlations between substantive personality traits and variables linked to response distortion, and is consistent with other complexities observed in the literature.

2. The item and general testing situation serve as cues for all three components, but other situational features are relevant to specific components. Examples regarding the primary component (T) are noted above. A major situational cue for both sd and im is the desirability of the given item. Job- or role-specific item desirability may be more influential. A situational factor particularly relevant to self-deception is the degree to which the testing situation presents a threat to the respondent's ego (or the respondent carries such threats from previous situations). Such conditions could activate ego protective mechanisms, leading to unconscious, overly desirable self-descriptions. Impression management can be activated by a number of situational factors. Most relevant in selection settings is the prospect of getting the desired job (i.e., "high stakes"). Warnings may function to deactivate traits underlying impression management, with the possible exception of risk taking, and use of subtle items may reduce opportunity for impression management (they may also reduce opportunity for expressing the targeted trait, thereby affecting T; Holden & Jackson, 1979).

3. A special situational feature, unique to each respondent and relevant to both self-deception and impression management, is the respondent's true standing on the targeted trait. Those with high true scores have less opportunity to distort in the positive direction (i.e., self-deceive or fake good), and those with low true scores have less opportunity to distort in the negative direction (i.e., self-deceive or fake bad). It follows that T will tend to correlate negatively with sd and im. This, in turn, suggests that observed mean differences in both can-fake and do-fake studies and correlations

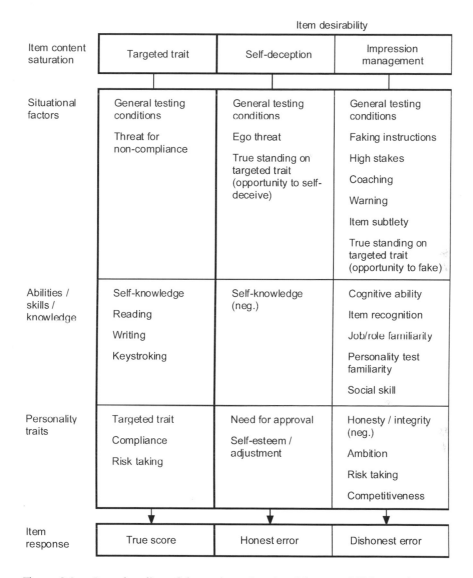

	Item desirability		
Item content saturation	Targeted trait	Self-deception	Impression management
Situational factors	General testing conditions Threat for non-compliance	General testing conditions Ego threat True standing on targeted trait (opportunity to self-deceive)	General testing conditions Faking instructions High stakes Coaching Warning Item subtlety True standing on targeted trait (opportunity to fake)
Abilities / skills / knowledge	Self-knowledge Reading Writing Keystroking	Self-knowledge (neg.)	Cognitive ability Item recognition Job/role familiarity Personality test familiarity Social skill
Personality traits	Targeted trait Compliance Risk taking	Need for approval Self-esteem / adjustment	Honesty / integrity (neg.) Ambition Risk taking Competitiveness
Item response	True score	Honest error	Dishonest error

Figure 3.3.   Complete lists of the various situational factors, abilities, and nontargeted traits.

between substantive trait scores and separate indices of response distortion underestimate the fakability of the substantive trait measures. More accurate estimates may be obtained using respondents with low T on the targeted trait. Estimating T, in light of unconscious self-deception bias, is not straightforward. Nonetheless, the role of ceiling (and floor) effects owing to respondents' true scores bears closer conceptual and empirical attention in response distortion research.

4.   Abilities (knowledge, skills) involved in responding to a personality item also vary by response component. As noted above, reading and writing are relevant to T (actually, to all 3 components). Self-knowledge is relevant to both T and sd, although in opposite ways: high self-knowledge contributes to T and reduces sd. Impression management brings other ability-related constructs into play. Cognitive ability will improve accuracy in judgments of job requirements and perceptions of item content and desirability. Familiarity with the job (e.g., from experience in similar jobs) would also offer an advantage in faking, as would familiarity with personality testing procedures. Social skill may further inflate the im component, as per Hogan and Hogan (1995).

5.   Personality traits other than the targeted dimension can be expressed in item responses, and the nontargeted traits may vary across the three response components. As noted above, the general testing situation offers opportunities to express compliance. A threat for noncompliance (e.g., withholding research credit in student samples) could activate risk taking in the form of nonpurposeful (i.e., random) responding. Nontargeted traits relevant to self-deception include need for approval and self-esteem: respondents high on such traits are more likely to unconsciously overestimate their true scores on scales targeting other dimensions (e.g., conscientiousness). Impression management can result as expressions of several nontargeted traits, including ambition, competitiveness, dishonesty, and risk taking. Snell et al. (1999) offer a few more: Machiavellianism, manipulativeness, delinquency, and locus of control.

6.   As per Figure 3.2, situational factors serve to activate (or deactivate) any of the noted ability-related characteristics and targeted as well as nontargeted personality traits. Perhaps the most salient of such interactions occurs in applicant settings, where the high stakes of getting a job (or not) can activate multiple traits and abilities. Warnings may activate risk taking, in that those higher on risk taking may be more likely to try to "beat the test" in the face of warnings. Coaching may be more effective for those high on cognitive ability and motivated by ambition (e.g., leading them to seek out coaching). The potential interplay among the multiple situa-

tional, ability, and trait variables defines the complexity involved in self-description using personality tests.

Applications of the Interactive Model to Relevant Findings

The model helps explain notable findings in the response distortion literature. Mean differences between straight-take and instructed faking conditions (in "can-fake" studies) result primarily from a situational main effect operating on the im component. The effect may be stronger, however, for those higher on cognitive ability as such individuals are likely to be better able to identify desirable items and choose appropriate responses. Results regarding the role of ability in faking have been mixed (Stanush, 1997). Moderating effects of item subtlety (i.e., subtle items create greater demand for ability) and compliance (i.e., ability will correlate with faking only in those who comply with instructions) seem worthy of investigation (Snell et al., 1999). The weaker mean differences between applicants and incumbents (in "do-fake" studies) may result from the replacement of faking instructions, exerting a strong situational main effect, with individual difference variables such as ambition and (low) honesty, whose influence is more weakly interactive.

Correlations between social desirability and personality variables tend to vary as a function of the situation (Elliot, Lawty-Jones, & Jackson, 1996; Hough, 1998; Rosse et al., 1998; Vasiloupolos et al., 2000). In general, traits judged as desirable (emotional stability, conscientiousness, and agreeableness) correlate more strongly with desirability in applicant than in incumbent settings, consistent with the view that activation of traits underlying social desirability varies by situational demands. Averaging across samples from Barrick and Mount (1996), impression management correlates stronger than self-deception does with four of the five primary dimensions (all but Extraversion) in applicants. The overall pattern of results suggests that measures of desirable personality traits are more susceptible to desirability bias under conditions expected to activate such bias. That impression management appears to be especially sensitive to applicant conditions further supports the need to examine response distortion effects from an interactionist perspective like that proposed here and by Snell et al. (1999).

Schmit and Ryan's (1993) ideal-employee factor, particular to applicants, may have emerged because T variance specific to each targeted trait (i.e., each person has a different T score per trait) was overwhelmed by im variance due to common sources (e.g., cognitive ability, ambition, and low honesty), activated by the high stakes of the applicant setting. Thus, in employment screening situations, all personality items can be cues for presenting a favorable impression, and respondents who are able and predisposed to do so will score higher on the items as a set, yielding a dominant ideal-employee factor.

Ones et al. (1993) reported mean validities of .40 and .29 for integrity tests in predicting job performance based on applicant and incumbent samples, respectively. The authors did not attempt to explain the difference, nor why it is opposite to expectations given that applicants have greater motivation to distort on integrity tests than do incumbents. Few constructs cut closer to the core of faking than integrity and honesty. The proposed model raises the possibility that integrity test scores in applicant settings capture other sources of variance besides integrity per se. Cognitive ability and ambition may be good candidates, as both are likely to be activated in applicant settings and both are linked to performance in most if not all jobs.

A critical implication of this interpretation is that integrity tests can work as selection tools for reasons other than their supposed construct validity. This ought to raise alarms as construct validity is the sine qua non of sound psychological measurement, and use of tests with uncertain construct validity is difficult to defend. The argument, moreover, bears extension to measures of any desirable trait, such as facets of conscientiousness, emotional stability, and agreeableness. It may help account for the noted discrepancy between results suggesting, on the one hand, negative effects of response distortion on construct validity and selection decisions, and, on the other, results suggesting no loss of criterion validity. Specifically, we suggest that explained variance in job performance lost due to degradation of construct validity associated with response distortion may be compensated by assessment of nontargeted abilities and traits that are also, inadvertently, related to performance.

Our argument bears directly on the effects of response distortion on selection decisions. Ignoring construct validity, shifting rank orders on the targeted trait matters little if the faker's lower standing on the targeted trait is compensated by higher standing on other dimensions that are, nonetheless, linked to job performance. Problems arise, however, when the traits underlying faking are in fact unrelated to job performance. The catch is that lack of construct validity (i.e., understanding what a given scale is measuring) undermines detection of such problem cases. It is better to know what a test is measuring and defend it on conceptual grounds than to rely blindly on empirical results capitalizing on biases that just happen to predict job performance. Our position, in essence, is that criterion validity serves construct validity, not the other way around, which is fundamental to psychological measurement as a scientific endeavor (Guion, 1998).

For those who treat deliberate score inflation on self-report personality scales as potentially relevant to predicting job performance (e.g., Hogan

& Hogan, 1995), our model clarifies several fundamental problems with that approach. First, although score inflation above T on a designated trait (e.g., Prudence) may contribute to predicting job performance independently of T, the source of inflation is ambiguous. It may be due to multiple nontargeted traits, abilities, and their interactions. Settling on any one (e.g., social skill) is premature without evidence to identify it to the exclusion of other possible sources.

Second, there is no assurance that a given score inflation is related to job performance. Two respondents might overestimate T equally, one in a way that contributes to performance on the job (e.g., as an interaction between ambition and abilities relevant to that job), and the other in a way that is unrelated or perhaps even detrimental to performance (e.g., narcissistic self-serving bias). Score inflation per se is insufficient to infer predictive advantage.

Third, the model offers grounds for expecting that the traits and abilities activated during screening may be different from those activated on the job. The biggest difference, noted earlier, is that applicant settings are high stakes situations. As such, they are likely to evoke maximal performance and, accordingly, personality score inflation. The actual work setting may include job demands qualitatively similar to those encountered during screening (e.g., to perform well), but they are likely to be much weaker. It is far from clear that someone motivated to make a good impression on a personality test when applying for a job will be as highly motivated to make a good impression after being hired, especially over intervals on which most performance criteria are based (i.e., 6 to 12 months).

A fourth problem, one that plagues all response distortion research, is that a high score on a desirable trait leaves ambiguous the combination of T, sd, and im components. Uncritical acceptance of sd and im as possible contributors to validity encourages uncritical blending of those two response components with T on the targeted substantive dimension. The implications for construct validity are severe, as we no longer know what our trait measures are assessing and, moreover, can no longer justify on rational grounds our predictions of performance, whether based on job analysis or mere speculation.

The proposed model keeps our heads above the sand and takes us in the opposite direction, toward clarifying the sources of variance in responses to any single personality test item and, by aggregation, variance in scale scores. It does this with the aim of distinguishing between variance that is related to job performance and variance that is unrelated. If variance independent of T is related to performance, then we might capitalize by development of separate construct-valid measures (e.g., social skill), thereby improving prediction and, moreover, understanding of job performance. If variance independent of T is unrelated to performance, then reducing that variance, preemptively or remedially, can be expected to improve the validity of our predictions based on T. The challenge made evident by our model is that identifying sd and im components and whether or not they are related to job performance is a daunting task, complicated by the simultaneous influences of multiple traits and abilities activated by distinct but sometimes overlapping situational forces.

Future Research Priorities

Nothing is more challenging and fundamental to the study of response distortion than distinguishing substantive from error variance in personality assessment: three people can score highly on a measure of conscientiousness, one because she is truly conscientious, another because he merely believes he is conscientious, and the last because of a deliberate attempt to present a favorable conscientious impression. Distinguishing between the second and third cases is made possible by comparing within each respondent scores obtained under straight-take (i.e., honest) versus typical applicant conditions (e.g., Griffith et al., in press). Distinguishing the first case from the others, however, requires knowledge of the respondent's true standing on the targeted dimension, for which we are forced by practicality to rely on self-reports made fallible by the very distortion we seek to identify. The search for T seems a futile pursuit. It may prove informative, however, to consider the conditions that might allow a reasonable inference of T based on realistic, albeit impractical, methods.

First, we should seek as a starting point the person's honest self-report using a standardized personality questionnaire, one that has been developed explicitly to minimize opportunity for response distortion through careful, construct-oriented scale development (e.g., the PRF; cf. Jackson, 1971). Second, we would arrange for the individual to be observed under a variety of real-life conditions over weeks and even months, serving different roles in interactions with realistic and representatively diverse data, people, and things, and combinations thereof. Third, the individual's responses to such varied stimuli would be monitored by multiple trained assessors with follow-up clinical interviews to allow detection of the more subtle clues to the individual's personality. Fourth, we would interview others who are closely familiar with the person, his friends, family, coworkers, clients, classmates, roommates, and other associates who have had reasonable opportunity to observe the assessee under assorted conditions. Fifth, we would gather all this information, compile it, and review it individually and as a team, while restricting as far we are able our own biases and value judgments. Finally, we would average across observers, situations, and behaviors, predetermined as indicative of targeted traits (perhaps in light of situational demands), to arrive at a reasoned estimate of the person's true personality profile. (Interesting and potentially important variations include adjustments for situational features, e.g., home versus work, which are ignored here for simplicity's sake.)

Despite the fallibility of each facet of this research program, the enterprise as a whole would be expected to offer as close a rendering as possible of T for each targeted trait. Taking the difference between these estimates of T and those provided by way of self-report (at step 1 in the above or using another self-report measure of interest) would offer arguably the best possible estimate of self-deception response bias associated with self-report. Only by applying the method to a reasonably sized and heterogeneous sample of individuals could we learn how important self-knowledge truly is in the validity of self-report measures. Only then could we assess the true relationship between self-deception and substantive traits and the impact of trait standing on opportunity to distort (i.e., as per ceiling and floor effects discussed earlier). Adding item and trait desirability ratings would allow further clarification of self-deception processes, and we might also better understand the relationship between self-deception and impression management. In short, the key to the most pressing issues facing distortion researchers, it seems, is finding valid measures of T and it behooves us to give greater thought to how this might be accomplished.

The described research regimen is offered as an ideal both to demonstrate the degree of challenge facing researchers seeking answers to key questions about response distortion and to set out a standard against which more practical pursuits of T might bear comparison in terms of rigor and likely validity. Alternatives worth considering include projective tests, conditional reasoning, clinical interviews, and, perhaps closest to the ideal described above, assessment centers. Faking researchers are urged to consider valid assessment of T as a difficult but worthwhile goal, as there seems little hope of resolving critical issues without it.

Other research needs, less critical but more tangible, include the following: (1) correlations between self-report personality tests and separate measures of self-deception and impression management under both applicant and incumbent conditions, preferably using a within-subjects design; (2) differences between personality scale scores from honest versus applicant conditions using a within-subjects design to allow individual assessment of "direct" impression management (i.e., as difference scores) for correlation with cognitive ability, ambition, competitiveness, self-esteem, need for approval, self-deception, questionnaire-based impression management, and other constructs postulated to underlie applicant faking; (3) measurement of situational activators of traits and abilities expected to underlie response distortion, for use in testing situational moderator effects and possibly reducing unwanted scale variance; (4) effects of perceived fakability of personality tests on applicant fairness reactions; and (5) direct comparisons among preventive and remedial approaches to dealing with response distortion.

Conclusions

We began this chapter by observing that lying, cheating, and faking (and related, less corrupt qualities, such as hiding one's faults) are as natural as self-preservation, and that personality tests invite self-promoting distortions, whether deliberate or unconscious. Our review of a broad cross-section of the response distortion literature leads us to conclude, contrary to some (e.g., Hough & Ones, 2001; Ones & Viswesvaran, 1998; Ones et al., 1996; Hogan & Hogan, 1995), that our widely shared beliefs about faking (e.g., Rees & Metcalf, 2003) are essentially accurate, that people can and do distort their responses to self-report personality tests, and that, when it happens, we ought to be concerned about it. Evidence supporting the "red herring" position (Ones et al., 1996) is based on a relatively small part of the bigger picture and even advocates of that position seem unconvinced, offering as they do palliative warnings and detection systems. We do not seek to diminish the power of those findings (although there may be grounds for doing so; Rosse et al., 1998). Rather, we seek to explain why they are at odds with other lines of evidence—and our instincts—collectively asserting that response distortion is not a red herring, that there is more to learn about self-report bias and its role in the prediction of work behavior.

No amount of research on the topic, nor situational and data manipulation, is likely to eliminate socially desirable responding, as desirability is inherent in the constructs we use to describe each another and ourselves. It is eminently reasonable, however, to expect to clarify our understanding of how traits, abilities, and situations interact to produce distortions in self-reports, so that we might reduce the unwanted effects of desirability bias thereby improving personality test validity. Making a good thing better is an honorable pursuit. The proposed model, we hope, will serve as a useful guide in this challenging, venerable, and worthwhile task.

REFERENCES

Abrahams, N. M., Neuman, I., & Githens, W. H. (1971). Faking vocational interests: Simulated versus real life motivation. *Personnel Psychology, 24,* 5-12.

Alliger, G. M., Lilienfield, S. I., & Mitchell, K. E. (1996). The susceptibility of overt and covert integrity tests to coaching and faking. *Psychological Science, 7*, 32-39.

Anastasi, A., & Urbina, S. (1997). *Psychological testing.* Upper Saddle River, NJ: Prentice Hall.

Arthur, W., Woehr, D. J., & Graziano, W. G. (2000). Personality testing in employment settings: Problems and issues in the application of typical selection practices. *Personnel Review, 30*, 657-676

Barrick, M. R., & Mount, M. K. (1996). Effects of impression management and self-deception on the predictive validity of personality constructs. *Journal of Applied Psychology, 81*, 261-272.

Becker, T. E., & Colquitt, A. L. (1992). Potential versus actual faking of a biodata form: An analysis along several dimensions of item type. *Personnel* Psychology, *45*, 389-406.

Bentler, P. M., Jackson, D. N., & Messick, S. (1971). Identification of content and style: A two-dimensional interpretation of acquiescence. *Psychological Bulletin, 76*, 186-204.

Block, J. (1965). *The challenge of response sets: Unconfounding meaning, acquiescence, and social desirability in the MMPI.* East Norwalk, CT: Appleton-Century-Crofts.

Bowers, K. S. (1973). Situationism in psychology: An analysis and a critique. *Psychological Review, 80*, 307-336.

Christiansen, N. D., Goffin, R. D., Johnston, N. G., & Rothstein, M. G. (1994). Correcting the 16PF for faking: Effects on criterion-related validity and individual hiring decisions. *Personnel Psychology, 47*, 847-860.

Costa, P. T. (1996). Work and personality: Use of the NEO-PI-R in industrial/organizational psychology. *Applied Psychology: An International Review, 45*, 225-241.

Crowne, D. P., & Marlowe, D. A. (1964). *The approval motive.* New York: Wiley.

Dahlstrom, W. G., Welsh, G. S., & Dahlstrom, L. E. (1972). *An MMPI handbook: Clinical interpretation.* Oxford, England: University of Minnesota Press.

Dahlstrom, W. G., Welsh, G. S., & Dahlstrom, L. E. (1975). *An MMPI handbook Vol. II: Research applications.* Minneapolis: University of Minnesota Press.

Damarin, F., & Messick, S. (1965). *Response styles as personality variables: A theoretical integration.* (Educational Testing Service RB 65-10). Princeton, NJ: Educational Testing Service.

Dunnette, S., Koun, S., & Barber, P. J. (1981). Social desirability in the Eysenck Personality Inventory. *British Journal of Psychology, 72*, 19-26.

Dunnette, M. D., McCartney, J., Carlson, H. C., & Kirchner, W. K. (1962). A study of faking behavior on a forced-choice self-description checklist. *Personnel Psychology, 15*, 13-24.

Dwight, S. A., & Donovan, J. J. (1998). *Warning: Proceed with caution when warning applicants not to dissimulate.* Paper presented at the 13th annual conference of the Society for Industrial and Organizational Psychology, Dallas, TX.

Edwards, A. L. (1953). The relationship between the judged desirability of a trait and the probability that the trait will be endorsed. *Journal of Applied Psychology, 37*, 90-93.

Edwards, A. L. (1957). Social desirability and probability of endorsement of items in the interpersonal check. *Journal of Abnormal & Social psychology, 55*, 394-396.

Ekehammar, B. (1974). Interactionism in personality from a historical perspective. *Psychological Bulletin, 81*, 1026-1048.

Elliot, S., Lawty-Jones, M., & Jackson, C. (1996). Effect of dissimulation on self-report and objective measures of personality. *Personality and Individual Differences, 21*, 335-343.

Endler, N. S., & Magnusson, D. (1976). Toward an interactional psychology of personality. *Psychological Bulletin, 83*, 959-974.

Epstein, S., & O'Brien, E. J. (1985). The person-situation debate in historical and current perspective. *Psychological Bulletin, 98*, 513-537.

Eysenck, S. B., Eysenck, H. J., & Shaw, L. (1974). The modification of personality and lie scale scores by special "honesty" instructions. *British Journal of Social & clinical Psychology, 13*, 41-50.

Furnham, A. (1984). Lay conceptions of neuroticism. *Personality and Individual Differences, 5*, 95-103.

Furnham, A. (1986). Response bias, social desirability and dissimulation. *Personality and Individual Differences, 7*, 385-406.

Furnham, A. (1990a). Faking personality questionnaires: Fabricating different profiles for different purposes. *Current Psychology, 9*, 46-56.

Furnham, A. (1990b). The fakeability of the 16PF, Myers-Briggs and FIRO-B personality measures. *Personality and Individual Differences, 11*, 711-716.

Furnham, A., & Craig, S. (1987). Fakeability and correlates of the perception and preference inventory. *Personality and Individual Differences, 8*, 459-470.

Gillis, J. R., Rogers, R., & Dickes, S. E. (1990). The detection of faking bad response styles on the MMPI. *Canadian Journal of Behavioural Science, 22*, 408-416.

Gough, H. G. (1975). *Manual for the California Psychological Inventory*. Palo Alto, CA: Consulting Psychologists Press.

Griffith, R. L., Chmielowski, T. S., & Yoshita, Y. (in press). Do applicants fake? An Examination of the Frequency of Applicant Faking Behavior. *Personnel Review*.

Griffith, R. L., Snell, A. F., McDaniel, M. A., & Frei, R. L. (1998). *Social desirability and construct validity: A method bias analysis of self deceptive enhancement and impression management*. Paper presented at the 13th annual conference of the Society for Industrial and Organizational Psychology, Dallas TX.

Guion, R. M. (1998). *Assessment, measurement, and prediction for personnel decisions*. Mahwah, NJ: Erlbaum.

Hartshorne, H., & May, M. A. (1928). *Studies in the nature of character, Vol. 1. Studies in deceit*. New York: Macmillan.

Haymaker, J. C., & Erwin, F. W. (1980). *Investigation of applicant responses and falsification detection procedures for the Military Applicant Profile* (Final Project Report, Work Unit No. DA644520). Alexandria, VA: U.S. Army Research Institute for the Behavioral and Social Sciences.

Helmes, E. (2000). The role of social desirability in the assessment of personality constructs. In R. D. Goffin & E. Helmes (Eds.), *Problems and solutions in human assessment* (pp. 21-40). Boston: Kluwer.

Helmes, E., & Holden, R. R. (2003). The construct of social desirability: One or two dimensions? *Personality and Individual Differences, 34*, 1015-1023.

Hinrichsen, J. J., Gryll, S., Bradley, L. A., & Katahn, M. (1975). Effects of impression management efforts on FIRO-B profiles. *Journal of Consulting and Clinical Psychology, 43*, 269–279.

Hogan, R., & Hogan, J. (1995). *Hogan personality inventory manual* (2nd ed.). Tulsa, OK: Hogan Assessment Systems.

Hogan, R., Hogan, J., & Roberts, B. W. (1996). Personality measurement and employment decisions: Questions and answers. *American Psychologist, 51*, 469-477.

Hogan, R., & Nicholson, R. A. (1988). The meaning of personality test scores. *American Psychologist, 43*, 621-626.

Holden, R. R., & Jackson, D. N. (1979). Item subtlety and face validity in personality assessment. *Journal of consulting and Clinical Psychology, 47*, 459-468.

Holden, R. R., Kroner, D. G., Fekken, G. C., & Popham, S. M. (1992), A model of personality test item response dissimulation. *Journal of Personality and Social Psychology, 63*, 272-279

Hough, L. M. (1998). Effects of intentional distortion in personality measurement and evaluation of suggested palliatives. *Human Performance, 11*, 209-244.

Hough, L. M., Eaton, N. K., Dunnette, M. D., Kamp, J. D., & McCloy, R. A. (1990). Criterion-related validities of personality constructs and the effect of response distortion on those validities. *Journal of Applied Psychology, 75*, 581-595.

Hough, L. M., & Furnham, A. (2003). Use of personality variables in work settings. In W. C. Borman, D. R. Ilgen, & R. J. Klimoski (Eds.), *Handbook of psychology: Industrial and organizational psychology* (Vol. 12, pp. 131-169). Hoboken, NJ: Wiley.

Hough, L. M., & Ones, D. S. (2001). The structure, measurement, validity, and use of personality variables in industrial, work, and organizational psychology. In N. Anderson, D. Ones, H. K. Sinangil, & C. Viswesvaran (Eds.), *Handbook of industrial, work, and organizational psychology* (Vol. 1, pp. 241-267). London: Sage.

Hough, L. M., & Schneider, R. J. (1996). Personality traits taxonomies & applications in organizations. In K. R. Murphy (Ed.), *Individual differences & behavior in organizations* (pp. 31-88). San Francisco: Jossey-Bass.

Hunter, J. E., & Schmidt, F. L. (2004). *Methods of meta-analysis: Correcting error and bias in research findings*. Thousand Oaks, CA: Sage.

Jackson, D. N. (1970). A sequential system for personality scale development. In C. D. Spielberger (Ed.), *Current topics in clinical and community psychology* (Vol. 2, pp. 61-96). New York: Academic Press.

Jackson, D. N. (1971). The dynamics of structured personality tests. *Psychological Review, 78*, 229-248.

Jackson, D. N. (1984). *Personality research form manual* (3rd ed.). Port Huron, MI: Research Psychologists Press.

Jackson, D. N. (1994). *Jackson Personality Inventory—Revised manual*. Port Huron, MI: Sigma Assessment Systems.

Jackson, D. N., & Helmes, E. (1979). Personality structure and the circumplex. *Journal of Personality and Social Psychology, 37*, 2278-2285.

Jackson, D. N., Wroblewski, V. R., & Ashton, M. C. (2000). The impact of faking on employment tests: Does forced-choice offer a solution? *Human Performance, 13*, 371-388.

Kirton, M.J. (1991). Faking personality questionnaires: A response to Furnham. *Current Psychology, 10*, 315-318.

Kluger, A. N., Reilly, R. R., & Russell, C. J. (1991). Faking biodata tests: are option-keyed instruments more resistant. *Journal of Applied Psychology, 76*, 889-896.

Krahe, B. (1989). Faking personality profiles on a standard personality inventory, *Personality and Individual Differences, 10*, 437-443.

Krueger, J. (1998). Enhancement bias in descriptions of self and others. *Personality and Social Psychology Bulletin, 24*, 505-516.

Leary, M. R., & Kowalski, R. M. (1990). Impression management: A literature review and two-component model. *Psychological Bulletin, 107*, 34-47.

Mahar, D., Cologon, J., & Duck, J. (1995). Response strategies when faking personality questionnaires in a vocational selection setting. *Personality and Individual Differences, 18*, 605-609.

McCrae, R. R., & Costa, P. T. (1983). Social desirability scales: More substance than style. *Journal of Consulting and Clinical Psychology, 51*, 882-888.

McFarland, L. A., & Ryan, A. M. (2000). Variance in faking across noncognitive measures. *Journal of Applied Psychology, 85*, 812-821.

Messick, S., & Jackson, D. N. (1961). Acquiescence and desirability as response determinants on the MMPI. *Educational and Psychological Measurement, 21*, 771-790.

Morf, M. E., & Jackson, D. N. (1972). An analysis of two response styles: True responding and item endorsement. *Educational and Psychological measurement, 32*, 329-353.

Mueller-Hanson, R., Heggestad, E. D., & Thorton, G. C. (2003). Faking and selection: Considering the use of personality from select-in and select-out perspectives. *Journal of Applied Psychology, 88*, 348-355.

Norman, W. T. (1963). Personality measurement, faking, and detection: An assessment method for use in personnel selection. *Journal of Applied Psychology, 47*, 225-241.

Ones, D. S., & Viswesvaran, C. (1998). The effects of social desirability and faking on personality and integrity assessment for personnel selection. *Human Performance, 11*, 245-269.

Ones, D. S., Viswesvaran, C., & Reiss, A. D. (1996). Role of social desirability in personality testing for personnel selection: The red herring. *Journal of Applied Psychology, 81*, 660-679.

Ones, D. S., Viswesvaran, C., & Schmidt, F. L. (1993). Comprehensive meta-analysis of integrity test validities: Findings and implications for personnel selection and theories of job performance. *Journal of Applied Psychology, 78*, 679-703.

Paulhus, D. L. (1986). Self-deception and impression management in test responses. In A. Angleitner, & J.S. Wiggins (Eds.), *Personality assessment via questionnaires: Current issues in theory and measurement* (pp. 143-165). New York: Springer-Verlag.

Paulhus, D. L. (1991). Measurement and control of response bias. In J. P. Robinson, P. R. Shaver, & L. S. Wrightsman (Eds.), *Measures of personality and social psychological attitudes* (pp. 1-17). New York: Academic Press.

Paulhus, D. L. (1998). *Manual for the Balanced Inventory of Desirable Responding (BIDR-7).* Toronto, Ontario, Canada: Multi-Health Systems.

Paulhus, D. L. (2002). Socially desirable responding: The evolution of a construct. In H. I. Braun, D. N. Jackson, & D. E. Wiley (Eds.), *The role of constructs in psychological and educational measurement* (pp. 49-69). Mahwah, NJ: Erlbaum.

Paulhus, D. L., & John, O. (1998). Egoistic and moralistic biases in self-perception: The interplay of self-deceptive styles with basic traits and motives. *Journal of Personality, 66,* 1025-1060.

Pervin, L. A. (1985). Personality: Current controversies, issues, and directions. *Annual Review of Psychology, 36,* 83-114.

Rees, C. J., & Metcalfe, B. (2003). The faking of personality questionnaire results: Who's kidding whom? *Journal of Managerial Psychology, 18,* 156-165.

Robie, C., Curtin, P. J., Foster, C., Philips, H. L., Zbylut, M., & Tetrick, L. E. (2000). The effect of coaching on the utility of response latencies in detecting fakers on a personality measure. *Canadian Journal of Behavioural Science, 32,* 226-233.

Rosse, J. G., Stecher, M. D., Miller, J. L., & Levin, R. A. (1998). The impact of response distortion on preemployment personality testing and hiring decisions. *Journal of Applied Psychology, 83,* 634-644.

Ryan, A. M., & Sackett, P. R. (1987). Pre-employment honesty testing: Fakability, reactions of test takers, and company image. *Journal of Business and Psychology, 1,* 248-256.

Sackheim, H. A., & Gur, R. C. (1978). Self-deception, self-confrontation, and consciousness. In G. E. Schwartz & D. Shapiro (Eds.), *Consciousness and self-regulation: Advances in research and theory* (Vol. 2, pp. 139-197). New York: Plenum.

Schmit, M. J., & Ryan, A. M. (1993). The Big Five in personnel selection: Factor structure in applicant and nonapplicant populations. *Journal of Applied Psychology, 78,* 966-974.

Schmit, M. J., Ryan, A. M., Stierwalt, S. L., & Powell, A. B. (1995). Frame-of-reference effects on personality scale scores and criterion-related validity. *Journal of Applied Psychology, 80,* 607-620.

Schwab, D. P. (1971). Issues in response distortion studies of personality inventories: A critique and replicated study. *Personnel Psychology, 24,* 637-647.

Schwab, D. P. (1974). Comments on "Re-examination of the fakability of the Gordon Personal Inventory and profile: A reply to Schwab." *Psychological Reports, 34,* 316-318.

Seisdedos, N. (1993). Personnel selection, questionnaires, and motivational distortion: An intelligent attitude of adaptation. In H. Schuler, J. L. Farr, & L. James, & M. Smith (Eds.), *Personnel selection and assessment: Individual and organizational perspectives.* Hillsdale, NJ: Erlbaum.

Smith, D. B., & Ellingson, J. E. (2002). Substance versus style: A new look at social desirability in motivating contexts. *Journal of Applied Psychology, 87,* 211-219.

Smith, D. B., & Robie, C. (2004). The implications of impression management for personality research in organizations. In B. Schneider & D. B. Smith (Eds.), *Personality and organizations* (pp. 111-138). Mahway, NJ: Erlbaum.

Snell, A. F., Sydell, E. J., & Luecke, S. B. (1999). Towards a theory of applicant faking: Integrating studies of deception. *Human Resource Management Review, 9,* 219-242.

Snyder, M., & Ickes, W. (1985). Personality and social behavior. In G. Lindzey & E. Aronson (Eds.), *The handbook of social psychology* (Vol. 11, pp. 883-947). New York: Random House.

Stanush, P. L. (1997). *Factors that influence the susceptibility of self-report inventories to distortion: A meta-analytic investigation.* Unpublished doctoral dissertation, Texas A&M University.

Stark, S., Chernyshenko, O. S., Chan, K., Lee, W. C., & Drasgow, F. (2001). Effects of the testing situation on item responding: Cause for concern. *Journal of Applied Psychology, 86,* 943-953.

Taylor, S. E., & Brown, J. D. (1988). Illusion and well-being: A social psychological perspective on mental health. *Psychological Bulletin, 103,* 193-210.

Tellegen, A. (1982). *Brief manual for the Differential Personality Questionnaire.* Unpublished manuscript, University of Minnesota.

Tett, R. P., & Burnett, D. D. (2003). A personality trait-based interactionist model of job performance. *Journal of Applied Psychology, 88,* 500-517.

Thornton, G. C., & Gierasch, P. F. (1980). Fakability of an empirically derived selection instrument. *Journal of Personality Assessment, 44,* 48-51.

Trent, T. T., Atwater, D. C., & Abrahams, N. M. (1986, April). Biographical screening of military applicants: Experimental assessment of item response distortion. In G. E. Lee (Ed.), *Proceedings of the tenth annual symposium of psychology in the department of defense* (pp. 96-100). Colorado Springs, CO: U.S. Air Force Academy, Department of Behavioral Sciences and leadership.

Vasilopoulos, N. L., Reilly, R., & Leaman, J. A. (2000). The influence of job familiarity and impression management on self-report measure scale scores and response latencies. *Journal of Applied Psychology, 85,* 50-64.

Viswesvaran, C., & Ones, D. S. (1999). Meta-analyses of fakability estimates: Implications for personality measurement. *Educational and Psychological Measurement, 59,* 197-210.

Walker, C. B. (1985, February). *The fakability of the Army's Military Applicant Profile (MAP).* Paper presented at the annual conference of the Human Resources Management and Organizational Behavior Association, Denver, CO.

Waters, L. K. (1965). A note on the "fakability" of forced-choice scales. *Personnel Psychology, 18,* 187-191.

Weiss, H. M., & Adler, S. (1984). Personality and organizational behavior. *Research in Organizational Behavior, 6,* 1-50.

Windle, C. (1954). Test-retest effect on personality questionnaires. *Educational and Psychological Measurement, 14,* 617-633.

Zerbe, W. J., & Paulhus, D. L. (1987). Socially desirable responding in organizational behavior: A reconception. *Academy of Management Review, 12,* 250-264.

Zickar, M. J., & Drasgow, F. (1996). Detecting faking on a personality instrument using appropriateness measurement. *Applied Psychological Measurement, 20*, 71-87.

Zickar, M. J., & Robie, C. (1999). Modeling faking good on personality items: An item-level analysis. *Journal of Applied Psychology, 84*, 551-563.

CHAPTER 4

ASSESSING RESPONSE DISTORTION IN PERSONALITY TESTS

A Review of Research Designs and Analytic Strategies

Jessica Mesmer-Magnus and Chockalingam Viswesvaran

The 1990s brought resurgence in the use of personality to explain organizational behavior (Hough & Ones, 2001). The increased use of personality measures in personnel selection can be attributed to the publication of well-developed taxonomies of personality traits (e.g., the Big Five; Hogan, Hogan, & Roberts, 1996) and to the meta-analytic cumulation (Hunter & Schmidt, 2004) of the extant literature using these taxonomies. Several meta-analyses have examined the validity of various personality scales for predicting a wide range of organizational behaviors (Barrick & Mount, 1991; Hurtz & Donovan, 2000; Salgado, 1997; Tett, Jackson, & Rothstein, 1991). Even though the exact magnitude and the utility of such magnitudes have been debated (cf. Murphy & Dzieweczynski, 2005; Ones, Viswesvaran, & Dilchert, 2005), there is a strong consensus that an indi-

A Closer Examination of Applicant Faking Behavior, 85–113
Copyright © 2006 by Information Age Publishing
All rights of reproduction in any form reserved.

vidual's personality plays a critical role in predicting and explaining behavior within an organizational setting. Specifically, substantial evidence suggests that personality variables correlate with a number of performance measures of importance to organizations (e.g., job performance, training performance, absenteeism, organizational commitment, etc.; Barrick & Mount, 1991; Ones, Viswesvaran, & Schmidt, 1993; Tett et al., 1991).

Due to the increasing use of personality tests in high stakes organizational decision making (e.g., personnel selection), concerns have been raised about the potential for test-takers to fake their responses, thus potentially reducing the predictive validity of the test. Specifically, concerns about the effects of faking for personnel selection center around whether (1) faking impacts the criterion-related or construct validity of personality scales, and (2) whether faking affects the quality of personnel selection decisions (Mueller-Hanson, Heggstad, & Thornton, 2003). Faking can be a problem both in written tests of personality and with other modes of personality assessments (e.g., clinical interviews). In this chapter, we focus solely on research related to the fakability of written personality tests.

The extant literature has, at times, referenced faking as impression management, socially desirable responding, lying, response distortion, and dissimulation (Viswesvaran & Ones, 1999). Although these terms have been used synonymously, conceptual distinctions are made between them (Paulhus, 1984). For example, Paulhus and Reid (1991) argue social desirability consists of two factors: self-deception and impression management. Impression management implies the respondent wishes to alter responses to be consistent with the expectations of others. However, self-deception reflects skewed responses indicative of the respondent's inaccurate self-concept. Similarly, distinctions are made between motivated faking (where the test-taker knows they are faking) and self-delusion (where the test-taker does not know they are faking or providing untruthful answers). In addition to examining the intentionality of response distortion (e.g., whether distortion is intentional or more a function of social desirability bias), researchers argue it is also important to consider the degree or extent to which the respondent has faked (Levin, 1995; Rosse, Stecher, Miller, & Levin, 1998). Regardless of these theoretical distinctions, the research designs and analytic strategies employed are mostly similar.

Investigations of faking in personality assessments have primarily focused on three key issues: the feasibility (Can respondents fake?), prevalence (Do they fake?), and consequences of faking (What are the effects to, for example, the criterion-related or construct validity of these tests?). In this chapter, we address the data collection, research design, and analytic strategies commonly employed to address these three questions. As long as fakability is a constant across individuals, many of the concerns of faking in organizational decision-making become moot or mitigated (i.e., if all applicants equally "fake-good" their responses to a personality test, then the quality of selection decisions based on these test scores should not be adversely impacted; see Douglas, McDaniel, & Snell, 1996). It is only when individuals differ in their levels of faking that such response distortion becomes an issue. Thus, in the final section of this chapter, we discuss the extent to which individuals differ, whether such differences are related to organizational behavior, and whether the popular scales used to capture faking (e.g., lie scales, social desirability scales) are useful in organizational decision-making. We review four prevalent scales developed to identify and assess respondent faking: two popular, commercially-available, personality instruments (the validity scales of the Minnesota Multiphasic Personality Inventory and the California Personality Inventory) and two independent scales (the Balanced Inventory of Desirable Responding and the Marlow-Crowne Social Desirability Scale).

Data Collection Strategies in Faking Research

What are the prevalence, feasibility, and consequences of respondent faking of personality instruments? To address these questions, researchers have primarily employed a between-subjects design or a within-subjects design. In the between-subjects design, two random groups of participants are formed. Responses are obtained from the two groups such that the testing environment in one group is designed to induce faking whereas in the other group more honest responses are encouraged (or at least a hypothesis is made that more accurate responses are provided by the participants). However, to the extent that social desirability response bias is always in operation, truly honest responding may not be within the willful control of the respondent (cf. Robie, Mueller, & Campion, 2001; Zickar, Gibby, & Robie, 2004). In the within-subjects design, the same participants take the personality test twice: once when the testing environment is designed to induce faking while in the other administration of the test more honest responses are encouraged.

In lieu of one of the above strategies, researchers have also opted to utilize only one group of test-takers, and instead of two test administrations under different experimental conditions, individual differences in faking are assessed and statistically controlled (e.g., social desirability scales, lie scales, validity indices, statistical indices of faking, etc.). This data collection strategy assumes a trait perspective of faking, wherein faking is conceptualized as an individual difference variable rather than a pattern of responding that may be contextually induced (Stark, Chernyshenko, Chan, Lee, & Drasgow, 2001). We will discuss these "trait" strategies in the second and third sections of this chapter.

Within- Versus Between-Subjects Designs

A key advantage of the within-subjects design is that, given the same number of participants, the within-subjects design has greater statistical power (there are twice the number of observations as compared with the between-subjects design; Campbell & Stanley, 1963). Within-subjects designs also facilitate an assessment of potential individual differences in the ability and motivation to fake (i.e., the trait perspective; McFarland & Ryan, 2000; Mersman & Schultz, 1998; Stark et al., 2001). Empirical results reported in the extant literature generally suggest that the degree of change in a test score as a function of instructions to fake is higher in within-subjects than in between-subjects designs. Viswesvaran and Ones (1999) report an average value for faking estimates from within-subjects design that is almost three times that reported for between-subjects design.

Within-subjects designs, however, are potentially liable to several threats to validity (i.e., testing, history, maturation; Cook & Campbell, 1979). For example, in within-subjects designs, respondents are asked to take the same test two times, raising the possibility that testing effects, rather than faking, explain differences in scores between administrations. Hausknecht, Trevor, and Farr (2002) report a significant pretesting effect for ability tests, and some have argued such pretesting effects may be even more pronounced for noncognitive (e.g., personality) assessments (e.g., Kelley, Jacobs, & Farr, 1994; Kulik, Kulik, & Bangert, 1984). This raises the possibility that purportedly faked responses may be more a function of testing effects than of response distortion. To address these concerns, more complicated designs have been introduced. For example, researchers may add a control group that also takes the test twice, but both times respondents are encouraged (via instruction or contextual clues) to generate honest responses. To assess interaction between pretesting and instructions given, a four-group design has been introduced (cf. Braver & Braver, 1988; Huck & Sandler, 1973, Solomon, 1949). Unfortunately, such designs are typically only feasible in laboratory settings (rather than in an organizational setting) using student samples (rather than actual job applicants or employees) thus raising concerns regarding generalizability. Thus, the more complicated designs, while elegant, typically lack ecological validity (Mook, 1983).

Inducing Faking Using Instructional Sets

Instructional sets are commonly used to create an environment conducive to respondent faking. In a between-subjects design, one group of participants is asked to take a personality test and respond honestly; the second group takes the same test, but is encouraged to fake their responses (either via explicit statements to fake or through instructions "to pretend" or "to appear desirable"). In a within-subjects design, the same group of respondents takes a personality test twice, once with instructions to fake and once without such instruction. The experimenter might request the respondent to "fake good" (increase the desirability of responses) or to "fake bad" (distort so as to look more negative). Similarly, researchers have issued warnings (regarding the feasibility of detecting faking or potential consequences of faking) and compared responses under this condition with a "no warning" condition (cf. Dwight & Donovan, 2003; Griffith, Frei, Snell, Hamill, & Wheeler, 1997; Hough, 1998). Most empirical research has focused on the feasibility and prevalence of respondents faking-good, since instances of faking-bad in an organizational context (e.g., applying for a job) are thought to be relatively rare (cf. Viswesvaran & Ones, 1999). Further, in organizational contexts, faking-bad likely results from respondents attempting to fake-good, but misunderstanding the response scale. A meta-analysis of empirical studies employing either a between-subjects or a within-subjects design revealed that when instructed to fake, individuals can change their scores. In studies employing a between-subjects design, respondents changed their scores by almost half a standard deviation. In studies using a within-subjects design, individuals instructed to fake changed their scores up to one and a half standard deviations (Viswesvaran & Ones, 1999).

Assessing Faking Under Naturally Occurring Conditions

A potential disadvantage of instructional sets is that they may give rise to demand characteristics (cf. Griffith et al., 1997; see also White & Moss, 1995). Instead of issuing instructions regarding whether or not to fake, many researchers have opted to employ more naturally occurring test administrations to assess faking (e.g., comparing applicant versus employee responses, anonymous versus nonanonymous responses, face-to-face versus computer administrations, etc.). For example, Hough, Eaton, Dunnette, Kamp, and McCloy (1990) compared the responses of individuals applying for army jobs to those who are already employed in these jobs. They found that although a similar group of respondents were able to distort their responses to a personality inventory when instructed to do so, there was little evidence that actual applicants (who, arguably, would be motivated to fake) faked their responses. Conversely, using a similar design (applicant versus nonapplicant), Rosse et al. (1998) found higher levels of faking among job applicants than nonapplicants. Further, they identified individual differences in faking tendency and found that applicant faking had negative implications for the quality of personnel selection decisions. Researchers have also combined comparisons of applicant and nonapplicant responses with differing instructions.

Even though it is more common for researchers to employ a between-subjects design in such studies (applicant versus nonapplicant), there is no reason why researchers cannot compare applicant scores with the scores obtained by the same individuals once they have been hired (a within-subjects comparison). This strategy assumes that differences between applicants and employees are due to applicant motivation to fake; another explanation, however, is that personality change may be a function of job experience (i.e., maturation; cf. Hough et al., 1990). The Attraction-Selection-Attrition (ASA) framework (Schneider, 1987) suggests attrition effects within an organization operate to ensure the maintenance of certain personality-types in organizations. Applicant-employee differences may be due to a number of (mutually overlapping) reasons: (1) there is a greater degree of response distortion by applicants (presumably attributable to the desire to garner an offer of employment), (2) true personality differences exist that result from experience on the job (e.g., socialization; Schneider, 1987), (3) employees have a greater knowledge of responses their organization would regard as socially desirable, (4) a greater degree of range restriction present in employee as compared with applicant samples (e.g., Rosse et al., 1998; Schneider, 1987), or (5) employees are less motivated or attentive than applicants during test-taking, potentially introducing error from careless responding, etc. Teasing response distortion from applicant-employee differences is difficult given these conditions and potential explanations. It may be that all that can be gleaned from identifying applicant-employee differences is that researchers must create different norms for personality scales used with applicant and employee samples. Obtaining the second measure of personality even before they have been hired (cf. Griffith, Rutkowski, Gujar, Yoshita, & Steelman, 2005) circumvents some of the problems but still leaves the possibility that the second response could be different from the responses given as applicants due to careless responding (or at least not taken the second test as seriously as the first under selection conditions).

A related research strategy is to compare the personality scores of applicant cohorts assessed under different labor market conditions (cf. Ones & Viswesvaran, 2001). The premise of this design is that under tight labor market conditions, when the unemployment rates are high, applicants will have a greater motivation to fake-good (so they may increase their likelihood of garnering an offer of employment) than when more job opportunities are available. This strategy facilitates an applicant-applicant comparison, which voids concerns that organizational experience or lack of motivation accounts for differences in faking.

Beyond applicant-employee or applicant-applicant comparisons, researchers have compared responses obtained from face-to face interactions (between test-taker and administrator) with scores obtained from computerized assessments, using both between- and within-subjects designs (e.g., Potosky & Bobko, 1997; Richman, Kiesler, Weisband, & Drasgow, 1999). The premise of these comparisons is that there will be less faking in the computerized administrations, because the perceived demand for socially desirable responding will be lower (than in a face-to-face interaction). Potosky and Bobko (1997) assessed socially desirable responding using scales administered by computer and those administered via paper-and-pencil. They found no difference in the level of response distortion across measures; they concluded each was similarly fakable. Richman et al. (1999) meta-analytically cumulated the results of comparisons between computerized and face-to-face interactions. They found that computerized test administrations resulted in less response distortion than did tests and interviews administered via paper-and-pencil or face-to-face. Note, however, that any difference between computerized administration and face-to-face traditional assessments could be a function of other variables, like the ease of presentation with computers. Despite research showing the measurement equivalence of computerized and traditional paper and pencil assessments of personality (e.g., Bartram & Brown, 2004; Lievens & Harris, 2003), equating the score differences between the two modes of administration to faking is a leap of faith.

Researchers have also compared score differences in anonymous and nonanonymous administrations (Dwight & Donovan, 2003; Rimland, 1962). Although between-subjects designs are commonly employed to investigate the effects of anonymity on response distortion, within-subjects designs can be used when the focus is only on the mean group difference (although such a strategy will lose the richness of assessing individual differences in fakability). Similar to comparing the mean scores from computer and traditional assessments, there is a leap of faith in attributing the differences between the anonymous and nonanonymous administrations to fakability. There is always the potential for more careless responding in anonymous administrations, for instance.

Other Techniques Employed to Assess Faking

Other techniques have been employed to assess the feasibility and incidence of faking. For example, attempts have been made to compare personality scores obtained under normal conditions to those obtained from personality assessments containing disguised purpose items. Such scales contain more subtle, less transparent items, and are thought to be more resistant to faking than counterpart measures with more obvious items (e.g., Becker & Colquitt, 1992; Kluger, Reilly, & Russell, 1991). Mueller-Hanson et al. (2003) have promised monetary rewards for faked (higher/lower) scores. James et al. (2005) has proposed the use of conditional reasoning to address the potential concerns of faking in personality assessments. Donovan, Dwight, and Hurtz (2003) proposed the use of a randomized response technique, whereby individual survey responses cannot be easily interpreted, thus reducing respondent tendency to respond in a socially desirable manner. Researchers (e.g., Martin, Bowen, & Hunt, 2002) have compared responses to scales where test-takers choose among answer options like "most like me" and "least like me." Forced choice scales are used in attempts to reduce faking (e.g., Maher, 1959). Several test cues have been hypothesized as addressing and/or estimating the potential faking in personality scores (Hough, 1998).

In sum, the data collection research designs used in the extant literature to assess faking can be broadly divided into those employing a within-subjects design and those opting for a between-subjects design. This classification can be utilized with different samples and techniques to produce a variety of data collection modes for faking research. Each mode has strengths and weaknesses with regard to the potential inferences that might be drawn regarding faking.

Some of the experimental combinations (e.g., use of student samples in random assignment with instructional sets) can produce a study with high internal validity to address the question whether individuals "can" fake personality tests. Collecting external outcomes on these groups can also give evidence of the effects of faking on the criterion-related validity and construct validity of personality tests. Unfortunately, little information is available regarding the external validity of these findings for the use of personality tests in organizations. It is difficult to address whether applicants fake during personnel selection (see also Griffith, Chmielowske, & Yoshita, in press, for a review). Research designed to collect data comparing applicants and employees under different experimental conditions (anonymous versus nonanonymous, computer versus traditional, etc.), are affected by confounding effects that preclude definitive conclusions about faking in high-stakes testing. In short, there seems to be no single logically tight data collection strategy that can be employed by researchers to simultaneously assess the incidence, feasibility, and consequences of faking in organizational settings. Rather, each strategy has strengths and weaknesses; triangulation of strategies will yield more valid results. Next, we turn to a review of analytic techniques employed in faking research.

Analytic Strategies Used in Faking Research

A number of analytic strategies have been proffered for assessing response distortion in faking research: (1) computing differences between groups of "honest" and "faked" scores (using effect size, point-biserial correlations), (2) controlling for social desirability response bias, (3) identifying incidence of faking using item response theory (IRT) models, and (4) examining response latencies as a means of isolating faked responses. Below, we review each analytic strategy.

The Role of Effect Sizes and Point-Biserial Correlations in Faking Research

To determine whether individuals can fake their responses in a personality test, researchers have computed the difference in mean responses between two groups, one group likely and one group unlikely to have faked responses. Specifically, researchers compute the effect size (the difference in group means divided by the standard deviation across responses; Cohen, 1977). This effect size is mathematically equivalent to computing the point-biserial correlation between group membership (e.g., honest responders or fakers) and personality scores. Statistical significance tests (t-tests) can also be conducted to test for significance of score differences between groups; these are easily converted to effect sizes (see Hunter & Schmidt, 2004, for computational formulae).

Other metrics have been developed to assess the effects of bias beyond the mean bias (cf. Kluger & Colella, 1993). Some researchers have argued in favor of assessing the effects of faking on the variability in scores, and perhaps more importantly, assessing whether faking increases true or error variance (Snell & McDaniel, 1998). Observed variance is the sum of error variance and true variance; score reliability is the ratio of true variance to error variance (cf. classical test theory; Nunnally, 1978). Thus, it is possible to obtain richer information about the effects of faking when variance and differences in reliability (across different administrations of personality tests- whether with a between- or within-subjects design) are examined in concert with mean differences. Beyond means and variances, researchers can also test the distributional overlap in personality scores under faking and without faking (Pedhazur & Schmelkin, 1991).

Statistically Controlling for Individual Differences in Faking

While the above analytic strategies are designed to address the question of whether individuals *can* fake in personality tests, until recently few solid suggestions were available regarding data collection or analytic strategies appropriate for assessing whether individuals actually *do* fake. A growing body of research purports a number of data collection strategies may be of use (e.g., self-report measures, mean-level differences, within-subjects designs conducted in field or laboratory settings; Donovan & Dwight, 2003; McDaniel, Douglas, & Snell, 1997; Stokes, Hogan, & Snell, 1993; Griffith, Chmielowski et al., in press). More extensive guidance is, however, available regarding useful analytic strategies for examining the effects of faking on criterion-related and construct validity of personality scores.

Researchers have compared the criterion-related validity under faking and honest response conditions (e.g., Hough et al., 1990). More commonly used is the strategy of measuring individual differences in faking (e.g., tendency toward socially desirable responding) and partialing out the effects of these individual differences from the correlation between predictor (i.e., personality test) and criterion (e.g., performance appraisal scores). Meta-analytic cumulation of these studies found that the reduction in the criterion-related validity due to partialing out social desirability scores is negligible (Ones, Viswesvaran, & Reiss, 1996). This finding has raised concerns about the robustness of the correlation coefficient to detect faking (Zickar & Drasgow, 1996). Rosse et al. (1998) found that although negligible reductions in correlation coefficients (between personality scores and criterion—the criterion-related validity) result from the statistical removal of variance associated with social desirability, the rank ordering of the individuals based on personality test scores is affected. When rank ordering is used for personnel selection, statistical control for social desirability will impact hiring decisions, especially in select-in conditions (i.e., using the personality test to determine a rank order of individuals to be considered for hire rather than for determining who will be eliminated from consideration—select-out) and in low selection ratio contexts (i.e., very competitive; Mueller-Hanson et al., 2003). In fact, Griffith, Rutkowsk et al. (2005) presents troubling data that suggests that social desirability and faking (as assessed with situational inducements) are negatively related.

The impact of social desirability on the construct validity of personality assessments has also been assessed. Personality test scores obtained under different faking contexts have been factor analyzed and multiple group factor analyses have tested the factor equivalence (cf. Griffith et al., 1997). The measurement equivalence under faking and honest response conditions has been explored using factor analysis (cf. Montag & Comrey, 1990; Schmit & Ryan, 1993). Researchers have modeled faking as a latent construct and tested whether the addition of this method factor adds to the fit of the model to data over a model that employs only substantive factors (i.e., does not consider individual differences in socially desirable responding). Ellingson, Sackett, and Hough (1999) have argued that the factor structure is affected due to faking; however, subsequent research casts doubt on the generalizability of such conclusions to real-world employment settings (Ellingson, Smith, & Sackett, 2001). The correlations between social desirability scale scores and personality constructs have been presented as evidence that social desirability scales measure a substantive personality variable (Hogan & Nicholson, 1988), although individual differences in social desirability do not predict criteria such as job performance (Hough & Ones, 2001).

Using Item Response Theory in Faking Research

Researchers have used item response theory (IRT) to model social desirability in personality assessments (cf. Zickar & Robie, 1999). When using IRT, researchers have the option of using one-, two- or three-parameter models. The one-parameter model assumes all items are of equal discriminability and guessing potential, differing only in difficulty level (i.e., strength of the trait). The two-parameter model also assumes equal guessing potential. Given the differences in obvious/transparent versus subtle/less transparent items, we think it would be more appropriate to use the three-parameter model in applying IRT models to faking research. Further, researchers have the option of dichotomizing the responses or employing polychotomous models. The richness of personality assessments is curtailed when responses to continuous items are dichotomized. Since most personality items measure an attitude that is more likely to be continuous rather than an either/or proposition, polychotomous models are more appropriate. However, three-parameter polychotomous models place greater demands on sample size than could be impossible in most organizational research.

IRT models can be used to flag invalid responses using appropriateness indices (Zickar & Drasgow, 1996) or to check for differential item functioning across groups. Stark et al. (2001) tested for differential item functioning of personality items between a group of 1,135 job applicants and a group of 1,023 nonapplicants who had taken the fifth edition of the 16PF and found that a substantial number of items exhibited differential item functioning. This is in contrast to the results of multigroup factor analyses that found substantial evidence of construct validity under faking conditions (cf. Ellingson et al., 2001). According to Stark et al., IRT models are superior to factor analyses because they make no assumptions about the underlying trait distributions; factor analyses assume both multivariate normality and a linear relationship between the items and the underlying factor.

Some caution has been raised regarding the use of IRT in faking research. Researchers (cf. Jensen, 1980 for a summary; see also Hunter, Schmidt, & Hunter, 1979) have found no evidence of bias in intelligence test items with different analyses (e.g., cleary regression model), but bias has been detected with differential item functioning. Hunter and Schmidt

(1996) surmise that differential item functioning in the absence of external bias in total scores is possible only if (1) there is differential item functioning in many items but the direction of favorability varies across items for groups such that they all cancel out in the total score, or (2) the individual items are more unreliable than the total score (such that DIF favoring one or other group is due to measurement error).

There is another caveat in using IRT models with personality items. IRT analyses assume local independence (that a response to an item is not influenced by preceding items). However, ample empirical evidence suggests individuals' responses to survey items are indeed influenced by preceding items (Bradburn, 1982; Knowles, 1988; Schuman & Presser, 1981; Schwarz & Sudman, 1992, for reviews). Tourangeau (1987) explained that the cognitive processes involved in responding to test items require that (1) the respondent understands the question (i.e., construct a schema), (2) the respondent recalls relevant beliefs, feelings, and opinions, (3) the respondent considers the recollected information and forms a judgment, and (4) the respondent chooses a response. So, consider a measure of the Big Five. The respondent is asked a variety of questions meant to assess conscientiousness, agreeableness, emotional stability, openness, and extraversion. Although individual items are meant to assess one of these five personality variables, as the individual considers each item, recollections relevant to other dimensions likely become salient, undoubtedly impacting answers to subsequent items. Therefore, when multiple personality items are presented, responses to one item will be influenced by the cognitive processes involved with responding to prior items. Consider an item that addresses conscientiousness of the individual. In responding to items on agreeableness the respondent would have remembered how s/he was reliable to friends, a recollection which will now affect the response to the item on conscientiousness. This is referred to as context effect in survey literature, and is a violation of the local independence assumption. More important, there could be individual differences in field dependence that could complicate the inferences from individual items.

Using Response Latencies to Identify Faking

Response latencies have been measured and linked to faking under the assumption that lying takes time (e.g., Holden, Wood, & Tomashewski, 2001). That is, individuals will take more time to respond to an item when they are trying to fake it since they will have to identify their own response as well as the desirable response, compare the two, and then make a choice. Researchers have assessed response latencies after controlling for item complexity, length, format, and so forth, and have confirmed that when respondents have been instructed to fake, they take longer to answer items (Brunetti, Schlottmann, Scott, & Hollrah, 1998). One criticism of the use of response latencies to detect faking is based on the premise when individuals try to fake, they merely provide socially desirable responses without giving much thought about identifying their "true" response. Therefore, they need only identify one response—the socially desirable response—rather than two responses, their true response and the more the desirable one.

Using Social Desirability Scales in Faking Research

Social desirability scales have played an important role in the data collection and analytic strategies employed in faking research. Social desirability scales were designed to assess individual differences in faking. To the extent all individuals engage in the same level of faking, each individual's score will be raised by a constant number. If so, norms may have to be revised as well as cut-scores and criterion-referenced interpretations. However, the relative rank ordering of individuals will remain constant and several organizational decisions can be made based on such distorted scores (Guion, 1998; Ones et al., 1996). Unfortunately, there is ample empirical evidence that there are individual differences in both ability and motivation to fake (cf. McFarland & Ryan, 2000).

Thus, almost all commercially sold personality tests include several scales to measure individual differences in and different aspects of faking. In fact, technical manuals of personality tests describe several indices that can be formed by combining the different scales. In the next section, we describe the faking indices utilized in two commercially available personality tests, the Minnesota Multiphasic Personality Inventory and the California Personality Inventory, and two independent scales popular to faking research, the Balanced Inventory of Desirable Responding and the Marlowe-Crowne Social Desirability Scale.

Commonly Used Scales in Faking Research

Understanding the potential for respondent faking in personality tests, and the importance of accurate responses (especially within clinical populations) to evaluation and prediction, the developers of Minnesota Multiphasic Personality Inventory (MMPI) and the California Personality Inventory (CPI) embedded scales within these tests to aid test administrators in the identification of inaccurate, incomplete, or invalid responses. Given the clear relevance and prevalence of these scales in faking research, we review each of the validity scales and associated indices embedded within these commercial tests. In addition, we review two independent scales commonly used to identify and statistically control for socially desirable responding in faking research. Specifically, we discuss the Balanced Inventory of Desirable Responding (BIDR) developed by Paulhus (1984) and the Marlowe-Crowne Social Desirability Scale (MC-SDS; Crowne & Marlowe, 1960) that are widely used in faking research.

Minnesota Multiphasic Personality Inventory

The Minnesota Multiphasic Personality Inventory was initially developed in the 1940s to assess individual emotional adjustment and attitude toward test taking. Currently, the MMPI is the most widely used clinical personality inventory, with over 10,000 published research references (Boccaccini & Brodsky, 1999; Groth-Marnat, 1999). While clinical personality instruments are inappropriate in an applicant setting due to concerns raised by the Americans With Disabilities Act, the MMPI is frequently used in postconditional offers of law enforcement and security personnel. Thus an examination of the MMPI validity scales addresses applicant faking, albeit in a different context. In this section, we describe the scales used in the MMPI to detect the invalidity of responses. We refer primarily to the MMPI-2 and MMPI-A versions. The original MMPI was composed of 13 scales, 10 clinical scales and 3 validity scales. The MMPI was the first personality instrument to embed a scale to detect invalid response patterns.

The most current version of the MMPI provides 22 different indices to assess whether an individual is attempting to distort the responses. These different indices are summarized in Table 4.1. These indices are based on the eight validity scales embedded within the MMPI: Cannot say (?), Variable Response Inconsistency (VRIN), True Response Inconsistency (TRIN), Infrequency (F-scale), Fb (F-Back), Lie (L), Correction (K), and Superlative Self-Presentation (S) scales.

The Cannot Say (?) scale captures the number of unanswered items left blank by the respondent. A high number (typically 30 or more) indicates an invalid profile as insufficient information has been made available to facilitate meaningful inferences. The VRIN and TRIN are indicators of the test-taker's propensity to respond to items in ways that are inconsistent or contradictory with previous responses. These scales use pairs of items and assess the extent of agreement. The VRIN scale contains items that should both either be answered true or be answered false. Inconsistency indicates the profile is of questionable validity. Item pairs in the TRIN scale are made of items that are opposite in content. If a test-taker answers false to both items, then a point is subtracted and if the test-taker answers true to both items in a pair, a point is added. Thus, a high score indicates an acquiescence bias (tendency to agree) and a low score suggests a nay-saying tendency (tendency to disagree).

The F-scale identifies respondents who have selected a number of responses that are reflective of atypical or nonconventional thinking

Table 4.1.   MMPI—Guide to the Interpretation of Validity Indices

Validity Index	Manual Guideline	Possible Reasons
Cannot Say Score > 29	May be invalid	Reading difficulties, severe psychopathology, lack of insight, uncooperative, obsessive
Cannot say Score between 11 & 29	Some scales may be invalid	Selective item omission
VRIN > 79	Profile invalid	Reading difficulties, confusion, intentional random responding, error in recording responses
VRIN between 65-79 (inclusive)	Profile is valid; however, it is characterized by some inconsistent responding	Carelessness, occasional loss of concentration
TRIN > 79T	Profile is invalid	Acquiescence response set
TRIN between 65T-79T (inclusive)	Profile is valid; however, it is characterized by some acquiescence	Partial acquiescence response set
TRIN > 79F	Profile is invalid	Nonacquiescent response set
TRIN between 65F-79F (inclusive)	Profile is valid; however, it is characterized by some nonacquiescence	Partial nonacquiescent response set
F-Scale > 79	May be invalid	Random/fixed responding, severe psychopathology, faking bad
F-Scale < 40	May be defensive	
F-Scale between 65-79 (inclusive)	Maybe exaggerated, but likely is valid	Exaggeration of existing problems
Fb-Scale > 89	May be invalid	Random/fixed responding, severe psychopathology, faking bad, change in responding
L-Scale > 79	Likely invalid	Faking good, pervasive acquiescence
L-Scale between 70-79 (inclusive)	May be invalid	Moderate faking good, moderate nonacquiescence
L-Scale between 65-69 (inclusive)	Questionably valid	Overly positive self-presentation
L-Scale between 60-64 (inclusive)	Likely valid	Unsophisticated defensiveness
K-Scale > 74	May be invalid	Faking good, pervasive nonacquiescence
K-Scale between 65-74 (inclusive)	May be invalid	Moderate defensiveness, moderate non acquiescence
K-Scale < 40	May be invalid	Faking bad, pervasive acquiescence
S-Scale > 74	May be invalid	Faking good, pervasive nonacquiescence
S-Scale between 70 & 74 (inclusive)	May be invalid	Moderate defensiveness, moderate nonacquiescence

within a normal population. These items were infrequently endorsed by the norm sample. Thus, high scores on the F-scale suggest the respondent is potentially engaging in random responding. In more recent versions of the MMPI, additional scales were added, lengthening the inventory. The original F-scale assessed random responding only for the first 370 items. To check whether such random responding occurs towards the end of the inventory, the F_B scale was created. Substantial difference between the F-scale and the F_B scale indicates the possibility of respondent fatigue.

The L-scale consists of 15 items and assesses the extent to which the respondent is attempting to present an overly (unrealistically) positive image of himself. In other words, this scale purports to identify individuals engaging in blatant or unsophisticated faking. The K-scale attempts to identify less blatant forms of faking. While a high score on the L-scale indicates the respondent has painted an unrealistically positive picture of himself, the K-scale identifies more sophisticated efforts at faking, wherein the respondent creates an overly positive (though not necessarily impossible) image. Finally, the S-scale employed a content-oriented strategy (most MMPI scales are empirically derived by contrast group analysis) to identify a group of extremely defensive respondents.

It is important to note that triggering one validity index does not automatically disqualify a respondent. The manual merely suggests caution should be used when interpreting the content scale scores (Butcher Graham, Ben-Porath, Tellegen, Dahlstrom, & Kaemmer, 2001). Nevertheless, the interpretation of scores in a selection or any other high stakes context becomes problematic (Sackett, Schmitt, Ellingson, & Kabin, 2001). In addition to these 22 validity indices reported in Table 4.1, the MMPI manual (Butcher et al., 2001) also presents several combinations of these validity scores to aid in the interpretation of content scores. For example, an S-Scale T-Score greater than 74 may indicate faking good or pervasive nonacquiescence. An S-Scale T-Score greater than 74 and a TRIN score greater than 79F narrows the interpretation to one of pervasive non acquiescence.

California Psychological Inventory

The California Psychological Inventory (CPI) is a paper-and-pencil, self-administered personality inventory intended for use within the normal adult population. The inventory is comprised of 434 true-false statements regarding an individual's typical behavior patterns, feelings, opinions, and ethical and social attitudes. While the MMPI was developed in clinical settings (though used in different settings later), the CPI has been normed with the normal adult population and has been translated into 40 languages and used in many contexts (Gough & Bradley, 1996). The CPI is frequently used for career development, personnel selection, and prediction of antisocial behavior (Groth-Marnat, 1999). Like the MMPI, the CPI has embedded scales to detect faking. Gough (1987) estimated that in large-scale career development testing situations approximately 1.7% of profiles will be invalid (.6% due to respondents faking good, .4% due to respondents faking bad, and .7% due to random responding; Groth-Marnat, 1999). It should be noted that the percentage of invalid profiles potentially generated in applicant (rather than career development) conditions may be quite different.

Three validity scales are embedded within the CPI to detect response distortion or invalid responding: the Good Impression scale, the Communality scale, and the Sense of Well-Being scale. The Good Impression scale purports to identify respondents "faking good." Specifically, it reflects the extent to which a respondent is concerned with making a favorable impression. Items assess goal-orientation, feelings of personal failure, antisocial tendencies, socially desirable behaviors, and harmonious relationships with others.

Random answering is detected by low scores on the Communality scale or by misalignment of test answers within the test booklet. Specifically, 95% of the items contained in the Communality scale are answered consistently. Respondents deviating from these patterns are thought to be randomly answering. Communality scale items assess good socialization, conformity, optimism, denial of neurotic traits, and conventionality of attitudes (Gough, 1987; Groth-Marnat, 1999).

The Sense of Well-Being scale was developed to identify respondents "faking bad." High scorers on this scale over-emphasize the extent to which they are enterprising, energetic, and have a high sense of security, and de-emphasize their worries and negative self-evaluations. High scorers on this scale may be faking good. Low scorers, on the other hand, emphasize diminished health and related that they have extensive difficulty meeting daily demands (Groth-Marnat, 1999). Importantly, Gough & Bradley (1986) admit that, at times, it may be difficult to differentiate between a profile that reflects exceptional psychological adjustment and one indicative of faking (cf. Gough, 1987).

Research suggests the CPI validity scales are successful in detecting more general forms of respondent faking, wherein the respondent attempts to create an overall more positive or more negative impression (e.g., "fake-good" or "fake-bad"; Dicken, 1960; Sandal & Enresen, 2002). However, in scenarios wherein an individual attempts to distort a specific attribute of personality, these scales have had somewhat less success in detecting the distortion. For example, Montross, Neas, Smith, and Hensley (1988) instructed volunteers to role-play a higher level of gender identification than their "honest profiles" indicated (males were instructed to fake higher masculinity; females were instructed to fake higher femininity). Neither the Good Impression nor the Sense of Well-Being scales were able to detect this form of faking.

Balanced Inventory of Desirable Responding

The Balanced Inventory of Desirable Responding (BIDR) was developed by Paulhus (1984) to assess tendency for respondents to engage in socially desirable responding. Factor analytic studies of social desirability have supported the partitioning of socially desirable response styles into two clusters: Alpha (Block, 1965) and Gamma (Wiggins, 1964). Alpha, the general adjustment factor of the MMPI, represents the tendency to give desirable self-reports on measures of personality (Edwards, 1957). Paulhus (1984) labeled this component of social desirability "self-deception." Gamma, the factor associated with "propagandistic bias" (Damarin & Messick, 1965), represents the distortion of self-descriptions to be consistent with those expected by a specific audience. Paulhus (1984) labeled this factor "impression management." The BIDR assesses both components of socially desirable response styles (self-deception and impression management). Self-deception has been termed an "egoistic bias"; impression management a "moralistic bias" (Paulhus & John, 1998). Specifically, self-deception is characterized by an overconfidence in one's social and intellectual abilities; impression management by claims of a heightened conformity to societal and moral norms and a rejection of deviant impulses (Paulhus & John, 1998; Peterson et al., 2003).

Paulhus (1986) and Paulhus and Reid (1991) further differentiated the self-deception response style into two components: (1) one component relates to the attribution of positive attributes (self-deceptive enhancement) and (2) the other component relates to the denial of negative attributes (self-deceptive denial). Subsequent factor analyses of these items confirmed a two-factor model for self-deception, with an intercorrelation between factors ranging from .10 to .19 (Kroner & Weekes, 1996; Paulhus & Reid, 1991; Pauls & Stemmler, 2003; Roth, Harris, & Snyder, 1988).

Since its creation, seven versions of the BIDR have been published. The most current form of the BIDR (BIDR-7; Paulhus & John, 1998) consists of three subscales, each comprised of 20-items: Self-Deceptive Enhancement (e.g., "I am a completely rational person."), Impression Management (e.g., "I never read sexy books or magazines."), and Self-Deceptive Denial (e.g., "Once I've made up my mind, other people can seldom change my opinion."). The Self-Deceptive Enhancement and Denial scales assess the two forms of self-deception; the Impression Management scale assesses the respondents' tendency to fake-good (i.e., to consciously exaggerate good qualities while simultaneously playing down less-desirable/bad qualities). Individuals who score high on the self-deception subscales tend to ignore evidence that implies their abilities are less impressive than they believe or that their moral/ethical standards are lower than they believe (Peterson et al., 2003).

Responses to the BIDR are made on a 7-point Likert-type scale, ranging from not true to very true. Only extreme responses (i.e., 6 and 7, or 1 and 2 for reverse-scored items) are scored (Paulhus & Reid, 1991). Reports suggest the BIDR has a high degree of internal consistency (Coefficient alpha = .83) and adequate reliability (test-retest; .65 -.69; Paulhus & Reid, 1991). In a study by Pauls and Stemmler (2003), scores on the BIDR and a measure of the Big Five were correlated. Self-ratings of self-deception correlated with neuroticism ($r = -.39$) and extraversion ($r = .36$; Pauls & Stemmler, 2003); impression management correlated with agreeableness ($r = .30$) and conscientiousness ($r = .41$). The BIDR did not significantly correlate with openness to experience. Researchers have reported that scores on the self-deception subscales are negatively correlated with depression ($r = -.49$; Holden, Starzyk, McLeod, & Edwards, 2000) and are useful for predicting performance in "caring" professions (e.g., health care, social work; Pauls & Stemmler, 2003). Conversely, scores on the impression management subscale are negatively related to antisocial behavior and low impulse control ($r = -.49$; Holden et al., 2000), and are useful for predicting performance in executive-level positions (Pauls & Stemmler, 2003).

Marlowe-Crowne Social Desirability Scale

Another instrument widely used to identify respondent faking is the Marlowe-Crowne Social Desirability Scale (MC-SDS; Crowne & Marlowe, 1960). Beretvas, Meyers, and Leite (2002) identified over 1,000 articles and dissertations that had employed the MC-SDS since its inception. The second independent social desirability scale developed, the MC-SDS was published in 1960 (Crowne & Marlowe, 1960). (The first such scale, the Edwards Social Desirability Scale, used items similar to those contained in the MMPI and was therefore highly associated with psychopathology; Leite & Beretvas, 2005). The MC-SDS was created specifically to avoid an association with psychopathology. The full version of the MC-SDS is comprised of 33-items. Respondents answer items as being either true or false of their typical behaviors, attitudes, or actions. An example item is "I never hesitate to go out of my way to help someone in trouble." Crowne and Marlowe (1960) reported high internal scale reliability (Kuder-Richardson, .88) and a high test-retest reliability of .89.

The MC-SDS assumes that socially desirable response bias represents a single, latent construct (Leite & Beretvas, 2005). The single-factor structure of the full 33-item scale has not been confirmed, but confirmatory factor analyses of short-forms of the MC-SDS (e.g., Reynolds', 1982,

Forms A, B, C and Strahan and Gerbasi's, 1972, Forms X1 and X2) have yielded good fits with single-factor models. Some authors argue a two-factor structure (similar to the preliminary versions of the BIDR) of the MC-SDS is more appropriate: wherein one factor indicates the attribution of socially desirable qualities and the other factors indicates the denial of socially undesirable characteristics (Ramanaiah, Schill, & Lock-Sing, 1977). Scores on the MC-SDS are correlated with neuroticism, extraversion and openness to experience (McCrae & Costa, 1983).

Researchers often collect information relevant to socially desirable response tendencies (using, for example, the BIDR and MC-SDS) in order to control for the deleterious effects of such response distortion on the predictive validity of personality and other noncognitive measures for organizational decision-making. McCrae and Costa (1983) assessed the feasibility of regarding social desirability scales as indices of invalid responding. They concluded that although items may be characterized as high or low in social desirability, there is little evidence to suggest that individuals attend to the social desirability of items when making responses. Further, when personality-performance correlations were corrected for scores on the MC-SDS, validity coefficients actually decreased in most cases. They caution against the practice of correcting personality test scores for faking by covarying out the effects of social desirability. This conclusion has received widespread support in the extant literature (i.e., Christiansen, Goffin, Johnston, & Rothstein, 1994; Leite & Beretvas, 2005; Ones et al., 1996).

In sum, a number of scales are available to researchers interested in examining social desirability bias and identifying respondent faking. We reviewed four inventories widely used in faking research: two commercial tests with embedded validity indices (the MMPI and the CPI) and two independent social desirability scales (the MCSDS and the BIDR). It is widely accepted that these scales are successful in identifying response patterns consistent with socially desirable responding (e.g., Dicken, 1960; Sandal & Endresen, 2002; Sivec et al., 1995; Wetter, Baer, Berry, Robinson, & Sumpter, 1993). However, researchers have suggested social desirability should be examined as a substantive personality variable rather than assessed solely for statistical control purposes (e.g., McCrae & Costa, 1983).

Summary and Conclusions

There is little debate that personality is a valuable predictor of outcomes important to organizational performance (e.g., job and training performance, job involvement, job satisfaction, organizational commitment, absenteeism, etc.; e.g., Barrick & Mount, 1991; Ones et al., 1993;

Tett et al., 1991). Despite increased optimism regarding the use of personality tests for personnel selection (Hough & Oswald, 2000) to concerns regarding the incidence, prevalence, and consequences of response distortion in these measures continue to be raised (Luther & Thornton, 1999; Mueller-Hanson et al., 2003; Zickar, Rosse, Levin, & Hulin, 1996).

Many research attempts have tried to answer key questions about applicant faking: (1) Can applicants fake their responses to personality measures?, (2) Do applicants fake?, and (3) What are the consequences of faking to the quality of selection decisions made based on these personality measures? These investigations have used a number of research strategies (i.e., within-subjects designs and between-subjects designs, and comparisons between applicants and nonapplicants, anonymous and nonanonymous responses, warning versus no-warning conditions, computer versus face-to-face administrations) and a variety of analytic strategies (i.e., effect size or point-biserial computations, models obtained using item response theory, examination of response latencies, and statistical control for social desirability).

An abundance of evidence supports the conclusion that respondents who are instructed to fake are able to distort their responses to personality inventories (Donovan et al., 2003; Dunnett, Koun, & Barber, 1981; Furnham & Craig, 1987; Hinrichsen, Gryll, Bradley, & Katahn, 1975; Hough et al., 1990; Schwab, 1971; Thornton & Gierasch, 1980; Viswesvaran & Ones, 1999). If applicants fake, the impact of such faking to the predictive validity of personality measures for personnel selection appears to be negligible (Barrick & Mount, 1991; Christiansen et al., 1994; Hough et al., 1990; Ones et al., 1996; Ones et al., 1993; Viswesvaran & Ones, 1999), however, some researchers have suggested that the impacts of faking may be larger when personality tests are used in a select-in approach rather than a select-out approach (Mueller-Hanson et al., 2003), as faking may alter the ranking of applicants, thus yielding less favorable selection decisions. However, just because an applicant can fake answers to personality assessments does not necessarily mean they will fake. The research question of whether applicants actually do fake is the most difficult to answer in that the designs are often fraught with threats to validity. However, when these techniques are triangulated (e.g., Self-reports, mean level differences, and within-subjects field designs) research conclusions lend support to the notion that faking does occur (cf. Donovan et al., 2003; Griffith, Chmielowski et al., in press; Stokes et al., 1993). Importantly, although it is clear that faking occurs, research suggests its prevalence is not nearly as high as once thought (Donovan et al., 2003; Griffith, Chmielowski, et al., in press; Ones & Viswesvaran, 1998; Rosse et al., 1998; Viswesvaran & Ones, 1999).

Finally, a number of techniques (e.g., the randomized response technique; Donovan et al., 2003) and scales are available to aid researchers in identifying incidence of response distortion and socially desirable responding in their research. While some scales are embedded within larger commercial personality inventories more commonly used within clinical populations (i.e., MMPI), Paulhus' Balanced Inventory of Desirable Responding and the Marlowe-Crowne Social Desirability Scale are frequently used in faking research and are easily incorporated into research related to organizational behavior. Concerns have been raised, however, about the efficacy of correcting for faking in personality instruments by statistically controlling for social desirability (cf. McCrae & Costa, 1983). Faking is an important question that is likely to generate substantial research in the coming years. By reviewing the different data collection and analytic strategies, this chapter summarizes the advantages and disadvantages of the different choices available to the researchers.

AUTHOR'S NOTE

Address correspondence regarding this manuscript to Jessica Mesmer-Magnus, Department of Management & Marketing, UNC-Wilmington, 601 South College Road, Wilmington, NC 28403; Email: magnusJ@uncw.edu or to Chockalingam Viswesvaran, Department of Psychology, Florida International University, Miami, FL 33199; Email: Vish@fiu.edu

REFERENCES

Barrick, M. R., & Mount, M. K. (1991). The Big Five personality dimensions and job performance: A meta-analysis. *Personnel Psychology, 44*, 1-26.

Bartram, D., & Brown, A. (2004). Online testing: Mode of administration and the stability of OPQ 32i scores. *International Journal of Selection and Assessment, 12*, 278-284.

Becker, T. E., & Colquitt, A. L. (1992). Potential versus actual faking of a biodata form: An analysis along several dimensions of item type. *Personnel Psychology, 45*, 389-406.

Beretvas, S. N., Meyers, J. L., & Leite, W. L. (2002). A reliability generalization study of the Marlowe-Crowne Social Desirability Scale. *Educational and Psychological Measurement, 64*(4), 570-589.

Block, J. (1965). *The challenge of response sets.* New York: Appleton.

Boccaccini, M. T., & Brodsky, S. L. (1999). Diagnostic test usage by forensic psychologists in emotional injury cases. *Professional Psychology: Research and Practice, 30*, 252-259.

Bradburn, N. (1982). Question wording effects in surveys. In R. Hogarth (Ed.), *Question framing and response consistency* (pp. 65-76). San Francisco: Jossey-Bass.

Braver, M. C. W., & Braver, S. L. (1988). Statistical treatment of the Solomon Four-Group Design: A meta-analytic approach. *Psychological Bulletin, 104*, 150-154.

Brunetti, D. G., Schlottmann, R. S., Scott, A. B., & Hollrah, J. L. (1998). Instructed faking and MMPI-2 response latencies: The potential for assessing response validity. *Journal of Clinical Psychology, 54*(4), 143-153.

Butcher, J. M., Graham, J. R., Ben-Porath, Y. S., Tellegen, A., Dahlstrom, W. G., & Kaemmer, B. (2001). *MMPI-2^{TM} (Minnesota Multiphasic Personality Inventory-2 ^{TM}): Manual for administration, scoring, and interpretation* (Rev. ed.). Minneapolis: University of Minnesota Press.

Campbell, D. T., & Stanley, J. C. (1963). Experimental and quasi-experimental designs for research in teaching. In N. L. Gage (Ed.), *Handbook of research on teaching* (pp. 171-246). Chicago: Rand McNally.

Christiansen, N. D., Goffin, R. D., Johnston, N. G., & Rothstein, M. G. (1994). Correcting the 16PF for faking: Effects on criterion-related validity and individual hiring decision. *Personnel Psychology, 47*, 847-860.

Cohen, J. (1977). Statistical power analysis for the behavioral sciences. New York: Academic Press.

Cook, T. D., & Campbell, D. T. (1979). *Quasi-experimentation: Design and analysis issues for field settings.* Boston: Houghton Mifflin.

Crowne, D. P., & Marlowe, D. (1960). A new scale of social desirability independent of psychopathology. *Journal of Consulting Psychology, 24*, 349-354.

Damarin, F., & Messick, S. (1965). *Response styles as personality variables: A theoretical integration.* Princeton, NJ: Educational Testing Service.

Dicken, C. F. (1960). Simulated patterns on the California Psychological Inventory. *Journal of Counseling Psychology, 7*(1), 24-31.

Donovan, J. J., Dwight, S. A., & Hurtz, G. M. (2003). An assessment of the prevalence, severity, and verifiability of entry-level applicant faking using the randomized response technique. *Human Performance, 16*(1), 81-106.

Douglas, E. F., McDaniel, M. A., & Snell, A. F. (1996, August). *The validity of non-cognitive measures decays when applicants fake.* Paper presented at the annual meeting of the Academy of Management, Cincinnati, OH.

Dunnett, S., Koun, S., & Barber, P. (1981). Social desirability in the Eysenck Personality Inventory. *British Journal of Psychology, 72*, 19-26.

Dwight, S. A., & Donovan, J. J. (2003). Do warnings not to fake reduce faking? *Human Performance, 16*(1), 1-23.

Edwards, A. L. (1957). *The social desirability variables in personality assessment and research.* New York: Dryden Press.

Ellingson, J. E., Sackett, P. R., & Hough, L. M. (1999). Social desirability corrections in personality measurement: Issues of applicant comparison and construct validity. *Journal of Applied Psychology, 84*, 215-224.

Ellingson, J. E., Smith, D. B., & Sackett, P. R. (2001). Investigating the influence of social desirability on personality factor structure. *Journal of Applied Psychology, 86*, 122-133.

Furnham, A., & Craig, S. (1987). Fakability and correlates of the perception and preference inventory. *Personality and Individual Differences, 8,* 459-470.

Gough, H. G. (1987). *California Psychological Inventory: Administrator's guide.* Palo Alto, CA: Consulting Psychologists Press.

Gough, H. G., & Bradley, P. (1986). *California Psychological Inventory: Manual* (3rd Ed.). Palo Alto, CA: Consulting Psychologists Press.

Griffith, R. L., Chmielowski, T., & Yoshita, Y. (in press). Do applicants fake? An Examination of the Frequency of Applicant Faking Behavior. *Personnel Review.*

Griffith, R. L., Frei, R. L., Snell, A. F., Hamill, L. S., & Wheeler, J. K. (1997, April). *Warnings versus no-warnings: Differential effects of method bias.* Paper presented at the 12th annual meeting of the Society for Industrial and Organizational Psychology, St. Louis, MD.

Griffith, R. L., Rutkowski, K., Gujar, A., Yoshita, Y. & Steelman, L. (2005). *Modeling applicant faking: New methods to examine an old problem.* Manuscript submitted for publication.

Groth-Marnat, G. (1999). *Handbook of Psychological Assessment* (3rd ed.). New York: Wiley.

Guion, R. M. (1998). *Assessment, measurement, and prediction for personnel decisions.* Mahwah, NJ: Erlbaum.

Hausknecht, J. P., Trevor, C. O., & Farr, J. L. (2002). Retaking ability tests in a selection setting: Implications for practice effects, training performance, and turnover. *Journal of Applied Psychology, 87,* 243-254.

Hinrichsen, J. J., Gryll, S. L., Bradley, L. A., & Katahn, M. (1975). Effects of impression management efforts on FIRO-B profiles. *Journal of Consulting and Clinical Psychology, 43,* 269.

Hogan, R., Hogan, J., & Roberts, B. W. (1996). Personality measurement and employment decisions. *American Psychologist, 51,* 469-477.

Hogan, R., & Nicholson, R. A. (1988). The meaning of personality test scores. *American Psychologist, 43,* 621-626.

Holden, R. R., Starzyk, K. B., McLeod, L. D., & Edwards, M. J. (2000). Comparison among the Holden Psychological Screening Inventory (HPSI), the Brief Symptom Inventory (BSI), and the Balanced Inventory of Desirable Responding (BIDR). *Assessment, 7,* 163-175.

Holden, R. R., Wood, L. L., & Tomashewski, L. (2001). Do response time limitations counteract the effect of faking on personality inventory validity? *Journal of Personality and Social Psychology, 81*(1), 160-169.

Hough, L. M. (1998). Effects of intentional distortion in personality measurement and evaluation of suggested palliatives. *Human Performance, 11,* 209-244.

Hough, L. M., Eaton, N. K., Dunnette, M. D., Kamp, J. D., & McCloy, R. A. (1990). Criterion-related validity of personality constructs and the effect of response distortion on those validities. *Journal of Applied Psychology, 75,* 581-595.

Hough, L. M., & Ones, D. S. (2001). The structure, measurement, validity, and use of personality variables in industrial, work, and organizational psychology. In N. Anderson, D. S. Ones, H. Sinangil, & C. Viswesvaran (Eds.), *Handbook of industrial, work, and organizational psychology* (Vol. 1, pp. 233-277), London: Sage.

Hough, L. M., & Oswald, F. L. (2000). Personnel selection: Looking toward the future- Remembering the past. *Annual Review of Psychology, 51*, 631-664.

Huck, S. W., & Sandler, H. M. (1973). A note on the Solomon 4-Group Design: Appropriate statistical analyses. *The Journal of Experimental Education, 42*, 54-55.

Hunter, J. E., & Schmidt, F. L. (2004). *Methods of meta-analysis: Correcting for error and bias in research findings.* Newbury Park, CA: Sage.

Hunter, J. E., & Schmidt, F. L. (1996). Intelligence and job performance: Economic and social implications. *Psychology, Public Policy and Law, 2*, 447-472.

Hunter, J. E., Schmidt, F. L., & Hunter, R. (1979). Differential validity of employment tests by race: A comprehensive review and analysis. *Psychological Bulletin, 86*, 721-735.

Hurtz, G. M., & Donovan, J. J. (2000). Personality and job performance: The Big Five revisited. *Journal of Applied Psychology, 85*, 869-879.

Jensen, A. R. (1980). *Bias in mental testing.* New York: Free Press.

James, L. R., McIntyre, M. D., Glisson, C. A., Green, P. D., Patton, T. W., LeBreton, J. M., et al. (2005) A conditional reasoning measure for aggression. *Organizational Research Methods, 8*, 69-99.

Kelley, P. L., Jacobs, R. R., & Farr, J. L. (1994). Effects of multiple administrations of the MMPI for employee screening. *Personnel Psychology, 47*, 575-591.

Kluger, A. N., & Colella, A. (1993). Beyond the mean bias: The effect of warning against faking on biodata item variances. *Personnel Psychology, 46*, 763-780.

Kluger, A. N., Reilly, R. R., & Russell, C. J. (1991). Faking biodata tests: Are option-keyed instruments more resistant? *Journal of Applied Psychology, 76*, 889-896.

Knowles, E. S. (1988). Item context effects on personality scales: Measuring changes the measure. *Journal of Personality and Social Psychology, 55*(2), 312-320.

Kroner, D. G., & Weekes, J. R. (1996). Balanced Inventory of Desirable Responding: Factor structure, reliability, and validity with an offender sample. *Personality and Individual Differences, 21*(3), 323-333.

Kulik, J. A., Kulik, C. C., & Bangert, R. L. (1984). Effects of practice on aptitude and achievement test scores. *American Educational Research Journal, 21*, 435-447.

Leite, W. L., & Beretvas, S. N. (2005). Validation of scores on the Marlowe-Crowne Social Desirability Scale and the Balanced Inventory of Desirable Responding. *Educational and Psychological Measurement, 65*(1), 140-154.

Levin, R. A. (1995, May). *Self-presentation, lies, and bullshit: The impact of impression management on employee selection.* Paper presented at the annual meeting of the Society for Industrial and Organizational Psychology, Orlando, FL.

Lievens, F., & Harris, M. M. (2003). Research on internet recruitment and testing: Current status and future directions. In I. Robertson & C. Cooper (Eds.), *The international review of industrial and organizational psychology.* Chichester, England: Wiley.

Luther, N. J., & Thornton, G. C., III (1999). Does faking on employment tests matter? *Employment Testing Law and Policy Reporter, 8*, 129-136.

Maher, H. (1959). Follow-up on the validity of a forced-choice study activity questionnaire in another setting. *Journal of Applied Psychology, 43,* 293-295.

Martin, B. A., Bowen, C. C., & Hunt, S. T. (2002). How effective are people at faking on personality questionnaires? *Personality and Individual Differences, 32,* 247-256.

McCrae, R. R., & Costa, P. T., Jr. (1983). Social desirability scales: More substance than style. *Journal of Consulting and Clinical Psychology, 51*(6), 882-888.

McDaniel, M. A., Douglas, E. F., & Snell, A. F. (1997, April). *A survey of deception among job seekers.* Paper presented at the 12th annual conference of the Society for Industrial and Organizational Psychology, St. Louis, MO.

McFarland, L. A., & Ryan, A. M. (2000). Variance in faking across noncognitive measures. *Journal of Applied Psychology, 85,* 812-821.

Mersman, J. L. & Shultz, K. S. (1998). Individual differences in the ability to fake on personality measures. *Personality and Individual Differences. 24,* 217-227.

Montag, L., Comrey, A. L. (1990). Stability of major personality factors under changing motivational conditions. *Journal of Social Behavior and Personality, 5,* 265-274.

Montross, J. F., Neas, F., Smith, C. L., & Hensley, J. H. (1988). The effects of role-playing high gender identification on the California Psychological Inventory. *Journal of Clinical Psychology, 44*(2), 160-164.

Mook, D. G. (1983). In defense of external invalidity. *American Psychologist, 38*(4), 379-387.

Mueller-Hanson, R., Heggstad, E. D., & Thornton, G. C. III (2003). Faking and selection: Considering the use of personality from select-in and select-out perspectives. *Journal of Applied Psychology, 88*(2), 348-355.

Murphy, K. R., & Dzieweczynski, J. L. (2005). Why don't measures of broad dimensions of personality perform better as predictors of job performance? *Human Performance, 18*(4), 343-357.

Nunnally, J. C. (1978). *Psychometric theory.* New York: McGraw-Hill.

Ones, D. S., & Viswesvaran, C. (1998). The effects of social desirability and faking on personality and integrity assessment for personnel selection. *Human Performance, 11*(2/3), 245-269.

Ones, D. S., & Viswesvaran, C. (2001, August). *Job applicant response distortion on personality scale scores: Labor market influences.* Paper presented at the 109th annual conference of the American Psychological Association, San Francisco, CA.

Ones, D. S., Viswesvaran, C., & Dilchert, S. (2005). Personality at work: Raising awareness and correcting misconceptions. *Human Performance, 18*(4), 389-404.

Ones, D. S., Viswesvaran, C., & Reiss, A. D. (1996). Role of social desirability in personality testing for personnel selection: The red herring. *Journal of Applied Psychology, 81,* 660-679.

Ones, D. S., Viswesvaran, C., & Schmidt, F. L. (1993). Comprehensive meta-analysis of integrity test validities: Findings and implications for personnel selection and theories of job performance [Monograph]. *Journal of Applied Psychology, 78,* 679-703.

Paulhus, D. L. (1984). Two-component models of socially desirable responding. *Journal of Personality & Social Psychology, 46,* 598-609.

Paulhus, D. L. (1986). Self-deception and impression management in test responses. In A. Angleitner & J. S. Wiggins (Eds.), *Personality assessment via questionnaire* (pp. 142 -165). New York: Springer.

Paulhus, D. L., & John, O. P. (1998). Egoistic and moralistic biases in self-perception: The interplay of self-deceptive styles with basic traits and motives. *Journal of Personality, 66*(6), 1025-1060.

Paulhus, D. L., & Reid, D. B. (1991). Enhancement and denial in socially desirable responding. *Journal of Personality and Social Psychology, 60,* 307-317.

Pauls, C. A., & Stemmler, G. (2003). Substance and bias in social desirability responding. *Personality and Individual Differences, 35,* 263-275.

Pedhazur, E. J., & Schmelkin, L. P. (1991). Measurement design and analysis: An integrated approach. Hillsdale, NJ: Erlbaum.

Peterson, J. B., DeYoung, C. G., Driver-Linn, E., Seguin, J. R., Higgins, D. M., Arseneault, L., et al. (2003). Self-deception and failure to modulate responses despite accruing evidence of error. *Journal of Research in Personality, 37,* 205-223.

Potosky, D., & Bobko, P. (1997). Computer versus paper and pencil administration mode and response distortion in noncognitive selection tests. *Journal of Applied Psychology, 82,* 293-299.

Ramanaiah, N. V., Schill, T., & Lock-Sing, L. (1977). A test of the hypotheses about the two-dimensional nature of the Marlowe-Crowne Social Desirability Scale. *Journal of Research in Personality, 11*(2), 251-259.

Reynolds, W. M. (1982). Development of reliable and valid short forms of the Marlowe-Crowne Social Desirability Scale. *Journal of Clinical Psychology, 38,* 119-125.

Richman, W. L., Kiesler, S., Weisband, S., & Drasgow, F. (1999). A meta-analytic study of social desirability distortion in computer-administered questionnaires, traditional questionnaires, and interviews. *Journal of Applied Psychology, 84,* 754-775.

Rimland, B. (1962). Personality test faking: Expressed willingness to fake as affected by anonymity and instructional set. *Educational and Psychological Measurement, 22,* 747-751.

Robie, C., Mueller, L. M., & Campion, J. E. (2001). Effects of motivational inducement on the psychometric properties of a cognitive ability test. *Journal of Business and Psychology, 16*(2), 177-189.

Rosse, J. G., Stecher, M. D., Miller, J. L., & Levin, R. A. (1998). The impact of response distortion on preemployment personality testing and hiring decisions. *Journal of Applied Psychology, 83,* 634-644.

Roth, D. L., Harris, R. N., & Snyder, C. R. (1988). An individual differences measure of attributive and repudiative tactics of favorable self-presentation. *Journal of Social and Clinical Psychology, 6*(2), 159-170.

Sackett, P. R., Schmitt, N., Ellingson, J. E., & Kabin, M. B. (2001). High stakes testing in employment, credentialling, and higher education: Prospects in a post-affirmative action world. *American Psychologist, 56,* 302-318.

Salgado, J. F. (1997). The five factor model of personality and job performance in the European community. *Journal of Applied Psychology, 82*, 30-43.

Sandal, G. M., & Endresen, I. M. (2002). The sensitivity of the CPI Good Impression Scale for detecting "faking good" among Norwegian students and job applicants. *International Journal of Selection and Assessment, 10*(4), 304-311.

Schwab, D. P. (1971). Issues in response distortion studies of personality inventories: A critique and replicated study. *Personnel Psychology, 24*, 637-647.

Schmit, M. J., & Ryan, A. M. (1993). The Big Five in personnel selection: Factor structure in applicant and nonapplicant populations. *Journal of Applied Psychology, 78*, 966-974.

Schneider, B. (1987). The people make the place. *Personnel Psychology, 40*(3), 437-453.

Schuman, H., & Presser, S. (1981). Questions and answers in attitude surveys: Experiments in question form, wording and context. New York: Academic Press.

Schwarz, N., & Sudman, S. (Eds.). (1992). *Context effects in social and psychological research.* New York: Springer-Verlag.

Sivec, H. J., Hilsenroth, M. J., & Lynn, S. J. (1995). Impact of simulating borderline personality disorder on the MMPI-2: A costs-benefit model employing base rates. *Journal of Personality Assessment, 64*, 295-311.

Snell, A. F., & McDaniel, M. A. (1998, April). *Faking: Getting data to answer the right question.* In M. McDaniel (Chair), Applicant faking with non-cognitive tests: Problems and solutions. Symposium conducted in the 13th annual conference of the Society for Industrial and Organizational Psychology, Dallas, TX.

Solomon, R. L. (1949). An extension of control group design. *Psychological Bulletin, 46*, 137-150.

Stark, S., Chernyshenko, O. S., Chan, K., Lee, W. C., & Drasgow, F. (2001). Effects of testing on item responding: Cause for concern. *Journal of Applied Psychology, 86*, 943-953.

Stokes, G. S., Hogan, J. B., & Snell, A. F. (1993). Comparability of incumbent and applicant samples for the development of biodata keys: The influence of social desirability. *Personnel Psychology, 46*, 739-762.

Strahan, R., & Gerbasi, K. C. (1972). Short, homogenous versions of the Marlowe-Crowne Social Desirability Scale. *Journal of Clinical Psychology, 28*(2), 191-193.

Tett, R. P., Jackson, D. N., & Rothstein, M. (1991). Personality measures as predictors of job performance: A meta-analytic review. *Personnel Psychology, 44*, 703-742.

Thornton, G. C. III, & Gierasch, P. F. (1980). Fakability of an empirically derived selection instrument. *Journal of Personality Assessment, 44*, 48-51.

Tourangeau, R. (1987). Attitude measurement: A cognitive perspective. In H. Hippler, N. Schwarz, & S. Sudman (Eds.), *Social information processing and survey methodology* (pp. 149-162). New York: Springer Verlag.

Viswesvaran, C., & Ones, D. S. (1999). Meta-analysis of fakability estimates: Implications for personality measurement. *Educational and Psychological Measurement, 54*, 197-210.

Wetter, M., Baer, R. A., Berry, D. T., Robinson, L. H., & Sumpter, J. (1993). MMPI-2 profiles of motivated fakers given specific symptom information: A comparison to matched patients. *Psychological Assessment, 5,* 317-323.

White, L. A., & Moss, M. C. (1995, May). *Factors influencing concurrent versus predictive validities of personality constructs: Impact of response distortion and item job content.* Paper presented at the annual meeting of the Society for Industrial and Organizational Psychology, San Diego, CA.

Wiggins, J. S. (1964). Convergences among stylistic response measures from objective personality tests. *Educational and Psychological Measurement, 24*(3), 551-562.

Zickar, M. J., & Drasgow, F. (1996). Detecting faking on a personality instrument using appropriateness measurement. *Applied Psychological Measurement, 20,* 71-87.

Zickar, M. J., Gibby, R. E., & Robie, C. (2004). Uncovering faking samples in applicant, incumbent, and experimental data sets: An application of mixed-model item response theory. *Organizational Research Methods, 7*(2), 168-190.

Zickar, M. J., & Robie, C. (1999). Modeling faking good on personality items: An item level analysis. *Journal of Applied psychology, 84,* 551-563.

Zickar, M., Rosse, J., Levin, R., & Hulin, C. L. (1996, April). *Modeling the effects of faking on personality scales.* Paper presented at the 11th annual conference of the Society for Industrial and Organizational Psychology, San Diego, CA.

CHAPTER 5

SENSITIVE OR SENSELESS

On the Use of Social Desirability Measures in Selection and Assessment

Gary N. Burns and Neil D. Christiansen

Although the use of personality tests to facilitate hiring decisions has increased over the past 2 decades, considerable skepticism still exists in the business world regarding the use of self-report measures in the areas of selection and assessment (Cook, 1993; Hoffman, 2001). In particular, the belief that personality measures can be easily faked remains the most widespread criticism of organizational personality assessment (e.g., Hogan & Hogan, 1992; Hogan, Hogan, & Roberts, 1996). Unlike ability or skill tests where there are verifiably correct answers, personality tests are self-descriptions in which it may be possible to choose a response calculated to obtain a valued outcome. Given this, it is hardly surprising that strategies aimed at combating applicant faking on personality tests continue to be developed and embraced by those who use personality tests for personnel decision making.

By far, most methods for dealing with applicant distortion of personality tests rely on measures of social desirability (SD) to identify those appli-

A Closer Examination of Applicant Faking Behavior, 115–150
Copyright © 2006 by Information Age Publishing
All rights of reproduction in any form reserved.

cants who have distorted responses the most. Sometimes also referred to as "lie" or "faking" scales, more than 80% of the commercial personality inventories include a SD scale and the vast majority of applied psychologists working in selection and assessment advocate their use when interpreting the results of personality tests (Goffin & Christiansen, 2003). Unfortunately, the psychometric training of most psychologists does not go into detail regarding how such scales were constructed, how they are intended to operate, or more importantly, whether research has supported their use. Moreover, the conceptualization of applicant faking as an attempt to respond in a way to create a favorable impression underlies almost all of the literature in the area, and SD measures are generally accepted as reliable and valid measures of faking behavior.

To be an intelligent consumer of research in this area, it is important to gain an understanding of SD, faking behavior, and when using SD scales in tandem with personality tests will be beneficial. To accomplish this, the nature of SD scales, their relationship with job performance, and their use in personnel selection and decision making is examined and reviewed. To tie these points together, an unpublished computational simulation examining different models of corrections is presented and discussed. The primary goal of this chapter is to provide psychologists a review of the use of SD scales in faking research and to highlight that the effectiveness, or ineffectiveness, of these scales is not a foregone conclusion. In doing so we draw attention to a number of unanswered questions to emphasize the research that is still needed on SD and SD scales.

Social Desirability Scales and What They Measure

Most measures of psychological constructs are developed by researchers interested in assessing individual differences relevant to some theory or outcome of interest. Based on an idea of what behaviors are construct relevant, items are written that concentrate either on the causes (sometimes called formative indicators) or effects (known as reflective or manifest indicators) of the construct. However, in the case of SD scales, the development of the measures often came first. Attempts to define the nature of the construct "social desirability," or even determine whether a stable individual difference was being measured, followed the development of the scales themselves. Use of the same term for whether SD measures represent response validity scales that gauge the level of a particular bias, or whether they are measures of a psychological construct, has created much confusion in that the two interpretations may be quite different.

SD scales share some common characteristics, such as gauging socially desirable responding using items in which the desirable response is relatively infrequent in a normative sample. In this context, socially desirable responding (SDR) refers to the general tendency to provide responses meant to make the respondent appear more favorable then they really are (Paulhus, 1991). Endorsing many responses that are both infrequent and desirable is considered an improbable event and higher scores are taken as an indication of the amount of positive distortion. Typical methods of SD scale construction include rationally developing items, empirically selecting items that differentiate groups from a faking manipulation, and having judges identify items that are highly desirable. Examples items include "I rarely feel resentful when I don't get my way" (Crowne & Marlowe, 1960) and "I never cover up my mistakes" (Paulhus, 1988).

Social Desirability Scales as Response Validity Scales

As early as the 1930s, concerns over the self-report nature of personality inventories and their susceptibility to motivated distortion had come to the forefront of the growing area of personality psychology (e.g., Bernreuter, 1933). Perhaps the first scale developed specifically for detecting such a positivity bias (or "defensiveness") when responding to a personality measure was in 1942 by Ruch, but the development of additional response validity scales quickly followed (Meehl & Hathaway, 1946). The L scale was utilized by the MMPI to detect blatant faking and was quickly replaced by the K scale, considered a more sophisticated or subtle detection instrument. A proliferation of SD scales followed throughout the next decade with as many as a dozen being used to detect a fake-good response bias (Wiggins, 1968). Up to this point, the concept of SD had largely been construed as a response bias, or more specifically, a response set that was a temporary reaction to situational demands (Tourangeau & Rasinksi, 1981). However, by the end of the 1950s researchers became interested in whether variability in SD scores might also represent a response style, a more stable influence on distortion than response sets (e.g., Crowne & Marlowe, 1960; Edwards, 1957).

As a response validity scale intended as an index of the amount of positive distortion, the research base overwhelmingly suggests that SD scales are sensitive to situational demands to respond in a desirable fashion. Viswesvaran and Ones (1999) meta-analyzed 26 studies that compared SD scores of a group told to respond honestly with one that had been given a fake good instructional set, showing that on average the faked SD scores were 1.06 standard deviations larger than the honest SD scores. In studies where the same individuals completed SD measures under both instructional sets (i.e., within-subject designs) the mean shift was larger ($d = 2.26$) as a result of carry-over effects that should probably be considered artifactual. Thus, SD scales have some validity as measures of the intended situational response bias.

Unfortunately, it is not clear how well this research generalizes to assessment of applicant faking in selection contexts. As Schmitt and Chan (1998) noted regarding the use of SD scales in the area of personality testing, "Evidence regarding the validity of the validity scales is not conclusive, at least not in the personnel selection area" (p. 161). A careful examination of the research in this area suggests that although somewhat sensitive to applicant faking, the relationship between faking and SD may not be very strong. For example, comparing incumbents and applicants for the same job, one study found that applicants (who likely have more motivation to fake good) scored 1.09 standard deviations higher than incumbents on a SD scale (Rosse, Stecher, Levin, & Miller, 1998). Rosse et al. further noted that the scores for applicants were similar to the norms for the SD scale reported for a fake good sample and that incumbents' scores were comparable to the norms reported for honest instructions (see Paulhus, 1988). Also supporting the interpretation that SD scores were related to applicant faking behavior, the mean difference for the SD scores was generally larger than that noted for the scales on a personality inventory (average $d = .65$).

Consider, however, that the mean shift in SD scores that results from instructing a group to fake good compared to one instructed to respond honestly is approximately one standard deviation (Viswesvaran & Ones, 1999) and that this is similar to the difference between what has been found comparing applicants with nonapplicants (Rosse et al., 1998). This means that approximately 20% of the variance in SD scores (based on $d = 1.0$) is explained by applicant distortion, amounting to a correlation of approximately .45. Because the results of laboratory faking studies have been taken to be an upper-bound or over-estimate of the faking behavior that takes place in real applicant samples (e.g., Smith & Ellingson, 2002; Viswesvaran & Ones, 1999), an upper-bound estimate of $r = .45$ may not be strong enough to consider SD a reliable and valid measure of applicant faking behavior. To the extent that this is true, the inference that faking does not affect validity, relying on studies that have partialed SD variance from applicants' trait scores or treated SD scores as a moderator of the validity of personality, may be questionable (e.g., Barrick & Mount, 1996; Hough, Eaton, Dunnette, Kamp, & McCloy, 1990; Ones, Viswesvaran, & Reiss, 1996). Conclusions based on such strategies with nonapplicant samples (comprised of individuals who have little motivation to fake) should be considered even more suspect, such as the majority of those samples contributing to the Ones et al. meta-analysis in which partialing was examined.

It should be noted that the above represents, at best, a crude estimate of the relationship between applicant SD scores and the amount of faking. Research that more concretely evaluates SD scales as indicators of applicant faking is sorely needed. For example, it is unclear how applicant SD scores are related to the amount of change in personality trait scores observed when individuals respond under normal assessment circumstances and when the trait scores are assessed as applicants. To avoid the strong carryover effects of within-subject designs in faking research, either a reasonably long period of time should separate responses to the personality measure or parallel measures of the same traits should be used. Furthermore, given the known problems with using difference scores (e.g., Edwards, 2002), using latent growth modeling to estimate change might be the best approach (although multiple-indicator models would likely be necessary as only two waves of data are used for change; see Raykov, 1999).

> **Unanswered Question #1:**   What is the relationship between applicants' social desirability scores and how much they alter their responses on a personality test as a result of the situational demands of the applicant context?

Social Desirability Scales as Measures of a Trait

By the beginning of the 1960s, researchers were interested in the possibility that scores on SD scales might reflect stable individual differences related to the personality of the test-taker. As a psychological construct, SD was defined by Edwards (1957) and Crowne and Marlowe (1960) as an individual's tendency to seek social approval or to act in a socially acceptable fashion. Building from this definition, Crowne and Marlowe (1964) later termed the concept simply "need for approval." They argued that individuals high in the need for approval have a desire to be liked but may lack the confidence, assertiveness, and social skills necessary to make the most of interpersonal situations (Schlenker, 1980).

One obstacle that emerged to this interpretation was that the correlation between scores on different SD scales was relatively low (e.g., Holden & Fekken, 1989; Stöber, 2001). Factor analyzing SD scales resulted in one factor that included Edwards SD scale and the MMPI K scale, a factor which was called *Alpha* (Block, 1965; Wiggins, 1964). A second factor, *Gamma*, also emerged with scales such as the MMPI Lie scale and a "positive malingering" scale (Wiggins, 1964). Researchers agreed that the *Gamma* factor represented motivated distortion and deliberate falsification, whereas the meaning of *Alpha* was debated (Damarin & Messick,

1965; Jackson & Messick, 1962). Work on the two factors of SDR was continued and refined by Sackeim and Gur (1978) in their development of the Other-Deception Questionnaire and the Self-Deception Questionnaire.

In the early 1980s, Paulhus conducted a series of experiments further defining the two components of SDR (Paulhus, 1984; Paulhus, 1986). Factor analyzing a host of different studies, Paulhus replicated a two factor structure. Most strongly associated with the first factor was the Other-Deception Questionnaire. Other scales clustered around it included MMPI Lie scale (Meehl & Hathaway, 1946), Wiggins SD scale (1959), and the Eysenck Personality Inventory Lie scale (Eysenck & Eysenck, 1964). In contrast, the strongest indicator of second factor was the Self-Deception Questionnaire with other factor markers including Edwards' SD scale (Edwards, 1957), MMPI K scale (Meehl & Hathaway, 1946), and the RD-16 (Schuessler, Hittle, & Cardascia, 1978). Paulhus also reported that two other common SD scales, the Marlowe-Crowne (Crowne & Marlowe, 1960) and the CPI Good Impression scale (Gough, 1957), loaded highly on both factors.

Building on this work, Paulhus (1984) argued that SDR could be viewed in terms of either self-deception or impression management. Self-deception (or deceptive self-positivity) refers to instances when respondents actually believe their positive self-reports, whereas instances of purposeful dissimulation are considered to be related to intentional attempts at impression management. Paulhus (1984) recommends that impression management (but not self-deception) be controlled in personality measures because this source of variance represents a conscious bias that may shift with any given situation whereas self-deception may reveal some degree of individuals' self-esteem. From this perspective, it is not particularly relevant whether the intentional distortion captured by impression management scales arises from situational demands as a response set or has a dispositional source in a response style in that both represent a source of distortion relevant to applied settings.

Although the results of research by Paulhus and his colleagues have indicated that impression management is more sensitive to situational factors than self-deception (Paulhus, 1984; Paulhus, Bruce, & Trapnell, 1995), it is unclear how relevant the distinction between the two different SD factors is in applicant contexts. For example, Christiansen (1997) assessed personality under normal instructions followed by SD under both honest and applicant instructions (i.e., half were provided with a description of a sales position and asked to respond as if applying for the job). In this study, the two SD factors were assessed using the Impression Management and Self-Deception scales of the Balanced Inventory of Desirable Responding (BIDR; Paulhus, 1984). Three aspects of the results were noteworthy: (a) the amount of mean shift in the two SD scales was similar, (b) in the applicant instructions condition factors representing the two SD scales correlated above .70, and (c) differences in the personality correlates across the two SD scales in the honest condition were markedly absent in the applicant instructions condition. However, given that such research is based only on simulated applicants and is probably related more to ability to distort than relevant motivational factors that affect applicants, more research is needed focusing on whether there are important differences in the two factors underlying SD scales in terms of relevance to applicant faking. Again, this would ideally be done relating applicants' scores on the two types of measures to the change observed in personality trait scores between a normal and applicant context.

> **Unanswered Question #2**: Does the relationship between applicants' SD scores and how much they have altered their response on a personality test depend on whether the SD scale primarily measures impression management or self-deception?

The notion that applicant faking of personality measures reflects individual differences in "impression management" in the sense that this term has been used in other areas such as social psychology is questionable. For example, it has been argued that applicant faking will not affect validity because those individuals who engage in impression management when answering personality questions in a favorable manner will tend to engage in impression management once on the job and engage in favorable behavior (Hogan et al., 1996; Smith & Robie, 2004). We believe that there is a fundamental difference in the motivation involved when completing a personality inventory as an applicant (i.e., to receive a valued outcome) and motivation involved when trying to engender positive social relations. Consider a college applicant who guesses at the right answer to a question on a college entrance examination requiring mathematical reasoning. Is the applicant attempting to manage the impression of admission officials throughout the country or trying to gain admission to the best school possible? Is this the same as when we might compliment a new acquaintance in the hope of becoming better friends? Although the instructions for completing a personality inventory may say that there is no right or wrong answer, the keying of the items argues otherwise. We suggest that attempts to construe faking behavior as either lying or impression management tactics may not only be unnecessarily evaluative but also off-base in terms of the cognitive and motivational factors involved. Rather, many applicants may construe their role as to discern the keyed response via any cues available just as they do on an ability test and that there is nothing (in their minds) dishonest about it.

Relationship between Social Desirability Scales and Substantive Personality

A considerable amount of research has examined the construct validity of SD scale scores in terms of relationships with well-understood personality traits when completed under normal assessment circumstances. For example, McCrae and Costa (1983) illustrated that SD was not only related to self-ratings of personality but to ratings of the participants by other individuals who should not be affected by the same self-serving biases. They argued that this was evidence that SD was a trait of substantive interest related to personality and not simply a contaminant of self-report personality measures. Ones et al. (1996) used meta-analysis to aggregate the personality correlates of SD scores across studies and found that after corrections SD scores were related to emotional stability ($\rho = .37$), conscientiousness ($\rho = .20$), agreeableness ($\rho = .14$), and extraversion ($\rho = .06$). The relationship of SD scales to an individual's personality when assessed without the additional situational demands to respond desirably present in applicant contexts has implications for interpreting scores as response validity scales and as indicators of the disposition of an applicant.

As an indicator of applicant faking, these relationships are problematic because there is no way to know whether applicants with high scores on social desirability scales have engaged in distortion or whether they legitimately have favourable trait elevations. As stated by McCrae and Costa (1983): "An individual who is in fact highly conscientious, well-adjusted, and cooperative would appear to be high in social desirability. Paradoxically, it is the most honest and upstanding citizen that these scales would lead us to accuse of lying!" (p. 883). The direction of these relationships is particularly pernicious because those with high SD scores are likely to be those with elevations on traits that tend to be positively correlated with job performance (see e.g., Barrick & Mount, 1991; Tett, Jackson, & Rothstein, 1991). It also suggests that valid trait variance will likely be removed when partialing or correcting trait scores. For all of these reasons, response validity scales should be relatively uncorrelated with true scores on the construct(s) of interest.

When considered as evidence of the validity of SD scores as indicators of a psychological construct, the direction of these relationships is also disconcerting because they are opposite of what would be expected from individuals who consistently need to seek approval or lack social confidence. That is, individuals who are *low* on emotional stability tend to seek social approval and those *low* on extraversion are those that lack social confidence (Costa & McCrae, 1992); one would therefore expect negative relationships with those traits rather than the positive ones observed. Thus, the theory of the construct being measured is not consistent with the data. Fur-

thermore, we are not aware of any evidence that individuals who score high
on SD scales tend to seek social approval across situations and most notably
absent are empirical data relating test scores to nontest behavior.

It is worth noting that the items used in SD scales differ in some ways
from those traditionally used to assess personality constructs. More specifically, items on SD scales ask about behavior not necessarily directly
related to the construct of interest (such as seeking approval), but focus on
areas where the responses might be biased by someone possessing the
trait of interest. For example, rather than saying "I worry a lot about what
others may think who do not really know me," example items from the
BIDR impression management items include "I always obey laws even if it
is unlikely I will get caught" and "When I hear people talking privately I
avoid listening." In the case of the example BIDR items, a point is scored
for extreme responses denying these undesirable behaviors. Not only are
there obvious relationships to other personality traits (conscientiousness
in the case of the first BIDR item and agreeableness for the second) but
on their face they would not seem directly related to those commonly
ascribed to SD as a psychological construct. Given this, it may not be surprising that when viewed as indicators of a single psychological trait SD
scales have difficulty with regard to construct validity.

In summary, it is clear that there is systematic variance to responses on
SD scales that suggest a response style that is related to personality. It is less
clear as to what the exact nature of that style is or whether a unitary construct is being assessed. It may very well be that it is a hodge-podge of other
traits based on the behavioral domains of the particular items included to
index the endorsement of desirable behaviors with low base rates.

> **Unanswered Question #3**:   Do SD scales assess a single con-
> struct (as opposed to a composite of personality traits)
> that predict nontest behavior consistent with the theory
> of the underlying construct?

Relationship Between Social Desirability Scales and Job Performance

At first glance, the literature relating SD scores to job performance
would seem straightforward. For example, Ones et al. (1996) meta-analyzed the results of 14 studies that included almost 10,000 combined participants and found the average correlation between SD scores and job
performance was .01. Similarly, Viswesvaran, Ones, and Hough (2001)
focused on the relationship between scores on impression management
scales and managerial performance. For over 20,000 managers the aver-

age relationship across studies was .04. Attempts to identify specific dimensions of performance (e.g., leadership, interpersonal relations) failed to find evidence that impression management scores had much implication for managerial success. It would therefore appear that scores on SD scales have little implication for predicting job success.

Two important caveats exist to qualify this inference. The first is that the participants from these samples were almost exclusively incumbents who already had jobs and presumably little reason to engage in faking behavior. Ample evidence exists that the construct validity of SD scores obtained from normal assessment situations differs from those obtained from applicants (cf. Hough, 1998; Rosse et al., 1998). Obviously, scores from the former lack the response to job-specific situational demands that are present in the latter and inform very little with regard to whether faking behavior might be related to job performance. To the extent that faking behavior reflects either social competence or a lack of integrity, it is easy to imagine that the correlates of job performance may depend on whether applicants or nonapplicants have provided responses.

The second caveat is that the above estimates of the relationship between SD scores and performance are across jobs. It has been shown that not only can such relationships depend on the position, but that for some positions scores may be positively correlated with job performance and for others negatively correlated (Tett, Jackson, Rothstein, & Reddon, 1999). In such cases, opposite relationships may cancel out when averaged. Inspecting the variability in effect sizes from Ones et al. (1996) reveals a fair amount of consistency (arguing against this hypothesis), but it should be recalled there were only 14 different positions included in the analysis. As an example, one could imagine that individuals who are candid despite strong demands to the contrary might perform better as police officers, resulting in a negative relationship with job success. Similarly, it is possible that those who are willing to say almost anything to get what they want might be more successful at telemarketing. Unfortunately, the research base is not currently able to inform us how likely it is SD scores may be relevant to the performance of particular positions. As shown below, the relationship between SD scores and job performance is one of the most critical variables for determining the effectiveness of the use of SD scales to enhance the validity of personality measures.

> **Unanswered Question #4:**   What is the relationship between social desirability scores obtained under normal assessment circumstances and how much applicants alter their responses on a personality test as a result of the situational demands of the applicant context?

Unanswered Question #5:   What is the relationship between
applicants' social desirability scores and job performance
and does this depend on the nature of the individual job?

Uses of Social Desirability Scales

Several strategies have been proposed for using SD scores in tandem
with the results of personality tests in the area of selection and assessment.
The assumptions and research relating to three types of specific uses will
be reviewed. First, the practice of partialing SD scores from trait scores is
reviewed along with other types of corrections to trait scores. This
includes both mechanical corrections as well as more subjective mental
adjustments that practitioners might apply to the interpretation of the
personality profile. The use of SD scores as a moderator of the personality-performance relationship is next considered. Finally, the use of SD
scores to set a cut-off and eliminate candidates solely on those scores is
reviewed.

Partialing or Correcting Based on Social Desirability Scores

The original use of SD scales in assessment involved attempts to
remove the effects of motivated distortion from personality scores by
regressing the trait scales onto the SD scores and computing a residual
score (Meehl & Hathaway, 1946). From this methodology, corrections
such as the K-correction for the MMPI were developed. Since this time little has changed; to researchers seeking unbiased measures and practitioners looking to hire the best applicants, SD has remained a variable of
interest and a viable option for combating distortion. In a survey of personality researchers Goffin and Christiansen (2003) found that 56% indicated that not only did they typically use a personality measure with a
faking scale but that they also typically corrected scores. Of their sample,
only 30% indicated that they would not correct personality scores even if
corrections were available. The remaining 14% indicated that they would
use corrections if their test of choice were so equipped. These results indicate that 70% of those surveyed were in favor of making corrections to
personality scores based on SD or other faking scales.

The difference between partialing and mechanical score corrections is
that partialing removes variance from trait scores based on the relationship in the specific sample, whereas mechanical corrections are intended
to approximate this based on relationships from a normative sample. For
example, Krug (1978) developed corrections for the Sixteen Personality

Factor Questionnaire, Fourth Edition (16PF4; Cattell, 1989; Cattell, Eber, & Tatsuoka, 1970) based on the regression equations from the normative data. Specifically, the corrections were constructed "based on the regressions of the distortion scales from the primary trait scales and represent an attempt to partial out distortion from the real trait variance" (Krug, 1978, p. 517). Such corrections take the form of conditional adjustments such that if SD scores are greater than X, A is subtracted from a primary trait. These are applied to each candidate such that higher SD elevations result in a lowering of scores on traits generally considered favorable and adding to those that are unfavorable (to reflect that candidates may have denied negative self-descriptions).

Typically, elevated SD scores result in an adjustment to many traits. In the case of the 16PF4, 10 of the 16 primary trait scales are altered by score corrections and so are all six of the second-order factors. Corrections can also affect the scores of a large number of candidates. For example, in a sample where half the assesses had completed the personality test as part of the assessment process for developmental purposes and half for selection purposes, relevant scores from over 70% of the sample were affected by the corrections (Christiansen, Goffin, Johnston, & Rothstein, 1994). This resulted in changes in the rank-order of over 85% of the candidates and markedly discrepant hiring decisions when compared to uncorrected scores. It would be an understatement to say that the effects of partialing or correcting SD scores can have an important effect on decisions based on the results of personality tests.

The underlying assumption in uses of SD scores to partial or mechanically correct trait scores is one of suppression, with faking behavior presumed to add variance to personality scores that will mask the relationship between personality and the criterion. In the classic psychometric literature on personnel selection, a suppressor is a variable that although not correlated with the criterion *is* correlated with the predictor measure and therefore contributes irrelevant variance. By measuring and removing this contaminant, a better estimate of candidates' true scores will emerge and better decisions will result. In the classic model the suppressor is uncorrelated with the criterion and partialing the suppressor (SD) from the other predictor (a personality score) has an equivalent effect on the validity of the predictor as including the suppressor in a multiple regression equation has on the multiple R (Conger & Jackson, 1972). Note that just as the multiple R cannot decrease in value when a predictor is added, partialing something unrelated to the criterion directly from predictor scores will not decrease their correlation with the criterion. However, when the supposed suppressor *is* related to the criterion of interest, partialing can have either a positive or negative effect on validity.

Research on the effects of partialing and correcting has generally failed to find beneficial effects, although there have been exceptions. For example, Kriedt, and Dawson (1961) examined the effects of partialing SD scores out of scores on the Gordon Personality Inventory and found that the validity was destroyed by the attempt to remove the response set. Christiansen et al. (1994) examined the effects of both mechanical corrections and partialing. In the case of corrections, the correlations of 7 out of 10 of the affected trait scales with job performance was smaller after correction as was the multiple correlation obtained when performance was regressed onto the second-order factors. Similarly, partialing SD from trait scores based on the sample-specific relationships also resulted in a decrease in validity rather than the expected increase. On the other hand, Barrick and Mount (1996) obtained mixed results across two studies when they used structural equation modeling to partial out impression management.

It is instructive to look at the relationships among the variables across these studies to understand why partialing generally failed to increase validity. For example, in the Kriedt and Dawson (1961) sample the SD measure was a better predictor of performance than any of the other sub-scales ($r = .47$). Similarly, in the Christiansen et al. (1994) study both 16PF response validity scales (fake good and fake bad) were positively related to performance (fake good $r = .16$ and fake bad $r = .09$). As a result any effort to remove their variance from personality removed valid trait variance related to the criterion. In the Barrick and Mount (1996) study, impression management was negatively related to turnover but positively related to conscientiousness, resulting in a decrease in validity after partialing. However, the results were different across the two samples of truck drivers when supervisors' ratings of performance were considered as the criterion. For sample 1, impression management was positively related to both conscientiousness ($r = .27$) and supervisors' ratings of performance ($r = .17$); partialing impression management resulted in a predictable decrease in validity of .04. In sample 2 however, the correlation between impression management and conscientiousness was positive ($r = .26$) but that with performance was negative ($r = -.07$). Partialing under these circumstances resulted in a corrected validity of .31, a .05 increase.

The effects of partialing and score corrections on validity depend on the correlations between applicants' personality scores, SD scores, and the criterion. In the most common situations in assessment and selection, in which the criterion is an indicator of job success and higher scores on the personality inventory are favorable, it can be presumed that the predictor-criterion relationship will be positive. Based on known relationships between SD and personality, along with the additional assumption that faking results in inflation of both SD and personality scores, it can be safely assumed that these relationships will also be positive. The only relationship left to vary across situations may be the most crucial: the relationship between applicants' SD scores and job performance. Unfortunately, as noted above, there are important unanswered questions about the nature of the boundary conditions of this relationship. What is clear is that when the SD-criterion relationship is positive, partialing and correcting will tend to result in hiring worse candidates. When the relationship is negative, partialing and corrections are likely to improve the quality of hires. In addition, to the extent that any of the key relationships depend on the job, a single set of mechanical corrections that are applied across jobs will necessarily be limited (Goffin & Christiansen, 2003).

A neglected area of study in examining the use of SD involves making subjective corrections to personality scores. In the area of individual assessment, mechanical methods of combining and correcting scores are not commonly utilized and a "clinical judgment" is often used to examine the entire profile of test scores (Ryan & Sackett, 1998). During this subjective decision-making process, an issue that an assessor may consider is whether self-presentation bias has impacted the responses of a candidate (Jeanneret & Silzer, 1998). Of the multiple cues that assessors could use to make this inference, the most salient is the presence of an SD scale. To our knowledge, only one study has examined the use of SD in subjective interpretation of personality profiles.

Specifically, to examine the effect of SD scores on the interpretation of personality profiles, Christiansen and Rozek (2004) presented practitioners in the area of selection and assessment with profiles of two candidates being considered for hire. The first candidate was high in cognitive ability but had average personality scores, whereas the second candidate had average cognitive ability scores but high personality scores. SD was manipulated across four conditions such that (a) no SD information was presented for either candidate, (b) the first candidate had a high SD score while the second had an average score, (c) the second candidate had a high SD score while the first had an average score, or (d) both candidates had high SD scores. The results of the study indicated that when applicants were presented with high SD scores they were rated as being less candid and sincere. The data also clearly indicated that candidates with higher SD scores were rated as less hirable than those with average scores. Finally, the researchers were able to isolate the process in which the SD scores were used in decision making, namely to give less weight to the personality test results relative to other assessments.

The results of Christiansen and Rozek (2004) indicate that SD scores may play a role in correcting personality scales in individual assessment even when no objective methods are utilized to adjust personality scores. Whether through a conscious decision or unconscious process, trained test-users judged candidates to be less hirable and less candid when SD scores were high. Unfortunately, there are many reasons to question whether the subjective adjustment of test scores based on SD will improve the quality of hiring decisions. Beyond those reasons discussed above for mechanical corrections, subjective corrections may face additional problems due to the nature of applicant distortion. For example, research has shown that the pattern of distorted responses on personality inventories depends on the job and the stereotypes associated with the job (e.g., Dicken, 1959; Furnham, 1990). These stereotypes, however, are not always accurate (Mahar, Cologon, & Duck, 1995), making it doubtful that a consistent adjustment could be used meaningfully across jobs or applicants.

Social Desirability as a Moderator of the Personality-Performance Relationship

The idea that SD scales may serve as a moderator of the criterion-related validity of applicant personality scores rests on the assumption that the trait scores of those with high SD scores will have lower validity than those with low SD scores. Indeed, one can argue that this is much closer to the intent of a *response validity* scale than an underlying model of suppression. It is also consistent with research that shows when some respondents have distorted their responses to a personality test that their scores rise to the top and are less valid. This has been shown in both simulations (Burns & Christiansen, 2004; Mueller-Hanson, Heggestad, & Thornton, 2003) and with actual applicants (Haaland, Christiansen, & Kaufman, 1999). To the extent that SD scales are sensitive to distortion that inflates trait scores of some applicants more than others, SD scores would be expected to moderate the personality-performance relationship.

To understand how the use of SD as a moderator relates to its use as a suppressor, or not using SD scores at all, one has to recall that conceptually, selection decisions are not made based on test scores (such as those from a personality inventory) per se, but rather are based on predicted performance scores. When SD scores are *not* used, predicted performance scores are just a linear transformation of trait scores that take into account the relationship with performance and inevitable regression toward the mean. When partialing or correcting using SD scores, the predicted performance scores are regressed toward their mean based on the actual or presumed relationship between trait and SD scores. The amount of adjustment does not depend on the deviation of the personality score from the mean, but is solely a function of elevation on SD scores. So whether an applicant's personality score is one quarter of a standard deviation above the mean or two standard deviations above, partialing or correcting would adjust it the same amount based on a high SD score. In fact, given a high SD score, mechanical correlation formulas would lower personality scores the same amount regardless of whether the personality scores were two standard deviations *above* or *below* the mean of the other applicants' scores, despite the improbability that the latter applicants had successfully distorted scores at all.

On the other hand, using SD scores as a moderator to compute predicted performance scores results in an adjustment to the ranking of applicants that depends upon the elevation of both the trait and SD scores. Just as personality and SD scores can be combined multiplicatively to form a continuous product from which decisions are made, a significant interaction also supports the use of multiple cut-score models that would eliminate applicants if their score were *above* a certain point on the SD scale but *below* a cut-off on the personality test score. Such noncompensatory models that remove applicants are discussed later in the next section because their use reflects a number of different assumptions.

In order to better understand the differences between approaches that use SD as a suppressor versus a moderator, consider the illustration in Figure 5.1. Three graphs depict the change in personality trait scores realized with both methods for a fairly typical selection situation. In this case, the relationships are based on correlations of .38 between applicants' personality scores and their SD scores. Separate adjustments are shown for low, average, and high SD scores across the three graphs; within each graph the change is shown for low, average, and high personality scores. Please note that the cross-product scores that represent use of SD as a moderator have been residualized on the personality and SD scores (based on the relationships between the simple effects and the interaction term).

Several distinctions across the two methods of adjustment are noteworthy. First, within each graph the change under the suppression model is constant across the range of personality scores: when SD is low the effect of adjustment is to increase all scores, when it is average they stay the same, and when SD is high all scores are lowered. In contrast, the effects of treating SD as a moderator and computing a (residualized) multiplicative composite are more complex and depend on elevation of both SD and personality scores. Generally speaking, when personality scores are high, treating SD as a moderator lowers the score and when personality scores are low the adjusted scores are increased. However, the amount of these changes is contingent upon the level of the SD scores.

It is also instructive to consider where the largest departures between the methods occur. Clearly, when scores are average on both measures the outcomes are similar. The greatest discrepancies are observed when scores on one measure are high and those on the other are low. That is, when

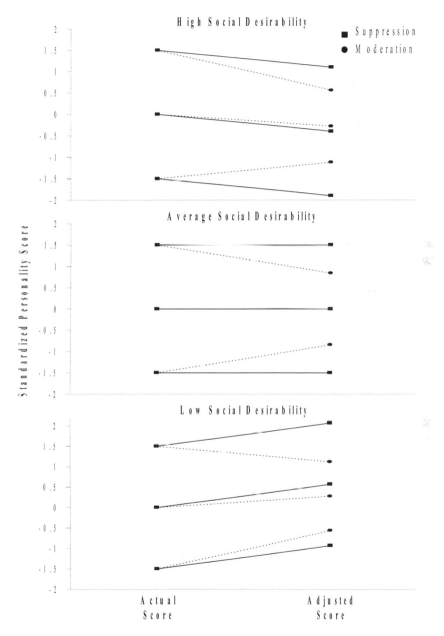

Note.   Low and High Social Desirability graphs were based on social desirability scores 1.5 standard deviations from the mean.

Figure 5.1.   Illustration of Suppression and Moderation Corrections on sTandardized Personality Scores.

personality is high and SD is low, as well as when personality is low and SD is high. Assuming a positive correlation between applicant personality and SD scores, most scores will *not* fall in these bivariate quadrants. This assumption will be true if the two sets of scores have a common cause (such as the motivation to distort); not surprisingly, research supports that this correlation tends to be larger in applicant samples than incumbent samples (Rosse et al., 1998). It follows then that as faking increases, this correlation will rise and the number of cases with large discrepancies in the outcomes of the methods will be reduced. As a result, the difference in effects obtained from the two methods of adjustment should decrease as the prevalence of faking increases.

Finally, it is important to note that the illustration does not show whether the changes move adjusted scores toward or away from actual performance. This is the essential question involved in whether validity will be enhanced, which is likely to depend not just on the relationship between applicants' personality and SD scores, but also with performance. Likewise, it may also depend on the amount and prevalence of applicant faking, as this may change all of the above parameters. To the extent that the SD scores are sensitive to the effects of faking it might be expected that any potential changes to validity will be augmented; when SD scores are relatively insensitive, potential changes to validity should be attenuated.

Unfortunately, research on SD as a moderator of the personality-performance relationship has been scarce. Yet despite this dearth of evidence, it has been adamantly declared that there is no SD moderation (Alliger & Dwight, 2000; Ones et al, 1996). The small amount of extant research on moderation suggests that in some cases the personality-performance relationship varies by SD.

For example, although not directly related to personnel selection, one of the first studies directly examining SD as a moderator was conducted by Ganster, Hennessey, and Luthans (1983). They examined the moderating role of SD in the prediction of a number of workplace variables, including organizational role characteristics and job attitudes. In an exploratory fashion, they examined the possible moderating effect of SD on 73 personality-outcome relationships and report significant interaction terms for only four possible combinations: need for achievement predicting promotion satisfaction, work satisfaction, and role conflict as well as for leader's structure predicting satisfaction with supervision.

Hough et al. (1990) were the first to examine SD as a moderator of the personality-performance relationship. Moving beyond the examination of the suppression model, Hough et al. examined whether SD scores moderated the criterion-related validity of a noncognitive instrument. To accomplish this they identified individuals with SD scores indicating that they were responding in an overly desirable fashion. The criterion-related validity for this group was then compared to the group without overly elevated SD profiles. Although the authors concluded that SDR does not moderate validity, careful inspection of the results suggest a much more mixed pattern. For example, in the prediction of effort and leadership 10 of the 11 scales showed a decrease when moving from the "accurate" group to the "overly desirable" group (only two of these decreases were "significant"). It is worth noting that the apparently small decrease in validity in the elevated SD group represents an average decrease of 26.9% of the variance explained in effort and leadership compared to the accurate group. Of the 33 reported relationships between personality scores and the criteria, 21 (64%) were smaller in the group with elevated SD scores, 10 were larger (30%), and two were similar. As mentioned previously, Hough et al. utilized a concurrent validation design where it is again open to question the extent to which faking behavior is likely to occur.

White, Young, and Rumsey (2001) examined the same personality inventory as Hough et al. (1990) using a predictive design and found stronger evidence of SD moderation. Dividing the sample into three levels of SDR (low, moderate, and high), White et al. report consistently lower validity coefficients among high SD respondents than among low and moderate respondents across seven subscales and four different criteria used in the military sample (field performance, leadership, delinquency, customer service). Specifically, validity was always lower in the high SD group than the low SD group, with the positive correlations found in the rest of the sample being observed to be consistently *negative* in the high SD group.

The presence of moderation was replicated in a concurrent validity design by White and Kilcullen (1998), although the results were not as strong. The authors examined the moderating effects of SD in four samples of military and civilian jobs: the noncognitive measure for each job consisted of biodata questionnaires that had been developed for each job. Their results indicated that SD played a large role in attenuating the validity of the biodata questionnaires. Specifically, high levels of SDR attenuated the prediction of delinquency among military recruits (low $r = .33$, moderate $r = .41$, high $r = -.12$) and the prediction of customer service among customer service employees (low $r = .22$, moderate $r = .23$, high $r = -.15$). Although not significant, SDR also appeared to attenuate the prediction of both special forces field performance (low $r = .31$, moderate $r = .22$, high $r = .05$) and leader job performance (low $r = .28$, moderate $r = .20$, high $r = -.05$). They conclude that it may be best to examine SD moderation by dividing the sample into three groups of SDR rather than the two observed in Hough et al. (1990).

The four studies reviewed here provide only a brief glimpse at the potential of SD to moderate the personality-performance relationship. Taken together they indicate that SD can moderate the relationship between variables, but that this moderation does not always occur. Relevant to personnel psychology, these studies indicated that the moderating effect of SD on criterion-related validity varies by motivation of the sample to distort. Three of the studies also suggest that moderation may be stronger in predictive designs than in concurrent designs, perhaps because applicants have more motivation to engage in SDR. Unfortunately, with only four studies examining the moderating effects of SDR on criterion-related validity, it is difficult to specify the conditions in which noncognitive measures will be affected.

> **Unanswered Question #6**:   To what extent does the relationship between applicants' scores on a personality test and job performance depend on the elevation of their social desirability scores?

> **Unanswered Question #7**:   When is it best to utilize SD as a moderator of the personality-performance relationship rather than a suppressor?

Eliminating Candidates With Elevated Social Desirability Scores

The third use for SD scales is as a component of a noncompensatory selection procedure in which applicants are removed from further consideration based on the elevation of their SD score. There are two assumptions that might underlie this use. First, it may be based on the belief that SD scores are negatively correlated with performance and that those above a threshold will be less successful if hired. It should be noted that even if this were the case, treating SD scores as another trait measure to be combined in a compensatory fashion would generally outperform such a noncompensatory approach. Be this as it may, with multiple predictors most organizational decision makers prefer a multiple cut score model (cf. Guion, 1998), perhaps because limitations to this practice are not well known. Second, it may be that using cut scores on both the trait and SD measures reflect the belief that a true multiplicative relationship exists, as discussed above. Of course, it could also reflect both assumptions. As one test consultant advises clients with regard to elevated SD scores: "Why would you want to hire someone who isn't honest?"

Although research on this use of SD scales is scarce, there is reason to believe that the procedure of removing applicants based on SD profiles is not an uncommon one. Personality test publishers who include response validity scales often include a warning that high SD respondents may have invalid personality scores (see Goffin & Christiansen, 2003). These warnings range from innocuous statements such as "Interpret with caution" to more severe warnings that the "Profile may be invalid" (Goffin & Christiansen, 2003). It also seems that test users put considerable weight into these warnings; discussions with more than a few consultants have left us with the sense that this method of utilizing SD scores is widely used.

Providing support for this anecdotal evidence, Rees and Metcalf (2003) surveyed individuals employed in personnel positions who lacked expertise in personality testing. Their results indicated that although over half the sample believed it was difficult to identify distorted profiles, they would still most likely reject any applicants identified as faking. Rather than directly eliminating applicants, Butcher, Morfitt, Rouse, and Holden (1997) readministered the test to those individuals with invalid profiles, informing them that the test had identified their profiles as invalid and instructing them to respond in a more honest manner. Applicants who still had invalid profiles with the second administration were assumed to be distorting their responses and were then rejected. Clearly, there could be ethical and legal issues involved in rejecting applicants, absent evidence that SD scores are job related.

Despite this, very little research has examined the effects of eliminating candidates based on SD scores. In fact, when Hough (1998) systematically examined this strategy she presented it as a new way of dealing with motivated distortion. Reverse scoring an Unlikely Virtues scale (a measure of SD), Hough utilized a .95 selection ratio as the first step in a multiple hurdle selection simulation. As a result, those individuals with the highest SD scores were removed from further consideration. Across three different applicant samples (one telecommunications and two law enforcement) Hough examined the effects of this procedure on the personality scale's correlation with SD and on the concurrent criterion-related validity. After eliminating applicants with high SD scores Hough found that the correlation between personality and the SD scale decreased in all three of the incumbent samples and in two of the applicant samples. Similarly, the criterion-related validity was minimally increased in two of the incumbent samples and attenuated in the third.

Schmitt and Oswald (in press) provide a more comprehensive examination of the effects of eliminating applicants by developing a computational model in which it was possible to vary some of the relevant parameters. As such, their simulation is a manipulation of the relationships between personality, a response validity scale such as SD, and a criterion. Taking SD profiles as an indicator of faking, they explored the effectiveness of removing cases under a variety of conditions (sample size, selection ratio, and proportion removed for suspected faking). Their chief conclusion was that given typical selection scenarios, the removal of high SD respondents has minimal impact on mean performance.

Examining the Schmitt and Oswald (in press) results more closely reveals that as applicants suspected of faking are eliminated the mean performance of selected applicants tends to decrease. Collapsing across all conditions the standardized mean performance of selected applicants was .304 when no one was removed, .290 when 5% were removed, .260 when 15% were removed, and .212 when 30% were removed. The driving force behind this trend is the SD-performance relationship; when the relationship between SD and performance is positive, eliminating those with high scores will tend to decrease the performance of those selected. Because Schmitt and Oswald varied this relationship from -.1 to .3, the average correlation across conditions was .075. This positive correlation drove the decrease in performance when high SD applicants were removed. Rather than collapsing, if one considers the results separately by condition, it is clear that the performance of selected applicants can *increase* by eliminating applicants when the SD scale is negatively correlated with performance.

In fact, it would seem that there is little new learned by this rather straightforward simulation. What is simulated is the effect of using SD scores as the first step in a multiple-hurdle strategy. The results follow directly from what is already known: (a) the effect on performance depends on the proportion screened out, such that when the selection ratio is quite high the use of an initial hurdle has little effect; (b) the direction of the change in performance depends upon the direction of the correlation between applicants' scores on that hurdle and the criterion. Thus, when SD scores are used to identify a relatively small number of potentially invalid cases and the validity of those scores is actually no worse than the rest of the cases, the net effect on performance is small and depends on whether applicants' SD scores are positively or negatively related to performance.

Modeling Uses of Social Desirability Scales

The effectiveness of corrections is often assumed to be a foregone conclusion: that corrections based on SD are ineffective in dealing with faking. As the review above indicates, research on the effectiveness of using SD scores is mixed at best. A better understanding of the complicated interplay between the factors is needed to inform when such practices are advisable and when they are folly. Such an exploration is ideally suited to computational modeling. Modeling faking provides a method of identifying which applicants are distorting and determining the extent of distortion while maintaining conditions similar to applicant settings. Existing computational models exploring faking (Schmitt & Oswald, in press; Zickar, Rosse, Levin, & Hulin, 1997) have ignored the complexity of the process of faking itself. To address this deficiency Burns and Christiansen (2005) developed a computational model to examine the uses of SD scales in selection decisions.

Modeling Faking

Moving beyond the relationships between variables examined in Schmitt and Oswald (in press), Burns and Christiansen (2005) focused on the process of faking as well as the sensitivity of SD to faking. Meta-analyses of the relationships between SD and other variables were examined to determine a set of parameters that represented estimates of the true correlations between personality, performance, and SD. Instead of manipulating these variables, they relied on the parameters provided by past research and focused on the effects that the amount of distortion (both proportion and mean shift) and the sensitivity of SD had on corrections.

The first step in modeling is to determine the assumptions about behavior that the model is to reflect. The authors identified the following assumptions to guide the development of their computational model of faking behavior: (1) personality is moderately related to job performance; (2) some applicants intentionally distort their responses to make a favorable impression; (3) applicants with favorable trait elevations distort less as they have less need and less room for score inflation even if the motivation were present; (4) those applicants that distort the personality scale will also distort the embedded SD measure; and (5) the amount that applicants distort is varied across both individuals and measures. Note that the third assumption is saying that faking will cause distorted scores to rise to the top of the score distribution and that faked responses will have worse validity than honest scores.

A summary of the parameters of the computational model is shown in Table 5.1. To examine the process of faking, the proportion of the sample distorting and the mean shift amongst distorted scores were manipulated. Estimates of the fakability of personality and SD were taken from Viswesvaran and Ones (1999) with the mean shift in personality caused by distortion manipulated to be .2, .6, or 1.0 and the mean shift in SD was specified as being 1.76 times greater than that observed for personality. From past examinations of the proportion of distorted scores (Donovan, Dwight, & Hurtz, 2003; Griffith, Chmielowski, & Yoshita, in press; McDaniel, Douglas, & Snell, 1997), the percentage of fakers was manipulated to be 25, 50, or 75%. To examine the effectiveness of corrections the sensitivity of SD scales to distortion was manipulated. Sensitivity (operationalized as the correlation between SD and the amount of distortion) was manipulated across four levels: .00, .20, .40, and .60.

Table 5.1.   Explanation of Simulation Imputs, Parameters, and Outputs

Parameter	Explanation
X	Scores on the predictor representing one or more traits on personality test taken without motivation to distort (i.e., honestly). Generated to have random normal distribution.
Y	Job performance scores generated to have random normal distribution.
SD	Scores on the predictor representing personality test taken without motivation to distort (i.e., honestly). Generated to have random normal distribution. Correlation between SD and d_X manipulated to be 0, .2, .4, or .6.
r_{x_y}	Criterion-related validity of honest personality test scores, set at .30.
r_{x_sd}	Correlation between honest personality and social desirability scores, set at .15.
r_{sd_y}	Correlation between honest social desirability scores and job performance, set at .04.
d_X	Amount of distortion of the personality test scores that have been faked, fixed such that applicant scores that are faked have a mean increase of .6 standard deviations. Variance is set at 75% of that observed in honest scores. The correlation between d_X and X is -.48.
d_{SD}	Amount of distortion of the social desirability scores that have been faked, fixed such that applicant scores that are faked have a mean increase of 1.06 standard deviations. Variance is set at 75% of that observed in honest scores. The correlation between d_{SD} and SD is 0. The correlation between d_{SD} and X is -.16. The correlation between d_{SD} and d_X is .36.
P	Proportion of honest scores that have d_x added to them. This was manipulated to be .25, .50, and .75, but must be interpreted in the context that approximately 20% of the d_X scores were negative. Therefore the net percent of applicant scores that were higher than their honest counterparts was approximately 20, 40, and 60%.
X_A	Applicant personality score where d_X has been added to the X score of those who have distorted and X has been used for those who did not.
SD_A	Applicant social desirability score where d_{SD} has been added to the SD score of those who have distorted and SD has been used for those who did not.

Effect of Faking on the Personality-Performance Relationship and Distribution of Scores

Before examining the effects of corrections across conditions, it is instructive to examine how the method of operationalizing faking in the Burns and Christiansen (2005) model affected the distribution of faked responses and criterion-validity across score ranges of the personality and SD measures. Several patterns were noteworthy. First, although the scores that would be affected by distortion were chosen randomly, fewer distorted scores than expected were found in the bottom of the score distributions and more distorted scores were found at the top of both distributions as the amount of distortion increased. Second, the validity of the applicant personality scores found at the top of the distribution tended to be worse than those at the bottom and in the overall sample, both in terms of the relationship with the honest personality scores as well as with performance. This effect also tended to be greater as the amount and proportion of distorted scores increased. These results are consistent with what has been found in actual applicant data (Haaland et al., 1999) as well as laboratory faking studies (Burns & Christiansen, 2004; Mueller-Hanson et al., 2003). This provides some support for the ecological validity of the computational model in terms of how the faking process was operationalized. It also underscores the complexity of how applicant faking affects personality measures, resulting in a deterioration of normally linear relationships.

Finally, the validity of applicant personality was examined across the score ranges of the applicant SD measure. Given that the applicant personality and SD scores are positively correlated and the validity of personality score was worse at the top of that distribution, it would be expected that the validity of personality scores might also be worse for those applicants with the highest SD scores. This was indeed the case and also tended to be more pronounced as the amount and proportion of distorted scores increased. It also set the stage for the possibility that including SD scores as a moderator of the personality-performance relationship might improve the prediction of performance.

Effect of Corrections on the Criterion-Related Validity

The next step was to examine the effect of corrections on the criterion-related validity. Rather than focus on just corrections based on the suppression model, Burns and Christiansen (2005) also examined corrections based on the moderation model. Overall, corrections resulted in a relatively small change in validity, ranging from -.011 to .043 with the average increase in validity being .011 (see Figure 5.2). Across most of the conditions the corrections based on moderation slightly outperformed corrections based on suppression. This trend reversed when both the amount of distortion and the proportion of distorted scores were at the very highest, with corrections based on suppression resulting in slightly superior validit.

The effectiveness of corrections, whether based on suppression or moderation, was largely driven by the model inputs. That is, as the amount of distortion, the proportion distorting, and the sensitivity of SD increased corrections were more effective at recovering lost validity. In addition, the proportion of distorted scores interacted with both the amount of distortion and the sensitivity of SD to produce enhanced validity for both types of corrections. The uncorrected and corrected validity coefficients across conditions are presented in Figure 5.2 and illustrate these main effects and interactions. Of these, it can be easily seen that the sensitivity of the SD scores to distortion was instrumental in how well corrections were able to improve validity.

In addition, a three-way interaction was observed for corrections based on moderation. As such, the effect of the proportion and sensitivity interaction decreased as the amount of distortion increased. This indicated that sensitivity of SD made less of a difference for corrections based on moderation when the majority of the sample were distorting a great deal. This can be seen in Figure 5.3.

The Cost of Faking and Utility of Corrections

In terms of effect sizes, the results revealed in the Burns and Christiansen (2005) simulation were small. After all, across all conditions the average effect in using SD to make corrections was just over a .01 enhancement of validity. Although such small enhancements might be easily dismissed, it must be remembered that they are typically gained at no cost in terms of the testing process. Utility analysis can therefore be used to illustrate the benefits of small changes in criterion related validity. Analyzing replications from a typical condition of the simulation (moderate amount of applicant distortion, 50% of the sample distorting, sensitivity of SD of .40), it was possible to examine both how much utility is lost due to faking and how much can be recovered by SD corrections. For this, we assume an organization that is screening 500 applicants for a general clerk position with a modest salary of $30,000 a year; SD_y was therefore estimated as 40% of the yearly income or $12,000 (Hunter & Schmidt, 1983; Schmidt & Hunter, 1983). As one component of a multi-hurdle sys-

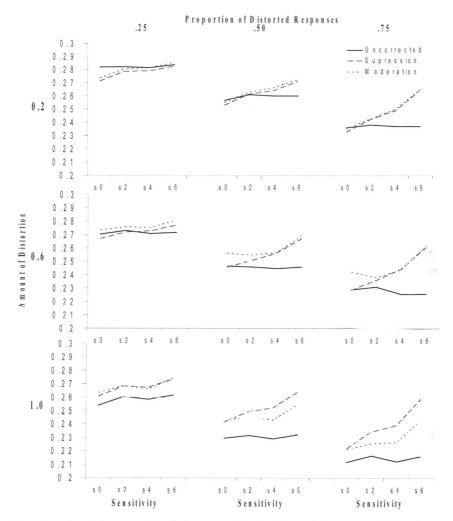

Note: Sensitivity is the correlation between social desirability and the amount of distortion: s0 refers to no correlation, s2 to a .20 correlation, s4 to a .40 correlation, and s6 to a .60 correlation.

Figure 5.2.   Uncorrected and Corrected Criterion-related Validity Coefficients.

tem, the organization intends to use the personality test to screen in 300 applicants and screen out 200 of them (i.e., the selection ratio is .60).

Before examining the utility recovered through corrections, it was possible to examine the actual cost of faking to the organization. In this condition, the validity of the personality scores was reduced from .30 to .248.

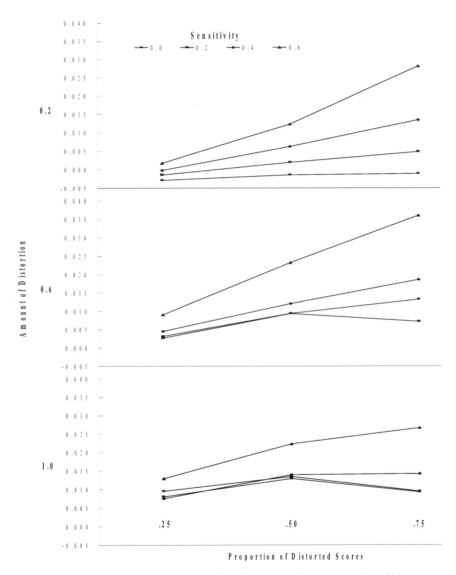

Figure 5.3.   Increase in Validity for moderation corrections across Conditions.

Based on this, the loss of utility due to applicant faking was approximately $422 per applicant hired or $126,600 overall. These are substantial losses to the utility of the personality test. Using SD as a suppressor would recover $21,600 ($72/applicant) or about 17% of the loss. On the other hand, using SD as a moderator would recover $36,000 ($120/applicant) or about 28% of the loss. With a higher paying job, the loss in utility per applicant due to faking and the recovered utility resulting from corrections would be commensurately higher.

Toward a Better Understanding of Social Desirability Scales

Can personality scales be used in personnel selection to improve the quality of hires? Does applicant faking pose a risk for personality scales in personnel selection? The answers to these questions are not mutually exclusive. Although a terminal problem that Guion and Gottier (1965) identified with the use of personality tests in this context was the over-reliance on incumbent samples (that may not resemble applicants in their motivation to distort), the vast majority of the studies that have been marshaled to support use of personality tests for selection continue to use a concurrent validation design. Whereas meta-analysis may have many advantages, aggregating the results of many such studies does not change this limitation of validation designs. What is clear is that concurrent designs can tell us whether personality *constructs* are job-related but very little about whether applicants' scores on personality tests are.

Our review suggests that it may be problematic to draw conclusions about the effects of faking based on less-than-perfect measures of faking behavior, namely SD scales. Unfortunately, that is precisely what has been done by an alarming number of well-known researchers. What is worse, the resulting assertions of the null that "faking does not matter" have not only been based on measures with limited sensitivity to faking, but have used a suppression model where it was not possible to have much improvement. Insult to the injury has occurred when the assertions have been based on nonapplicant samples with little motivation to distort (e.g., Ones et al., 1996; Ones & Viswesvaran, 1998).

The review has also highlighted a number of unanswered questions and areas where additional research is needed to better understand when SD is best suited as a response validity scale and when corrections based on SD are most appropriate. In particular, more research is needed on whether those who *normally* score high on SD scales are more likely to engage in applicant faking, the extent that those *applicants* with high SD scores are more likely to engage in applicant faking, and what the relationship between honest and applicant SD scores might be. Perhaps equally important, research is needed on the relationship between actual applicants' social desirability scores and job performance and whether this depends on the job in question.

Given the limited sensitivity to faking and the complications that arise from the substantive relationships between SD scores and personality constructs likely to be of interest in selection contexts, it may be that a refinement of these scales is necessary before they are more useful. It may be possible to identify items that show a greater difference between those who have faked and those who have not, at the same time as choosing items that are relatively uncorrelated with personality. This would increase sensitivity while reducing the correlation with the trait scores that decisions will be made on.

Another possibility to increase the overall sensitivity would be to use SD scales in tandem with other methods of identifying applicants who have distorted the most. For example, research has shown that faking spuriously increases the covariance among both SD and personality items and that this can be used to index the amount of faking (Chaney & Christiansen, 2004). It has been found that the validity of the personality scores from applicants identified with such a covariance index is worse than that observed in the overall sample or for those with elevated scores on a traditional SD scale (Christiansen, Robie, & Bly, 2005). Similarly, research has shown that including bogus tasks items in a training and experience section of an application may help identify those who engage in intentional distortion (Anderson, Warner, & Spencer, 1984; Pannone, 1984). If the goal were to correctly identify those applicants that have faked and avoid false accusations, looking at the convergence of these methods would probably be the most fruitful approach.

An issue not addressed in this chapter is whether faking represents a single strategy or whether there are different approaches employed by different applicants. Herein we have addressed whether applicants have faked or not, and by how much. There is growing evidence that faking may not a single, simple strategy of just choosing the response that is perceived to be most desirable for the job. For example, Zickar, Gibby, and Robie (2004) indicated that in an applicant sample three IRT classes were needed to fit applicant responses and Levin and Zickar (2002) discussed the difference between self-presentation, lies, and bullshit. Moreover, many applicants may be concerned over presenting a uniformly favorable profile, picking and choosing which type of negative behavioral tendencies to admit to. Evidence suggests the basis of these choices may be beliefs about which traits are job-relevant and that there are important individual differences in these belief systems (Christiansen, Burns, & Montgomery, 2005). Taken together, this suggests another important limitation to SD scales: they may capture only one faking strategy.

Another issue not addressed is under what circumstances corrections should be attempted at all, given that they are most useful when faking is severe. One possibility is suggested by the recurrent finding that the validity of applicant personality scores is worse at the top of the distribution. Although in many circumstances performance data may not be available, scores on another variable related to personality but not susceptible to self-report biases may be. Examples might be GPA or assessment center ratings. For example, if GPA were available from college records it could be observed whether the unstandardized regression coefficient associated with applicants' scores at the top of a conscientiousness distribution were similar to that from those at the bottom. If so, either a great amount of distortion did not occur or it probably did not meaningfully affect validity. If validity appears to be different at the top and bottom, corrections might be expected to be useful. Obviously, more research on practical techniques for identifying the circumstances where the most faking has occurred in order to more judiciously apply SD-based corrections is needed.

Conclusion

Although it is widely assumed that SD scales faithfully measure the amount an applicant has faked, serious questions remained about the construct validity of these scales. Clearly they have some sensitivity to applicant faking, but reasonable people might disagree whether it is currently strong enough to justify the inference that high scorers have misrepresented themselves and that their trait scores should be tampered with. This may be more of an ethical issue than a practical one, as even with limited sensitivity some gain in utility may be realized by using SD scores for corrections. However, when these corrections will be successful and which strategy will be most effective would appear to depend on a wide array of parameters, many of which are in need of more research before firm conclusions may be drawn.

REFERENCES

Alliger, G. M., & Dwight, S. A. (2000). A meta-analytic investigation of the susceptibility of integrity tests to faking and coaching. *Educational and Psychological Measurement, 60,* 59-72.

Anderson, C. D., Warner, J. L., & Spencer, C. C. (1984). Inflation bias in self-assessment examinations: Implications for valid employee selection. *Journal of Applied Psychology, 69,* 574-580.

Barrick, M. R., & Mount, M. K. (1991). The big five personality dimensions and job performance: A meta-analysis. *Personnel Psychology, 44,* 1-26.

Barrick, M. B., & Mount, M. K. (1996). Effects of impression management and self-deception on the predictive validity of personality constructs. *Journal of Applied Psychology, 42,* 261-272.

Bernreuter, R. G. (1933). Validity of personality inventory. *Personnel Journal, 11,* 383-386.

Block, J. (1965). *The challenge of response sets.* New York: Appleton-Century-Crofts.

Burns, G. N., & Christiansen, N. D. (2004, April). *Effects of faking on the linear construct relationships of personality test scores.* Paper presented at the 19th annual conference of the Society for Industrial and Organizational Psychology, Chicago

Burns, G. N., & Christiansen, N. D. (2005, April). *Use of social desirability as a suppressor versus moderator.* Paper presented at the 20th annual conference of the Society for Industrial and Organizational Psychology, Los Angeles, CA.

Butcher, J. N, Morfitt, R. C., Rouse, S. V., & Holden, R. R. (1997). Reducing MMPI-2 defensiveness: The effect of specialized instructions on retest validity in a job applicant sample. *Journal of Personality Assessment, 68,* 385-401.

Cattell, R. B. (1989). *The 16PF: Personality in depth.* Champaign, IL: IPAT.

Cattell, R. B., Eber, H. W., & Tatsuoka, M. M. (1970). *Handbook for the Sixteen Personality Factor Questionnaire.* Champaign, IL: IPAT.

Chaney, S. K., & Christiansen, N. D. (2004, April). *Disentangling applicant faking from personality: Using covariance to detect response distortion.* Paper presented at the annual conference of the Society for Industrial and Organizational Psychology held in Chicago.

Christiansen, N. D. (1997). *Sensitive or senseless? The usefulness of social desirability measures for making inferences concerning applicant response distortion.* Unpublished manuscript.

Christiansen, N. D., Burns, G. N., & Montgomery, G. E. (2005). Reconsidering forced-choice item formats for applicant personality assessment. *Human Performance, 18,* 267-307.

Christiansen, N. D., Goffin, R. D., Johnston, N. G., & Rothstein, M. G. (1994). Correcting the 16PF for faking: Effects on criterion-related validity and individual hiring decisions. *Personnel Psychology, 47,* 847-860.

Christiansen, N. D., Robie, C., & Bly, P. R. (2005, April). *Using covariance to detect applicant response distortion of personality measures.* Paper presented at the annual conference of the Society for Industrial and Organizational Psychology held in Los Angeles.

Christiansen, N. D., & Rozek, R. F. (2004, April). *Effects of socially desirable responding on hiring judgments.* Paper presented at the 19th annual conference of the Society for Industrial and Organizational Psychology, Chicago.

Conger, A. J., & Jackson, D. N. (1972). Suppressor variables, prediction, and the interpretation of psychological relationships. *Educational and Psychological Measurement, 32,* 579-599.

Cook, M. (1993). *Personnel selection and productivity* (Rev. ed.). New York: Wiley.

Costa, P. T., Jr., & McCrae, R. R. (1992). *The Revised NEO-PI/NEO-FFI manual supplement.* Psychological Assessment Resources: Odessa, FL.

Crowne, D. P., & Marlowe, D. (1960). A new scale of social desirability independent of psychopathology. *Journal of Consulting Psychology, 24,* 349-354.

Crowne, D. P., & Marlowe, D. (1964). *The approval motive.* New York: Wiley.

Damarin, F., & Messick, S. (1965). *Response styles as personality variables: A theoretical integration* (ETS RB 65-10). Princeton, NJ: Educational Testing Service.

Dicken, C. F. (1959). Simulated patterns of the Edwards Personal Preference Schedule. *Journal of Applied Psychology, 43*, 372-378.

Donovan, J. J., Dwight, S. A., & Hurtz, G. M. (2003). As assessment of the prevalence, severity, and verifiability of entry-level applicant faking using the randomized response technique. *Human Performance, 16*, 81-106.

Edwards, A. L. (1957). *The social desirability variable in personality assessment and research*. New York: Dryden.

Edwards, J.R. (2002). Alternatives to difference scores: Polynomial regression analysis and response surface methodology. In F. Drasgow & N. Schmitt (Eds.), *Measuring and analyzing behavior in organizations: Advances in measurement and data analysis* (pp. 350-400). San Francisco: Jossey-Bass/Pfeiffer.

Eysenck, H. J., & Eysenck, S. B. G. (1964). *The manual of the Eysenck Personality Inventory*. London: University of London Press.

Furnham, A. (1990). Faking personality questionnaires: Fabricating different profiles for different purposes. *Current Psychology: Research and Reviews, 9*, 46-55.

Ganster, D. C., Hennessey, H. W., & Luthans, F. (1983). Social desirability response effects: Three alternative models. *Academy of Management Journal, 26*, 321-331.

Goffin, R. D., & Christiansen, N. D. (2003). Correcting personality tests for faking: A review of popular personality tests and an initial survey of researchers. *International Journal of Selection and Assessment, 11*, 340-344.

Gough, H. G. (1957). *The California Psychological Inventory administrator's guide*. Pala Alto, CA: CPP.

Griffith, R. L., Chmielowski, T. S., & Yoshita, Y. (in press). Do applicants fake? An examination of the frequency of applicant faking behavior. *Personnel Review*.

Guion, R. M. (1998). *Assessment, measurement, and prediction for personnel decisions*. Mahwah, NJ: Erlbaum.

Guion, R. M., & Gottier, R. F. (1965). Validity of personality measures in personnel selection. *Personnel Psychology, 18*, 135-164.

Haaland, D., Christiansen, N. D., & Kaufmann, G. (1999, April). *Applicant distortion of personality measures in police selection: Reasons for optimism and caution*. Symposium conducted at the 14th annual meeting of the Society for Industrial and Organizational Psychology, Dallas, TX.

Hoffman, E. (2001). *Ace the corporate personality test*. New York: McGraw-Hill.

Hogan, R., & Hogan, J. (1992). *Hogan Personality Inventory manual*. Tulsa, OK: Hogan Assessment Systems.

Hogan, R., Hogan, J., & Roberts, B. W. (1996). Personality measurement and employment decisions: Questions and Answers. *American Psychologist, 51*, 469-477.

Holden, R. R., & Fekken, G. C. (1989). Three common social desirability scales: Friends, acquaintances, or strangers? *Journal of Research in Personality, 23*, 180-191.

Hough, L. M. (1998). Effects of intentional distortion in personality measurement and evaluation of suggested palliatives. *Human Performance, 11*, 209-244.

Hough, L. M., Eaton, N. K., Dunnette, M. D., Kamp, J. D., & McCloy, R. A. (1990). Criterion-related validities of personality constructs and the effect of response distortion on those validities [Monograph]. *Journal of Applied Psychology, 75,* 581-585.

Hunter, J. E., & Schmidt, F. L. (1983). Quantifying the effects of psychological interventions on employee job performance and work-force productivity. *American Psychologist, 38,* 473-478.

Jackson, D. N., & Messick, S. (1962). Response styles on the MMPI: Comparison of clinical and normal samples. *Journal of Abnormal & Social Psychology, 65,* 285-299.

Jeanneret, R., & Silzer, R. (1998) An overview of the individual psychological assessment. In R. Jeanneret & R. Silzer (Eds.), *Individual Psychological Assessment.* San Francisco, CA: Jossey-Bass.

Kriedt, P. H., & Dawson, R. I. (1961). Response set and the prediction of clerical job performance. *Journal of Applied Psychology, 45,* 175-178.

Krug, S. E. (1978). Further evidence on 16 PF distortion scales. *Journal of Personality Assessment, 42,* 513-518.

Levin, R. A., & Zickar, M. J. (2002). Investigating self-presentation, lies, and bullshit: Understanding faking and its effects on selection decisions using theory, field research, and simulation. In J. M. Brett & F. Drasgow (Eds.), *The psychology of work: Theoretically based empirical research* (pp. 253-276). Mahwah, NJ: Erlbaum.

Mahar, D., Cologon, J., & Duck, J. (1995). Response strategies when faking personality questionnaires in a vocational selection setting. *Personality and Individual Differences, 18,* 605-609.

McCrae, R. R., & Costa, P. T., Jr. (1983). Social desirability scales: More substance than style. *Journal of Consulting and Clinical Psychology, 51,* 882-888.

McDaniel, M. A., Douglas, E. F., & Snell, A. F. (1997, April). *A survey of deception among job seekers.* Paper presented at the 12th annual conference of the Society of Industrial and Organizational Psychology, St. Louis, MO.

Meehl, P. E., & Hathaway, S. R. (1946). The K Factor as a suppressor variable in the MMPI. *Journal of Applied Psychology, 30,* 526-564.

Mueller-Hanson, R., Heggestad, E. D., & Thornton, G. C. (2003). Faking and selection: Considering the use of personality from a select-in and select-out perspective. *Journal of Applied Psychology, 88,* 348-355.

Ones, D. S., & Viswesvaran, C. (1998). The effects of social desirability and faking on personality and integrity assessment for personnel selection. *Human Performance, 11,* 245-269.

Ones, D. S., Viswesvaran, C., & Reiss, A. D. (1996). Role of social desirability in personality testing for selection: The red herring. *Journal of Applied Psychology, 81,* 660-679.

Paulhus, D. L. (1984). Two-component models of socially desirable responding. *Journal of Personality and Social Psychology, 46,* 598-609.

Paulhus, D. L. (1986). Self-deception and impression management in test responses. In A. Angleitner & J. S. Wiggins (Eds.), *Personality assessment via questionnaire* (pp. 143-165). New York: Springer-Verlag.

Paulhus, D. L. (1988). *Assessing self-deception and impression management in self-reports: The Balanced Inventory of Desirable Responding.* Unpublished manuscript, University of British Columbia, Vancouver, Canada.

Paulhus, D. L. (1991). Measurement and control of response bias. In J. Robinson, P. Shaver, & L. Wrightsman (Eds.), *Measures of Personality and Social Psychological Attitudes* (pp. 17-59). San Diego, CA: Academic Press.

Paulhus, D. L., Bruce, M. N., & Trapnell, P. D. (1995). Effects of self-presentation on personality profiles and their structure. *Personality and Social Psychology Bulletin, 21*, 100-108.

Pannone, R. D. (1984). Predicting test performance: A content valid approach to screening applicants. *Personnel Psychology, 37*, 507-514.

Rees, C. J., & Metcalf, B. (2003). The faking of personality questionnaire results: Who's kidding whom? *Journal of Managerial Psychology, 18*, 156-165.

Raykov, T. (1999). Are simple change scores obsolete? An approach to studying correlates and predictors of change. *Applied Psychological Measurement, 23*, 120-126.

Rosse, J. G., Stecher, M. D., & Miller, J. L., & Levin, R. A. (1998). The impact of response distortion on preemployment testing and hiring decisions. *Journal of Applied Psychology, 83*, 634-644.

Ryan, A. M., & Sackett, P. R. (1998). Individual assessment: The research base. In R. Jeanneret & R. Silzer (Eds.), *Individual Psychological Assessment*. San Francisco: Jossey-Bass.

Sackeim, H. A., & Gur, R .C. (1978). Self-deceotpion, self-confrontation and consciousness. In G. E. Schwartz & D. Shapiro (Eds.), *Consciousness and self-regulation: Advances in research* (Vol. 2, pp. 139-197). New York: Plenum.

Schlenker, B. R. (1980). *Impression management: The self concept, social identity, and interpersonal relations*. Monterey, CA: Brooks/Cole.

Schmidt, F. L., & Hunter, J. E. (1983). Individual differences in productivity: An empirical test of estimates derived from studies of selection procedure utility. *Journal of Applied Psychology, 68*, 407-414.

Schmitt, N., & Chan, D. (1998). *Personnel selection: A theoretical approach*. Thousand Oaks, CA: Sage.

Schmitt, N., & Oswald, F. L. (in press). The impact of corrections for faking on the validity of noncognitive measures in selection settings. *Journal of Applied Psychology*.

Schuessler, K., Hittle, D., & Cardascia, J. (1978). Measuring responding desirability with attitude-opinion items. *Social Psychology, 41*, 224-235.

Smith, D. B., & Robie, C. (2004). The implications of impression management for personality research in organizations. In B. Schneider & D. B. Smith (Eds.), *Personality and organizations* (pp. 111-140). Mahwah, NJ: Erlbaum.

Smith, D. B., & Ellingson, J. E. (2002). Substance versus style: A new look at social desirability in motivating contexts. *Journal of Applied Psychology, 87*, 211-219.

Stöber, J. (2001). The Social Desirability Scale-17 (SDS-17): Convergent validity, discriminant validity, and relationship. *European Journal of Psychological assessment, 17*, 222-232.

Tett, R. P., Jackson, D. N., & Rothstein, M. G. (1991). Personality measures as predictors of job performance: A meta-analytic review. *Personnel Psychology, 44*, 703-742.

Tett, R. P., Jackson, D. N., Rothstein, M., & Reddon, J. R. (1999). Meta-analysis of bidirectional relations in personality-job performance research. *Human Performance*, *12*, 1-29.

Tourangeau, R., & Rasinksi, K. A. (1981). Cognitive processes underlying context effects I attitude measurement. *Psychological Bulletin*, *103*, 299-314.

Viswesvaran, C., & Ones, D. S. (1999). Meta-analyses of fakability estimates: Implications for personality measurement. *Educational and Psychological Measurement*, *59*, 197-210.

Viswesvaran, C., Ones, D. S., & Hough, L. M. (2001). Do impression management scales in personality inventories predict managerial job performance ratings? *International Journal of Selection and Assessment*, *9*, 277-289.

White, L. A., & Kilcullen, R. N. (1998, April). *How socially desirable responding affects the criterion-related validity of self-report measures.* Paper presented at the 13th annual conference of the Society for Industrial and Organizational Psychology, Dallas, TX.

White, L. A., Young, M. C., & Rumsey (2001). Assessment of background and life experiences (ABLE) implementation issues and related research. In J. P. Campbell & D. J. Knapps (Eds.), *Exploring the limits in personnel selection and classification* (pp. 526-558). Mahway, NJ: Earlbaum.

Wiggins, J. S. (1959) Interrelationships among MMPI measures of dissimulation under standard and social desirability instruction. *Journal of Consulting Psychology*, *23*, 419-427.

Wiggins, J. S. (1964). Convergences among stylistic response measures from objective personality tests. *Educational and Psychological Measurement*, *24*, 551-562.

Wiggins, J. S. (1968). Personality structure. *Annual Review of Psychology*, *19*, 293-350.

Zickar, M. J., Gibby, R. E., & Robie, C. (2004). Uncovering faking samples in applicant, incumbent, and experimental data sets: An application of mixed-model item response theory. *Organizational Research Methods*, *7*, 168-190.

Zickar, M. J., Rosse, J. G., Levin, R. A., & Hulin, C. L. (1997). *Modeling the effects of faking on personality tests.* Unpublished manuscript.

CHAPTER 6

APPLICANT FAKING BEHAVIOR

Teasing Apart the Influence of Situational Variance, Cognitive Biases, and Individual Differences

**Richard Griffith, Tina Malm, Andrew English,
Yukiko Yoshita and Abhishek Gujar**

There is now considerable evidence to suggest that a significant portion of applicants fake their responses to personality measures (Donovan, Dwight, & Hurtz, 2003; Donovan, Dwight, & Schneider, 2005; Griffith, Chmielowski, & Yoshita, in press; McDaniel, Douglas, & Snell, 1997; Rosse, Stecher, Miller, & Levin, 1998; Stokes, Hogan, & Snell, 1993). This support does little to ease the minds of practitioners who have had a long history of concern regarding the impact of faking on their selection decisions. Why are practitioners concerned? What is it about someone who fakes his or her score that HR professionals find undesirable? Those who design selection procedures never promised perfect prediction, and selection errors are made at a much higher rate than I/O psychologists would like to admit. But there is something about fakers that makes them stand out. Much of this anxiety regard-

A Closer Examination of Applicant Faking Behavior, 151–178

ing applicant faking stems from assumptions that we have about deception.

While both unconscious and conscious forms of faking behavior have been identified (e.g. Self-Deceptive Enhancement/Denial and Impression Management), a great deal of the discomfort can be attributed to faking behavior being seen as volitional. The fundamental attribution error may cause us to target the locus of the behavior within the individual. Applicants who faked lied to us. They deceived us. They tricked us into giving them a job. Our socialization regarding this kind of behavior is strong, and in general we associate deception with a lack of integrity. This categorization may cause those concerned with faking to extend their attitudes about deception to the employment setting. The secret fear of some employers is that fakers may not stop at the application stage with their dishonest behavior; they may steal, lie to customers, and conceal mistakes.

These assumptions regarding fakers remain largely untested, and there currently is little empirical evidence to support them (e.g. Rosse, Levin, & Nowicki, 1999). The truth is that little is known about the characteristics of the faker and how those characteristics, as well as situational influences, combine to form the phenomenon of faking. Deception is a complex activity, and applicant faking behavior is likely to have similarly complex dynamics. Much of the literature frames applicant faking behavior as a unitary construct, or at best divides it into simple conscious and unconscious influences. While parsimonious, these conceptualizations of faking have not proven useful, because they do not give us much information about the dynamics of the behavior or the underlying trait and situational influences. Recent research supports the notion that multiple forms of faking behavior may be occurring in an applicant setting (Zickar, Gibby, & Robie, 2004). Therefore it is reasonable to expect multiple influences on that behavior.

Comprehensive models of applicant faking behavior have been suggested (Douglas, McDaniel, & Snell, 1996; McFarland & Ryan, 2000; Snell & McDaniel, 1998; Snell, Sydell, & Lueke, 1999). These models recognize the dynamic nature of the phenomenon, and incorporate a variety of possible influences on applicant faking. Some of these factors, such as integrity and cognitive ability, are internal to the applicant. Other factors, such as the presence of warnings or the desirability of the job, may be a function of the situation. While faking is an individual behavior (Snell & McDaniel, 1998; Zickar & Robie, 1999), the influence of situational variance should not be ignored. The situation of applying for a job is highly constrained. We are socialized to put our "best foot forward" and to emphasize our strengths in an employment setting. Therefore, applicants may not be "faking" in the sense that they are intentionally attempting to deceive the employer, but may be answering in a fashion that they perceive to be consistent with the demands of the situation. Most likely the deciding factor of whether applicants will elevate their scores in an applicant setting will be an interaction of the situation and applicant characteristics. While these models offer the most complete explanation of applicant faking behavior to date, many elements have not been empirically tested in an integrated fashion.

This chapter explores these dynamics by examining theoretical models of potential influences on applicant faking behavior. More specifically, we attempted to tease apart the influence of situational variance, cognitive biases, and individual differences (ability, integrity, locus of control, and self-monitoring) on response distortion in the applicant setting. In addition we will discuss the influence of previously suggested faking-related constructs such as self-deceptive enhancement and impression management. We will also review the results of a number of recent studies we have conducted utilizing these variables that may shed light on the dynamics of applicant faking behavior. Our overall goal was to consolidate the findings of several integrated studies to develop an empirical nomological network around the phenomenon of applicant faking behavior. Our hope was that this nomological network may then lead to a better understanding of the interactions and processes that underlie faking behavior.

What is Faking?

While conceptually simple, in practice it has been extremely difficult to directly identify which individual or group of individuals faked their responses in an employment context. So before we set out to examine possible antecedents of faking behavior, we thought it best to clearly define the variable of interest, and to identify a method of measurement to best assess this behavior. The faking of personality measures used for personnel selection has been referred to in the literature as response distortion, impression management, social desirability, displaying unlikely virtues, and self enhancement (Hough, Eaton, Dunnette, Kamp, & McCloy, 1990; Hough & Paullin, 1994; Ones, Viswesvaran, & Korbin, 1995). This lack of clear terminology and a widely accepted operational definition of the phenomenon has slowed research efforts and often led researchers to inconclusive or contradictory results. We were particularly interested in directly assessing applicant faking behavior rather than using proxy measures such as social desirability or unlikely virtue scales.

We started our search for an operational definition of faking by examining the goals of personality measurement in the applicant setting. These goals, loosely based in latent trait theory, generally revolve around identifying traits that are associated with good job performance and then accurately assessing whether applicants do in fact possess high levels of the trait. However, applicants obviously have their own goals, which are not necessarily congruent with those of the organization. The primary goal of the applicant is to obtain employment. *Some* applicants may see the personality test as a chance to increase the odds of obtaining this goal and elevate their scores to be more competitive in the hiring process. Therefore they may report characteristics that they think potential employers find desirable. But we want to know who they really are by getting a sense of how they will behave, particularly in the employment context. We want to know their true scores.

Latent trait theory suggests that an individual's behaviors are influenced by constructs, and that individual differences on these constructs can be accurately measured. Central to this theory is the notion of the true score. A true score is the individual's actual level of the construct, but this level cannot be directly measured and must be inferred. In an applicant setting, individuals may answer noncognitive measures in a fashion that does not reflect this true score in order to increase their chances of being hired. Thus conceptually, the difference between the individual's true score and the score in the applicant setting is an appropriate measure of faking (Figure 6.1).

While seemingly uncomplicated, the operational conditions necessary for this definition of faking cannot be fully met. The true score is a theoretical abstraction, and direct measurement of a true score is not possible. In fact, some personality researchers would argue whether a true score, as we refer to it here, even exists (Hogan, 1991). Because true scores can never be known, measurement researchers have traditionally relied on alternative methods to derive the true score. Classical test theory offers an operational definition of the true score, and a method to estimate this score (Crocker & Algina, 1986). Multiple observations, either through repeated measurement of the construct, or in the weaker form of multiple items, are averaged to produce an estimate of the true score. Because of the imprecise nature of this estimate, the standard error of measurement (SEM) is used to form a confidence interval around the mean score, and the true score is assumed to reside within this interval (Traub, 1994). While identifying the true score is necessary to determine the amount of applicant faking using our definition, it may be possible to estimate faking using an estimate of the true score that can serve as a baseline.

Individuals must be assessed under two types of conditions in order to meet our conceptual definition of applicant faking behavior. First, they must complete the noncognitive measure while applying for a job. Under this condition respondents are under the impression that scores on the personality measure are used as part of the selection process, and may be

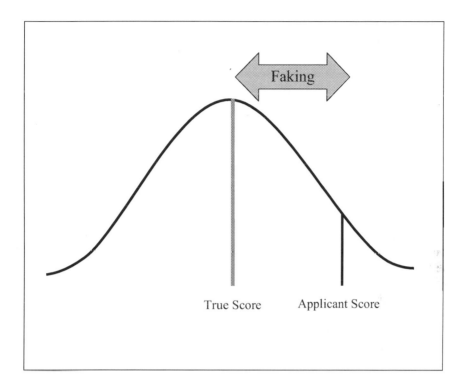

Figure 6.1.   Graphic of Applicant Faking.

motivated to represent themselves in a favorable light. Second, individuals must be directed to honestly complete a personality measure under pure research conditions to provide a baseline for our operational definition of faking. Here, respondents are not exposed to any incentives and are therefore not motivated to fake. This honest score can serve as a baseline score that approximates the true score. On a group level, the average scores of these conditions should differ from each other. Therefore we operationally defined faking as the difference between response conditions (examining the change between applicant scores and the honest scores).

Deconstructing Faking

Our primary interest was to examine variables that may help us explain the variance in our conceptualization of faking. Some applicants may choose to fake to a large extent, some moderately, and some not at all. We don't expect a single source of variance (such as social desirability) to account for this variation. Instead, we based much of our focus on the previous theoretical models suggested by McFarland and Ryan (2000) and Snell and McDaniel (1998). To that effect, we included both external influences as well as characteristics internal to the applicant.

External Influences and Situational Variance

Interactional psychology suggests that situations cause different people to behave similarly and similar people to behave differently. This psychological perspective is based on three basic assumptions (Tett & Gutterman, 2000). First, situational influences on behavior are mediated by how situations are perceived (Mischel, 1973). Second, people influence and are influenced by their environment. Third, in order for personality traits to be expressed, trait-relevant situations are required (Kenrick & Funder, 1988). Lord (1982) found that the best predictors of consistency were tailored to each individual's unique perceptions. In other words, our perceptions of a situation will drive our behavior in that specific situation.

One such specific situation is the applicant setting, which has built-in constraints that influence responding. It is possible that situational demands may be partially responsible for elevated scores. Perhaps it is not that applicants are "faking" (or responding dishonestly per se), but that they are responding in a manner dictated by the situation. The applicant environment is highly constrained. We are taught at a relatively young age the "rules" of applying for a job, and to portray a favorable initial impression on our prospective employer. We are also socialized to emphasize our strengths. Therefore, applicants may not be faking in the sense that they are attempting to deceive their prospective organization, but simply may be responding consistently with the demands of the situation. Therefore, the situation may be driving their responses more than their true score on the trait in question. In this case, the situation (employment setting) strongly influences responding, and a common pattern (inflated scores) may be identified. This pattern of responses may be outside of the conscious awareness of the applicant.

Many researchers have focused on this phenomenon, but under various titles including cross-situational consistency, trait activation, and frame of reference (FOR). An existing trait may not be recorded due to an irrelevant circumstance. Research has suggested that situational relevance should be considered when developing a personality instrument or experiment. Tett and Gutterman (2000) offer support for the notion of situational effects on behavior. They suggest that the behavioral expression of a trait depends on the arousal of that trait, by what they refer to as trait-related situational cues. In a review of contextual specificity, Mischel and Shoda (1995) suggested that a trait that is dependent on, or "activated" by, a particular situation, and should be measured only under that specific situation. This line of reasoning provides additional support for the use of personality measures that incorporate a work frame-of-reference (Schmit, Ryan, Stierwalt, & Powell, 1995).

FOR modifications usually consist of altering the instructional set to emphasize a specific situation or setting. Another method to adjust respondent FOR is to alter the stem of the item to reflect a specific setting such as school (Schmit et al., 1995) or work (Hunthausen, Truxillo, Bauer, & Hammer, 2003). Thus, an item that originally was phrased "I keep my surroundings organized" may be modified to read, "I keep my surroundings organized at work". Due to more consistent item endorsements across respondents, FOR measures have been shown to produce lower error variances (Robie, Schmit, Ryan, & Zickar, 2000). Previous studies also suggest that gains in both the reliability and criterion validity of Big Five measures can be made by incorporating a specific FOR into the instruments (Bing, Davidson, Whanger, & VanHook, 2004; Hunthausen et al., 2003; Schmit et al.). These benefits are likely the result of a more uniform interpretation by respondents, and are evidenced through higher reliabilities than noncontextual measures (Schmit et al.).

Previous research examining FOR also investigated the role of Socially Desirable Responding (SDR) as a factor in applicant responding. Schmit et al. (1995) suggested that utilizing a FOR may increase SDR because items were more transparent than generally framed items. The contextual information present in FOR items essentially provides the respondent a "blue print" on *how* to fake. The authors examined whether SDR was a primary influence on applicant responses, or if applicants were responding in a fashion consistent with self-presentation theory (Hogan, 1991). Both forms of responding would result in elevated scores, but would have differential effects on criterion validity. If applicants were responding in socially desirable fashion validities would be expected to decrease. However, if self-presentation theory was operating, transparent items would allow applicants to present themselves as they would normally in other social interactions, and validities should increase with an FOR modification. While the results of Schmit et al. implied that self-presentation theory had more influence on responses, the results of Bing et al. (2004) suggested SDR may still be a factor. In a replication of the Schmit et al. study, Bing found a different pattern in the ranking of validities across FOR conditions that suggested both self-presentation theory and SDR were influencing scores on noncognitive measures. However, no measure of socially desirable responding or applicant faking was included in the aforementioned studies, so the conclusions drawn regarding FOR and applicant faking must be inferred from the pattern of validities, and thus interpreted with caution. It should be noted that traditional measures of SDR have been heavily criticized as indicators of actual faking behavior (see Burns & Christiansen, in this volume).

It is plausible that adding a work FOR to a noncognitive selection measure may actually decrease the amount of applicant faking that occurs. A personality measure embedded in a work FOR may not necessarily decrease faking directly, but may reduce the opportunity to fake. Traditionally, opportunity to fake has been defined in relation to a respondent's true score. The lower the true score, the more opportunity the applicant has to elevate his or her score in the applicant setting. Because individuals are responding to items as they would behave at work, specific item responses may be constrained (especially negative behaviors like arguing, being unfriendly, untidiness and lack of effort). Since the work setting constrains certain negative behaviors and reinforces other positive behaviors, it is likely that individuals will have less opportunity to fake when responding to items in an FOR condition. When answering consistently with this work FOR, respondents will display high levels of positive characteristics and low levels of negative characteristics regardless of the presence of a motivated setting. Douglas et al. (1996) suggested a theoretical model of applicant faking which included the opportunity to fake, and may partially explain the reduction of applicant faking with an FOR instrument. Douglas et al. stated that those who have a high true score on the construct of interest may not be able to fake because of the ceiling of the measurement instrument. Thus even if an applicant with a high true score chose to fake, they would be restricted in the opportunity to raise their score significantly. When respondents complete an FOR measure their "true score" should, on average, be higher than a general frame of reference, leaving less room for elevation. Previous research has supported this pattern of responses. Schmit et al., (1995) demonstrated that the inclusion of an FOR modification in noncognitive measures significantly increases the respondent's mean scores in a positive fashion. We suspected that this elevation would result in smaller magnitudes of applicant faking, as well as smaller percentages of applicant fakers.

Cognitive Biases

Another potential source of faking variance is cognitive bias. While scores on a personality scale are seen to reflect an individual's general level on a trait, cognitive biases may change the way a person frames the perception of his or her level on that trait. Examples of common cognitive biases that can influence responses on noncognitive instruments are

nay saying (the tendency to respond negatively to an inquiry) and acquiescence (the tendency to answer in the affirmative). Additionally, attribution research has also established an inherent "self-serving bias" in
people's success and failure explanations. Due to our tendency to protect
and enhance our self-esteem, we tend to credit ourselves for success and
avoid blame for failure (Ross & Fletcher, 1985).

One cognitive bias that has received little attention in the response distortion literature is temporal bias. Previous research has demonstrated a
tendency for well-adjusted individuals to see the future as being better than
the past. For example, Wilson and Ross (2000) conducted a series of studies
that showed that people tend to describe their past selves less favorably
than their present selves, and that enhancement in the perception of the
present self can be illusory (i.e., in the absence of actual improvement). Positive personality traits are desirable attributes and individuals may wish to
demonstrate them in the appropriate context. Further, people tend to perceive more control over their future than their past, and hence would be
more prone to have more positive estimates of their future selves in comparison to their past selves (Robinson & Ryff, 1999). This is a "feel good"
type of projection, where anticipated improvements serve as conditions for
inspiration and self-enhancement. Buehler, Griffin, and Ross (1994) state
that people often tend to ignore reality in the face of negative past experiences, and remain overly optimistic.

In an employment setting, applicants may be less motivated to accurately self-assess themselves and hence temporal comparisons would be
prevalent. Applicants may find a future state (which may be illusory) more
appealing and desirable, and the individual may anticipate the acquisi-
tion of this future state (Wilson & Ross, 2000). Furthermore, people tend
to perceive more control over their future than their past (Robinson &
Ryff, 1999).

Thus, answering questions about future job performance, applicants
may not be overtly dishonest, but simply manifesting a "hard-wired" tendency to predict optimistic estimates future performance. This phenomenon could possibly explain differences found between applicant and
incumbent scores on personality inventories (Stokes et al., 1993). While
the situation may dictate a future frame of reference for applicants that
may inflate scores, incumbents may adopt a present frame of reference

Individual Differences

The focus on the internal characteristics of the individual is consistent
with fundamental attribution theory, in which behaviors tend to be associated with individual characteristics instead of the situation. A variety of

individual characteristics have been hypothesized to contribute to applicant faking behavior, including cognitive ability, integrity, and self-monitoring. Previous research has suggested that applicants must identify what construct is being measured, and choose the "correct" response for the specific job that the applicant is attempting to obtain (Douglas et al., 1996; McFarland & Ryan 2000). In order to make the correct responses, cognitive ability may be required.

Ability

Individuals high in cognitive ability tend to recognize and solve problems, especially ambiguous problems, more successfully than those low in ability. Personality measures may be characterized as "ill-defined" (Vernon, 1933), and often have various acceptable solutions. They are embedded in (and require) prior everyday experience, and they also require a certain amount of verbal ability in order to understand the underlying construct. High ability individuals may be better able to analytically apply problem solving strategies and verbal abilities by first detecting the underlying construct of an item, and figuring out the desirable answer. They might also understand the advantage of faking personality selection measures in order to obtain the job. Individuals low in cognitive ability may fail to do so, naively approaching the noncognitive measure, failing to see which personality characteristics an employer might find to be desirable for his or her employees. However, correctly identifying some desirable personality traits does not require a rocket scientist (unless you are selecting for the job of rocket scientist!). Single construct measures would be fairly easy, while job analysis-driven targeted profiles would require some problem-solving ability. Scale characteristics, such as elaborate warnings or forced choice format, may moderate this relationship considerably (Vasilopoulos, Cucina, & McElreath, 2005).

Research investigating whether cognitive ability influences applicant response distortion offers inconsistent findings. While, Anderson, Sison, and Wester (1984), Burkhart, Gynther, and Christian (1978), and Garry (1953) failed to find a significant relationship between the *ability* to fake noncognitive measures and intelligence, other researchers report significant findings. Alliger, Lilienfeld, and Mitchell (1996) reported that intelligence and the ability to fake an overt integrity test were positively correlated ($r = .36$, $p < .05$). Correlations between intelligence and the ability to fake the covert integrity measure were not significant, however. Nguyen (2001) found respondents' cognitive ability to predict faking behavior on the Work Judgment Survey, accounting for 3% of the variance in faked judgment scores. Furthermore, Mersman, and Shultz (1998) found that cognitive ability was positively related to the ability to fake measures of extraversion, conscientiousness, and stability. Finally,

Wrensen and Biderman (2005) also found the ability to fake to be significantly related to cognitive ability.

Only Austin, Hofer, Deary, and Eber (2000) investigated the relationship between cognitive ability and response distortion in an applicant setting, by investigating the actual amount individuals chose to fake. They
found positive correlations for a police applicant sample between cognitive ability and the personality traits of stability, dominance, enthusiasm,
extraversion, independence, and conscientiousness, which all represent
highly desirable traits. However, negative correlations were reported
between cognitive ability and the personality trait of apprehensiveness (an
undesirable trait). The authors suggested that cognitive ability may have
influenced those response choices. While the literature is somewhat
inconclusive in regards to the role of cognitive ability and applicant faking, we expected to see a positive relationship between those two variables.

Integrity

Previous research has suggested that integrity may be related to the
ability to fake. McFarland and Ryan (2000) reported that individuals with
higher integrity scores were less likely to increase their scores through faking. Other studies, however, found individuals scoring high on integrity
tests to be more apt to fake (Wrensen & Biderman, 2005).

Integrity is often considered synonymous with honesty, implying truthfulness, fairness, and refusal to engage in fraud or deceit. Douglas et al.
(1996) suggested that an individuals' behavior is strongly influenced by
their values. Individuals' moral or ethical values may be measured by
integrity measures, which assess job commitment, moral reasoning,
responsibility, and work ethics (Ones, Viswesvaran, & Schmidt, 1993).
Conceivably, individuals with high integrity will have high moral standards, and therefore would not distort their response because they believe
doing so is wrong. Research has demonstrated that individuals with high
integrity tend to be reserved, responsible, and moderate, while those with
low integrity tend to be impulsive, spontaneous, and emotional (Sackett &
Wanek, 1996).

According to McFarland and Ryan (2000), "research within social psychology has indicated that individuals who are inherently honest and
have high moral standards are less likely to attempt to control the images
they present to others because they believe that doing so is wrong" (p.
814). If integrity is a significant predictor of faking behavior, then there
may be some cause for concern if fakers are hired. Research has demonstrated that integrity tests are valid predictors of employee theft (Ones et
al. 1993; Sackett & Wanek, 1996; Shaffer & Schmidt, 1999). Criteria that
integrity test scores have been shown to predict include: likelihood of get-

ting caught stealing, supervisor ratings of employee's dishonesty, criminal history, and theft admissions. Other research has found that those applying to and working in high-theft stores have lower integrity scores than those applying to and working in low theft stores (Shaffer & Schmidt). We suspected that an applicant's honest integrity score would be a significant predictor of faking behavior.

Locus of Control

Locus of Control refers to the degree of control an individual feels that they have over life. When we started our theory development we tried to examine literature outside of the selection (and I/O in general) arena. We hoped that by examining similar behaviors and situations we might be able to refine previously suggested models. Our preliminary investigation of the ethical behavior and academic dishonesty literature kept pointing to locus of control as a possible predictor of applicant faking behavior. If generalizable to the applicant situation, this literature would suggest that individuals with an internal locus of control (the belief that there are consequences associated with behavior and that individuals have control of what happens to them) may be less likely to distort their responses. Conversely, individuals with an external locus of control (the belief that what happens to them is beyond their personal control) would be more susceptible to response distortion in an applicant setting. While initially appearing counterintuitive, this relationship is logical. Those individuals who feel little sense of control may exert influence when they feel they have an opportunity. This pattern has been supported in the business ethics and academic dishonesty literature.

Trevino and Youngblood (1990) found that locus of control influenced ethical decision-making. Individuals with internal locus of control exhibited more ethical decisions ($r = -0.42, p < 0.01$). In addition, locus of control has been empirically linked to academic dishonesty (Leming, 1980). Coleman and Mahaffey (2000) found evidence that suggests that those individuals with an internal locus of control are less tolerant of cheating than those with an external locus of control. The authors suggest that "people who believe the majority of their experiences are beyond their control are therefore more likely to gain some relief if they can assert some control over their outcome" (p. 129). This relationship may be extended to response distortion in an applicant setting. Individuals who feel more responsible for their own actions (internal) may be less likely to exhibit unethical behavior (faking or response distortion to get what they want). However, individuals with an external locus of control may fake as a way to cope with what they perceive as a lack of control over the selection process.

Self-Monitoring

Self-monitoring describes the extent to which an individual regulates his or her own behavior in social situations and interpersonal relations. Specific characteristics of high and low self-monitors can be related to faking behavior. First, responding to a personality measure represents a display of personality in a public situation. Since high self-monitors are very concerned about their public image, they tend to engage in certain forms of impression management in that they attempt to control the information about themselves that others receive (Gangestad & Snyder, 2000), and are more likely to present themselves in as favorable a light as possible—even if it means stretching the truth (Delery & Kacmar, 1998). They might also try to adapt their true personality style to one they believe the employer is looking for.

Second, whereas high self-monitors monitor, control, and mold their expressive behaviors to fit the situation, the self-presentation of low self-monitors is controlled from within by their affective states. Because of this tendency to maintain consistency between the way they perceive themselves and behavior displayed in public (Delery & Kacmar, 1998), they may also be less likely to fake personality measures, since this would represent great dissonance between attitudes and behavior.

McFarland and Ryan (2000) and Wrensen and Biderman (2005) failed to find a significant relationship between self-monitoring and the *ability* to fake. We expected, however, that self-monitoring would exhibit a significant relationship between the actual *amount* that individuals fake in an applicant setting. An employment setting is conceptually different from a research condition with participants instructed to fake personality measures. While we would expect an instructional set design to result in elevation independent of the level of self-monitoring, an applicant setting may trigger the desire of high self-monitors to leave a favorable impression on others (potential employers).

Social Desirability

Any examination of applicant faking and individual differences would be deficient if it did not include the construct of Socially Desirable Responding (SDR). SDR has been a strong focus of the investigation of applicant faking behavior, and often has been considering synonymous with faking (Ones, Viswesvaran, & Reiss, 1996). The problem with many scales that measure SDR is that it is not clear whether they actually reflect faking, or whether they reflect substantive personality traits that have real and meaningful relationships with other traits. SDR has been split into two independent factors, Self-Deceptive Enhancement (SDE), and Impression Management (IM). SDE has been characterized as an unconscious inflation of existing traits (Paulhus, 1984), while IM is characterized by intentional deception, and thus has been the focus of much research focusing on performance. These constructs have been used as "flags" for faking, statistical controls, and potential relevant sources of job performance (Ones et al.), but alone have not adequately unraveled applicant faking behavior.

Testing the Model

While theoretical models of faking have been proposed, collecting the data to test those models poses several challenges. First, data must be collected in an applicant setting. If our goal is to study actual applicant faking behavior, the instructional set approach to faking misses the mark, and has been criticized quite heavily in the literature (Snell & McDaniel, 1998). It captures the ability of individuals to fake, but not their actual propensity to fake in an applicant setting. Improvements have been made to this approach by instructing applicants to pretend they are applying for a job, or offering incentives. Even though there is merit to these approaches, we felt that they were deficient for the purposes of our research.

However, obtaining applicants in a field study introduced another set of problems. First was locating a partner organization that would have a large number of applicants, with a high rate of retention. We did not see this hurdle as insurmountable. However, the amount of administration time necessary to collect personality and antecedent data presented another barrier. To capture the complexity of the model, a variety of measures would need to be included in the model, and our projected total testing time quickly raised above a level that most organizations would support. Finally the issue of collecting an "honest" baseline score presented a problem. Even though these data could be collected after the applicant was hired, thereby reducing their motivation to fake, we thought employees may be reluctant to answer in a way inconsistent with their initial attempts. In a previous study we conducted (Griffith et al., in press) we did not find this effect, but it nonetheless remained a concern. In addition, we were interested in varying situational cues that might have an impact faking behavior, and when using a field study you are stuck with the situational characteristics that are given to you. Worse yet, these characteristics may differ across organizations (which might be necessary to collect a large amount of data) further confounding possible results.

To address these challenges, our current line of research adopted a hybrid approach, which allowed for the collection of a large amount of applicant data in a flexible framework. We designed a study that utilized deception to capture the applicant situation dynamic, but in a more controlled setting. In our series of studies, participants were informed before the experimenter's arrival that they would be participating in a scheduled research day. The experimenter portrayed himself as an employee for a local, university-based consulting firm. Participants were then told that while they were participating in research, they would also have the opportunity to apply for a student job at the consulting firm. Participants were given a job description and asked to read it carefully. This job description advertised the job as having attractive pay while offering flexible work hours and the ability to work from home or school. Following the description of the job, participants were given an applicant packet.

The applicant packet was comprised of a job application, the Wonderlic Personnel Test, and the NEO-FFI. Half of our subject pool completed a modified NEO-FFI, which contained a work frame of reference. This modification was intended to tease out the influence of situation variance on the applicants, and we expected it to result in a lower magnitude and prevalence of applicant faking. The other half of the subject pool completed the standard NEO-FFI, with no specific FOR. Once the respondents completed the instruments, participants were debriefed, and the experimenter told the participants that no job existed. A follow-up packet containing the NEO-FFI was then administered under an additional instructional set, which directed participants to answer the questionnaires as honestly as possible. Additionally, to assess antecedent influences on faking behavior participants were asked to honestly respond to the BIDR, Locus of Control, Integrity, and Self-Monitoring scales. The complete description of the study design can be found in Griffith, Rutkowski, Gujar, Yoshita, and Steelman (2005), and detailed results are presented in Griffith, Malm, English, Yoshita, Gujar, and Monnot (2004).

Given the experimental nature of the study, we first need to determine whether our simulated applicant setting was effective. To ensure that the manipulation was successful, a brief survey was randomly collected from a sub-sample of participants. Participants were asked to rate their interest in the job on a 5-point scale (1 = no interest and 5 = very interested). 79.9% of the participants rated their interest a 2 or greater (Mean= 3.43, Median= 4, Mode= 5, S.D. = 1.51). These results generally support the notion that our applicant setting would provide the faking variance that we were interested in studying. The means of the conditions also strongly support the viability of the manipulation.

Examining the means across applicant and honest conditions for both the general NEO-FFI and work FOR NEO-FFI groups (Table 6.1) indicates that means were higher in the applicant condition than in the hon-

Table 6.1.   Descriptive Statistics for General Applicant NEO-FFI

| | General NEO-FFI | | | | Work FOR NEO-FFI | | | |
| | Applicant Condition | | Honest Condition | | Applicant Condition | | Honest Condition | |
Measure	Mean	SD	Mean	SD	Mean	SD	Mean	SD
Neuroticism	26.92	8.38	32.43	8.81	25.04	7.27	28.28	8.22
Extraversion	44.93	6.89	40.69	7.53	44.21	6.61	41.85	7.58
Openness	41.34	6.12	39.11	6.00	39.41	5.56	38.46	6.21
Agreeableness	47.22	7.70	41.83	7.23	49.00	6.91	45.45	8.37
Conscientiousness	49.49	7.55	43.14	8.13	50.69	7.11	46.32	8.81

est condition for all traits (excluding neuroticism in which lower scores would be expected). This pattern was found for both the general and work FOR groups.

Situational Variance

In order to tease out the effects of situational variance, we compared the degree to which applicants faked two personality measures—one phrased in a general frame of reference, and one phrased in a work frame-of-reference. More specifically, while the general version of the personality measure did not specify any FOR, the work version of the personality measures was modified to reflect an employment setting. Because of the ambiguity found in general personality measures, it was expected that a work frame-of-reference would decrease applicant response distortion. That was in fact what we found. Our MANOVA results revealed significant mean level differences between the general and FOR group such that the participants appeared to fake the general NEO-FFI to a greater amount than the work FOR NEO-FFI (English, Griffith, Graseck, & Steelman, 2005).

We also expected to detect significant differences in the number of applicants identified as fakers between the general and work FOR conditions, such that there would be less fakers in the FOR condition. We expected to see a lower percentage of individuals faking in the FOR condition. In order to identify which applicants faked their responses, we used a methodology suggested by Griffith et al. (in press). We calculated the probability that an individual would score within a confidence interval established around the subjects' score on the NEO-FFI constructs under the honest condition. Individuals who scored above the upper bound of this confidence interval were categorized as fakers. The confidence interval estimate was calculated by multiplying the standard error of measurement (SEM) for the each of the 5 constructs by 1.96. An individual's true score should fall within this range with a 95% probability. We then calculated a standardized effect size (d) to examine the magnitude of faking in both the applicant and faking conditions.

Our analyses suggested that the FOR condition not only resulted in lower mean levels of faking, but that it significantly reduced the percentage of individuals who raised their scores for two of the five NEO-FFI constructs (See Table 6.2) Although the results were not statistically significant for all constructs, practical gains in the reduction of fakers were evident across the board.

Temporal Biases

In our research examining cognitive biases (Gujar & Griffith, 2006), we adopted the past, present, and future perception methodology of Robinson and Ryff (1999). This methodology was modified, by including past, present, and future versions of the Summated Conscientiousness Scale (SCS), a personality measure designed to measure an applicant's level of conscientiousness. This methodology allowed us to identify individuals that are prone to temporal self-deception by looking at the differences in scores on the past and future versions of the measure. We expected to see respondents' mean score on the future version of the scale to be significantly higher than respondents' mean score on the present and past versions of the noncognitive instrument. The results of our analysis supported these research hypotheses. We expected a relatively small effect size for these differences; however, the amount of change between these measurement conditions exceeded our estimations greatly, with the mean for the future version being significantly higher than the mean for the present version, which in turn was significantly higher than the mean for

Table 6.2. Percentage of Fakers in General and Work FOR NEO-FFI

Big Five Personality Traits	General NEO-FFI	Work FOR NEO-FFI	Pearson Chi-Square	Sig
Neuroticism	3.37%	7.30%	1.84	.13
Extraversion	26.97%	22.47%	1.45	.14
Openness to experience	22.47%	16.85%	2.37	.08
Agreeableness	29.21%	19.10%	6.40	.01
Conscientiousness	34.27%	24.72%	2.95	.05

the past version. Just by changing the verb tense, we observed an 11-point difference between past and future versions of the SCS (on a scale with 140 possible points). Thus the temporal bias accounted for almost a full standard deviation difference between past and future versions of the conscientiousness measure. This difference occurred for both the general and work FOR conditions.

Individual Differences

The results of the study offered some mixed support for all of our predictions, but in general, the hypotheses were better supported in the FOR condition (the correlations for all antecedent measures can be found in Tables 6.3 and 6.4.). Overall we found a stronger pattern of prediction for the individual difference variables in the FOR condition than the general condition. In the general condition, Integrity was moderately correlated with the amount of applicant faking on the conscientiousness measure, but failed to show a significant correlation with the other four constructs. Locus of control was significantly correlated with the amount of faking on neuroticism, extraversion, agreeableness, and conscientiousness in the general NEO-FFI condition.

In the work FOR condition, integrity predicted the amount of faking for all 5 constructs. Locus of control was also significantly correlated with neuroticism and the magnitude of correlations was similar to that found in the general condition. Neither cognitive ability nor Self-monitoring were significantly related to the amount of applicant faking in either the general or work condition. Finally, both Impression Management and Self-Deceptive Enhancement were not related to the amount of faking in the expected direction. In fact, for several constructs the correlation was opposite in sign! This is particularly troubling since many researchers and practitioners rely on measures of SDR to counter applicant faking.

Dissecting Applicant Faking Behavior

The goal of our current line of research is to uncover some of the individual characteristics, and situational influences that may be associated with applicant faking behavior. Previous conceptualizations of faking behavior often speak of the phenomenon in monolithic terms. Research has often framed faking as one thing and "fakers" as one type of person (mostly bad). Theoretical models have suggested that there are many influences to faking behavior, some of which we have included in a series of studies to begin to test these models. These factors may be present in

Table 6.3. Descriptive Statistics and Correlations between Amount of Faking NEO-FFI and Predictor Measures in General Condition

	N	M	SD	1	2	3	4	5	6	7	8	9	10	11
1. Amount of faking NEO neuroticism	178	-5.09	9.47	—										
2. Amount of faking NEO extraversion	178	4.13	7.82	-.63**	—									
3. Amount of faking NEO openness	178	1.72	6.57	-.30**	.36**	—								
4. Amount of faking NEO agreeableness	178	5.10	8.84	-.64**	-.68**	.34**	—							
5. Amount of faking NEO conscientiousness	178	6.01	8.74	-.68**	-.74**	.37**	-.71**	—						
6. Self-deceptive enhancement	158	61.39	7.14	.29**	-.07	.08	-.09	-.13	—					
7. Impression management	160	57.06	10.75	.20*	-.05	-.06	-.12	-.12	.41**	—				
8. Integrity	202	30.29	5.22	.09	-.02	.03	-.12	-.19*	.25**	.52**	—			
9. Locus of control	157	63.77	9.77	-.20*	.16*	-.05	.25**	-.29**	-.29**	-.15	-.26**	—		
10. Self-monitoring	158	61.32	7.14	-.11	.02	-.08	.05	.05	-.15	-.34**	-.15	.18*	—	
11. Ability WPT)	192	22.52	6.77	.05	-.08	-.12	-.13	.08	.04	.08	.05	-.05	-.10	—

Note: ** Correlation is significant at the 0.01 level (2-tailed)
* Correlation is significant at the 0.05 level (2-tailed)

Table 6.4. Descriptive Statistics and Correlations Between Amount of Faking NEO-FFI and Predictor Measures in Work Condition

	N	M	SD	1	2	3	4	5	6	7	8	9	10	11
1. Amount of faking NEO neuroticism	178	-2.38	7.74	—										
2. Amount of faking NEO extraversion	178	1.96	6.58	-.57**	—									
3. Amount of faking NEO openness	178	.71	5.35	-.33**	.32**	—								
4. Amount of faking NEO agreeableness	178	2.82	8.55	-.56**	.61**	.28**	—							
5. Amount of faking NEO conscientiousness	178	3.47	8.39	-.61**	.61**	.35**	.75**	—						
6. Self-deceptive enhancement	171	61.49	7.13	.16*	-.13	-.14	-.14	-.27**	—					
7. Impression management	172	58.63	11.12	.07	-.12	-.12	-.13	-.15*	.48**	—				
8. Integrity	237	30.79	5.14	.20**	-.18*	-.19*	-.23**	-.31**	.28**	.42**	—			
9. Locus of control	162	40.01	10.07	-.27**	.20*	-.18*	.28**	.28**	-.27**	-.17*	-.30**	—		
10. Self-monitoring	171	61.49	7.13	-.04	.06	.09	.05	-.10	-.11	-.26**	-.13	-.03	—	
11. Ability (WPT)	168	22.36	7.08	-.01	.00	.02	-.13	-.10	.20*	.18	.02	-.23*	-.07	—

Note: ** Correlation is significant at the 0.01 level (2-tailed)

* Correlation is significant at the 0.05 level (2-tailed)

different amounts in job applicants. Thus, even if two people have similar elevated scores, that elevation may have been influenced by different factors. Some of these antecedents may be associated with poor performance, while others may not. If those who elevate their scores in the applicant condition differ on the reasons driving that elevation, it is plausible that they may differ on important outcome variables. Therefore, it does not make sense to lump "fakers" into a single stereotyped category.

Even we fell into this trap when writing this chapter, often referring to fakers and nonfakers. The term itself may interfere with the understanding of the phenomenon because it is affectively loaded (faking equals lying to some). We believe that there are many reasons that applicants "fake" (some volitional and some not). Perhaps it might have been more accurate to identify fakers as "those who consciously or unconsciously elevated their scores in an applicant setting"; however it is considerably less parsimonious. In a research field flooded with divergent definitions and terms, we thought it best not to coin another. While the phenomenon of faking may be continuous, dichotomizing the variable does lend itself to a simpler description of the prevalence of the behavior. We, as well as other researchers such as Zickar et al. (2004), do believe that this continuous variable is comprised of distinct response categories. Levin and Zickar (2002) propose that the same elevation may be produced by those who are self-presenting, those who are outright lying, and those are just bullshitting. Faking behavior is dynamic, and may demonstrate considerable individual differences. Our data suggests that there are differences in the amount people fake. In fact some are so poor at faking that they actually lower their scores in the applicant condition. It stands to reason that there may be other key differences as well.

This chapter reviewed the results of a large-scale study (over a thousand "applicants") examining several factors suggested by theoretical models to contribute to applicant faking. Our research suggests that external factors, such as situational variance, as well as internal factors including integrity and locus of control are related to applicant faking behavior. Several variables that have traditionally been associated with faking behavior, such as ability, SDE and IM, did not demonstrate predictive relationships.

One goal of this research was to challenge previous conceptualizations of faking as a single entity (or at best two subentities: SDE and IM), and break down the faking variance into possible components of faking behavior. While in no way comprehensive, our results indicate that there are several stable determinants.

The situation has a powerful influence on our day-to-day behavior, and is no less influential in an applicant setting. When we utilized a work FOR condition to tease apart situational variance associated with faking we found that significantly less faking occurs. Given the recent findings regarding the increased reliability and validity of FOR measures, our results add additional weight to the argument that FOR measures are superior to general measures in an applicant setting. When we investigated the impact of temporal cognitive biases on the elevation of personality scores, we found that future representations of the self tended to be more positive than past representations. This finding may also yield applied benefits and suggests that bio-data type measures, which have a past frame-of-reference, may suffer less from the elevating effects of temporal cognitive biases.

While several forms of cognitive bias have previously been hypothesized to be related to applicant faking, little research has examined the role of temporal bias. The results of Gujar and Griffith (2006) suggest that the wording of the stem (past present, or future tense) may have an impact on the self-reporting of traits. When item stems were framed in a future tense significantly higher means were observed in comparison to present and past tense phrased items. Perhaps most surprisingly self-deceptive enhancement, a construct long associated with unconscious faking behavior, was not significantly correlated with this temporal effect.

We also tested several individual differences that have been suggested to influence faking behavior. As suggested by McFarland and Ryan (2000), integrity was negatively related to the amount of applicant faking. This study extends that finding by demonstrating the relationship in a hybrid motivated design. The variable of integrity demonstrated more consistent relationships across the big five constructs in the work FOR condition than the general condition. One possible explanation for this pattern of results is that in the general condition, situational variance may be a stronger contributor to faking behavior and its influence may mask that of integrity. In the work FOR, the variance in the interpretation of the situation may be reduced, and integrity then becomes a more reliable predictor. Locus of control was also a significant predictor of faking behavior for both general and work FOR conditions. This result is not surprising given the constructs' link to academic dishonesty. The employment testing situation is in many ways simply an extension of testing in school.

The constructs that have long been associated with applicant faking, Self Deceptive Enhancement and Impression Management, were not significantly positively correlated with observed levels of faking in our research, and in some instances were even *negatively* correlated. While this seems counterintuitive, these constructs have been criticized as poor proxies for faking behavior in the past (Christiansen, 1997). These findings further question their inclusion in a model of faking, and their use as "flags" for faking behavior. Caution with interpretation should be taken here as these constructs were measured in the honest condition and thus represent substantive variance, not just a response set (McCrae & Costa, 1983)

The Search for Meaningful Criterion Relationships

While we attempted to build a nomological net around the faking construct, we missed one key variable. One of the motivating factors for untangling the antecedents of applicant faking behavior is to begin to understand their impact on the prediction of performance. Different antecedents may ultimately result in different outcomes on the criterion of interest. Some fakers simply may be more susceptible to situational influences. Recent cross cultural research presented in this book supports the notion that there are measurable differences in the strength of situational cues for individuals. Would this type of faker be a poor performer? On the other hand the antecedents of integrity and locus of control were the most robust predictors of applicant faking behaviors. Both of these variables have been linked to negative outcomes by a large body of literature. Integrity may be the largest concern because of its association with theft and other counterproductive behavior.

Researchers have only recently begun to investigate the faking/performance relationship directly. Boyce (2005) did not find significant criterion related validity differences between scores collected in an applicant condition and scores collected in an honest condition. However, Boyce found significant negative relationships between d-scores (amount of faking) and performance ratings for the upper half of the distribution. Donovan et al. (2005) used a methodology similar to Griffith et al. (in press) to identify applicant fakers in a large applied sample. Although their results were not significant, when the job performance of fakers was compared to their honest counterparts, d effect sizes showed lower performance for the faking group.

We encourage additional research investigating criterion related validities and the performance levels of those identified as fakers. Determining the job performance of individuals who have faked has been difficult, mostly due to the difficulties in identifying faking behavior. If faking is influenced by different factors, it stands to reason that fakers may vary significantly on job performance. Future research should examine this often argued, but as of yet untested question. Are fakers really poor performers?

While still contested, evidence is mounting that a significant number of applicants do fake in employment situations (Donovan et al., 2003; Donovan et al., 2005; Griffith et al., in press; McDaniel et al., 1997; Rosse et al., 1998; Stokes et al., 1993). However, one of these applicants is not like the other. Applicant faking may be influenced by a variety of factors that may be present in differing amounts for different applicants. By untangling these sources of faking variance we may gain a better understanding of the nature of the phenomenon, and determine the impact of faking on the quality of employment decisions.

REFERENCES

Alliger, G. M., Lilienfeld, S. O., & Mitchell, K. E. (1996). The susceptibility of overt and covert integrity tests to coaching and faking. *Psychological Science, 7,* 32-39.

Anderson, H. N., Sison, G., & Wester, S. (1984). Intelligence and dissimulation on the personal orientation inventory. *Journal of Clinical Psychology, 40,* 1394-1398.

Austin, E. J., Hofer, S. M., Deary, I. J., & Eber, H. W. (2000). Interactions between intelligence and personality: Results from two large samples. *Personality and Individual Differences, 29,* 405-427.

Bing, M. N., Davidson, H. K., Whanger, J. C., & VanHook, J. B. (2004). Incremental validity of the frame of reference effect in personality scale scores: A replication and extension. *Journal of Applied Psychology, 89*(1), 150-157.

Boyce, A. (2005, April). *An investigation of faking: Its antecedents and impacts in applicant settings.* Paper presented at the 20th annual conference of the Society for Industrial/Organizational Psychology, Los Angeles.

Buehler, R., Griffin, D., & Ross, M. (1994). Exploring the "planning fallacy": Why people underestimate their task completion times. *Journal of Personality and Social Psychology, 67,* 366-381.

Burkhart, B. R., Gynther, M. D., & Christian, W. L. (1978). Psychological mindedness, intelligence, and item subtlety endorsement patterns on the MMPI. *Journal of Clinical Psychology, 34,* 76-79.

Christiansen, N. D. (1997). *Sensitive or senseless? The usefulness of social desirability measures for making inferences concerning applicant response distortion.* Unpublished manuscript.

Coleman, N., & Mahaffey, T. (2000). Business student ethics: Selected predictors of attitudes toward cheating. *Teaching Business Ethics, 4,* 121-136.

Crocker, L. M., & Algina, J. (1986). *Introduction to classical and modern test theory.* New York: Holt, Rinehart, and Winston

Delery, J. E., & Kacmar, K. M. (1998). The influence of applicant and interviewer characteristics on the use of impression management. *Journal of Applied Social Psychology, 28,* 1649-1669.

Donovan, J. J., Dwight, S. A., & Hurtz, G. M. (2003). An assessment of the prevalence, severity, and verifiability of entry-level applicant faking using the randomized response technique. *Human Performance, 16,* 81-106.

Donovan, J. J., Dwight, S. A., & Schneider, D. (2005, April). Prevalence and impact of faking in an organizational setting. In S. A. Dwight (Chair), *Faking it: Insights and remedies for applicant faking.* Symposium conducted at the 20th annual conference for the Society of Industrial and Organizational Psychology, Los Angeles.

Douglas, E. F., McDaniel, M. A., & Snell, A. F. (1996, August). *The validity of non-cognitive measures decays when applicants fake.* Paper presented at the meeting of the Academy of Management, Nashville, TN.

English, A., Griffith, R., Graseck, M., & Steelman, L. (2005). *Frame-of-reference, applicant faking, and the predictive validity of non-cognitive measures: A matter of context.* Manuscript submitted for publication.

Gangestad, S. W., & Snyder, M. (2000). Self-monitoring: Appraisal and reappraisal. *Psychological Bulletin, 126,* 530-555.

Garry, R. (1953). Individual differences in ability to fake vocational interests. *The Journal of Applied Psychology, 37,* 33-37.

Griffith, R. L., Malm, T., English, A., Yoshita, Y., Gujar, A., & Monnot, M. (2004). individual differences and applicant faking behavior: One of These applicants is not like the others. In N. D. Christiansen (Chair), *Beyond Social Desirability in Research on Applicant Response Distortion.* Symposium conducted at the 19th annual conference for the Society of Industrial and Organizational Psychology, Chicago.

Griffith, R. L., Rutkowski, K. A., Gujar, A., Yoshita, Y., & Steelman, L. A. (2005). *Modeling applicant faking behavior.* Manuscript submitted for publication.

Griffith, R. L., Chmielowski, T. S., & Yoshita, Y. (in press). Do applicants fake? An examination of the frequency of applicant faking behavior. *Personnel Review.*

Gujar, A., & Griffith, R. (2006). *The effect of temporal context of personality measures in personnel selection.* Paper presented at the 21st annual conference of the Society for Industrial and Organizational Psychology, Dallas, TX.

Hogan, R. (1991). Personality and personality measurement. In M. D. Dunnette & L. M. Hough (Eds.), *Handbook of industrial and organizational psychology* (Vol. 2, 2nd ed., pp. 873-919). Palo Alto, CA: Consulting Psychologists Press.

Hough, L. M., Eaton, N. K., Dunnette, M. D., Kamp, J. D., & McCloy, R. A. (1990). Criterion-related validities of personality constructs and the effect of response distortion on those validities. *Journal of Applied Psychology Monograph, 75*(5), 581-595.

Hough, L. M., & Paullin, C. (1994). Construct-oriented scale construction: The rational approach. In G. S. Stokes, M. D. Mumford, & W. A. Owens (Eds.), *Biodata handbook: Theory, research, and use of biographical information in selection and performance prediction.* Palo Alto, CA: Consulting Psychologist Press.

Hunthausen, J. M., Truxillo, D. M., Bauer, T. N., & Hammer, L. B. (2003). A field study of frame of reference effects on personality test validity. *Journal of Applied Psychology, 88,* 545-551.

Kenrick, D. T., & Funder, D. C. (1988). Profiting from controversy: Lessons from the person-situation debate. *American Psychologist, 43,* 23-34.

Leming, J. S. (1980). Cheating behavior, subject variables, and components of the internal-external scale under high and low risk conditions. *Journal of Educational Research, 74,* 83-87

Levin, R. A., & Zickar, M. J. (2002). Investigating self-presentation, lies, and bullshit: Understanding faking and its effects on hiring decisions using theory, field research, and simulation. In F. Drasgow & J. Brett (Eds.), *The new millennium of work* (pp. 253-276). San Francisco: Jossey-Bass.

Lord, C. G. (1982). Predicting behavioral consistency from an individual's perception of individual similarities. *Journal of Personality and Social Psychology, 42,* 1076-1088.

McCrae, R. R., & Costa, P. T., Jr. (1983). Social desirability scales: More substance than style. *Journal of Consulting and Clinical Psychology, 51,* 882-888.

McDaniel, M. A., Douglas, E. F., & Snell, A. F. (1997, April). *A survey of deception among job seekers.* Paper presented at the 12th annual conference of the Society for Industrial and Organizational Psychology, St. Louis, MO.

McFarland, L. A., & Ryan, A. M. (2000). Variance in faking across non-cognitive measures. *Journal of Applied Psychology, 85,* 812-821.

Mersman, J. L., & Shultz, K. S. (1998). Individual differences in the ability to fake on personality measures. *Personality and Individual Differences, 24*(2), 217-227.

Mischel, W. (1973). Toward a cognitive social learning reconceptualization of personality. *Psychological Review, 80,* 252-283.

Mischel, W., & Shoda, Y. (1995). A cognitive-affective system theory of personality: Reconceptualizing situations, dispositions, dynamics, and invariance in personality structure. *Psychology Review, 102,* 246-268.

Nguyen, N. T. (2001). Faking in situational judgment tests: An empirical investigation of the Work Judgment Survey. *Dissertation Abstracts International, 62,* 3109A.

Ones, D. S., Viswesvaran, C., & Korbin, W. P. (1995, May). *Meta-analyses of fakability estimates: Between-subjects versus within-subjects designs.* Paper presented at the 10th annual conference of the Society for Industrial and Organizational Psychology, Orlando, FL.

Ones, D. S., Viswesvaran, C., & Reiss, A. D. (1996). Role of social desirability in personality testing for personnel selection: The red herring. *Journal of Applied Psychology, 81,* 660-679.

Ones, D. S., Viswesvaran, C., & Schmidt, F. L. (1993). Comprehensive meta-analysis of integrity test validities: Findings and implications for personnel selection and theories of job performance. *Journal of Applied Psychology Monograph, 78*(4), 679-703.

Paulhus, D. L. (1984). Two component models of social desirable responding. *Journal of Personality and Social Psychology, 46,* 598-609.

Robie, C., Schmit, M. J., Ryan, A. M., & Zickar, M. J. (2000). Effects of item context specificity on the measurement equivalence of a personality inventory. *Organizational Research Methods, 3,* 348-365.

Robinson, M. D., & Ryff, C. D. (1999). The role of self-deception in perceptions of past, present, and future happiness. *Personality and Social Psychology Bulletin, 25*(5), 595-606.

Ross, M., & Fletcher, G. J. (1985). Attribution and social perception. In G. Lindzey & E. Aronson (Eds.), *Handbook of social psychology* (Vol. 2, pp. 73-122). New York: Random House.

Rosse, J. G., Levin, R. A., & Nowicki, M. D. (1999). Assessing the impact of faking on job performance and counter-productive job behaviors. Paper presented in Sackett, P. (Chair), *New empirical research on social desirability in personality measurement.* Symposium for the 14th annual meeting of the Society for Industrial and Organizational Psychology, Atlanta, GA.

Rosse, J. G., Stecher, M. D., Miller, J. L., & Levin, R. A. (1998). The Impact of response distortion on preemployment personality testing and hiring decisions. *Journal of Applied Psychology, 83,* 634-644.

Sackett, P. R., & Wanek, J. E. (1996). New developments in the use of measures of honesty, integrity, conscientiousness, dependability, trustworthiness, and reliability for personnel selection. Personnel Psychology, *49,* 787-829.

Schmit, M. J., Ryan, A. M., Stierwalt, S. L., & Powell, A. B. (1995). Frame-of-reference effects on personality sale scores and criterion related validity. *Journal of Applied Psychology, 80,* 607-620.

Shaffer, D. J., & Schmidt, R. A. (1999). Personality testing in employment. *Society for Human Resource Management Legal Report.* Retrieved November, 2005, from http://www.shrm.org/whitepapers/documents/61294.asp

Snell, A. F., & McDaniel, M. A. (1998, April). *Faking: Getting Data to Answer the Right Questions.* Paper presented at the 13th annual conference of the Society for Industrial and Organizational Psychology, Dallas, TX.

Snell, A. F., Sydell, E. J., & Lueke, S. B. (1999). Towards a theory of applicant faking: Integrating studies of deception. *Human Resource Management Review, 9,* 219-242.

Stokes, G. S., Hogan, J. B., & Snell, A. F. (1993). Comparability of incumbent and applicant samples for the development of biodata keys: The influence of social desirability. *Personnel Psychology, 46,* 739-762.

Tett, R. P., & Guterman, H. A. (2000). Situation trait relevance, trait expression, and cross-situational consistency: Testing a principle of trait activation. *Journal of Research in Personality, 34,* 397-423.

Traub, R. E. (1994). *MMSS reliability for the social sciences: Theory and applications.* Thousand Oaks, CA: Sage.

Trevino, L. K., & Youngblood, S. A. (1990). Bad apples in bad barrels: A causal analysis of ethical decision-making behavior. *Journal of Applied Psychology, 75*(4), 378-385.

Vernon, P. E. (1933). The biosocial nature of the personality trait. *Psychological Review, 40,* 533-548.

Wilson, A. E., & Ross, M. (2000). The frequency of temporal-self and social comparisons in people's personal appraisals. *Journal of Personality and Social Psychology, 78,* 928-942.

Wrensen, L., & Biderman, M. (2005, April). *Factors related to faking ability: A structural equation model application.* Interactive poster session presented at the 20th annual conference of the Society for Industrial and Organizational Psychology, Los Angeles.

Vasilopoulos, N. L., Cucina, J. M., & McElreath, J. M. (2005). Do warnings of response verification moderate the relationship between personality and cognitive ability? *Journal of Applied Psychology, 90*(2), 306-322.

Zickar, M. J., Gibby, R. E., & Robie, C. (2004). Uncovering faking samples in applicant, incumbent, and experimental data sets: An application of mixed-model item response theory. *Organizational Research Methods*, 7, 168-190.

Zickar, M. J., & Robie, C. (1999). Modeling faking good on personality items: An item-level analysis. *Journal of Applied Psychology, 84,* 551- 563.

CHAPTER 7

UNDERSTANDING RESPONSES TO PERSONALITY SELECTION MEASURES

A Conditional Model of the Applicant Reasoning Process

Andrea F. Snell and Chris D. Fluckinger

The persistent discourse on the fakability of personality selection methods has somewhat masked two points that we believe are critical for a new, and more successful, future for these controversial predictors of job performance. First, even when severely criticized by many influential researchers in the field (Campion, Dipboye, Hollenbeck, Murphy, & Schmitt, 2004), selection practitioners and their academic research partners are loathe to abandon them. Second, it is the inconsistent, and often disappointing, predictive validities of these measures that is at issue—not whether faking matters. From the perspective of these two points, we would like to add our voices to the rumbling growl in the research field that process models are long overdue. Specifically, we contend that an applicant response model which attempts to explicate the situational

A Closer Examination of Applicant Faking Behavior, 179–208
Copyright © 2006 by Information Age Publishing
All rights of reproduction in any form reserved.

influences on applicant responses, in contrast to a faking model, would help integrate the empirical literature and provide a theoretical structure for future investigations.

This model is designed with only one goal in mind: improvement of the predictive utility of personality selection measures. There are those who argue that until personality tests can predict more than 4-6% of the variance in job performance, they should be avoided in favor of more valid selection tools (Murphy & Dzieweczynski, 2005). The most effective method to silence this criticism is to increase the validities to a level commensurate with alternative predictors such as cognitive ability and situational judgment tests. As recently reviewed by several personality researchers (Barrick & Mount, 2005; Hogan, 2005a; Hough & Oswald, 2005), there are numerous reasons for continued use of personality selection measures. Personality dispositions and attributes *are* predictive of important criteria, add incremental validity to the prediction of job performance and are not burdened by adverse impact problems that plague many of the alternatives. A communality that unites the naysayers and the believers is that both would consider increased predictive validities for personality selection measures to be a good thing.

Before presenting our applicant response model, we present an overview of the two avenues of research which most directly address the predictive utility of personality selection measures. First, there is the nomological net that has emerged from the personality literature. For decades, personality research has been continually striving to produce a taxonomy of predictor-criterion relationships and evaluate whether modifications to the measured constructs will enhance the prediction of job performance criteria. Second, research which investigates applicant response distortion is based on the premise that if you can reduce faking on personality selection measures, their predictive validities will improve. Both bodies of literature lay the groundwork for the proposed applicant response model.

LESSONS LEARNED FROM THE PERSONALITY LITERATURE

The personality literature that is most relevant to our applicant response model is the research on the criterion-related validity of personality constructs as predictors of important organizational outcomes such as job performance. Several meta-analyses have demonstrated that personality characteristics, especially those related to the Big Five, are predictive of job performance (Barrick, Mount, & Judge, 2001; Hogan & Holland, 2003; Judge & Ilies, 2002). In spite of these empirical findings, the critics contend that the predictive validities of personality measures are not only inferior to other predictors but are too small to warrant their continued use (Murphy & Dzieweczynski, 2005).

In an effort to address those who reject the use of personality variables as predictors of job behaviors, several personality researchers have critically examined the inconsistent validities generated by personality selection measures. We briefly review the recent summaries of the research literature and their recommendations for future personality selection tests for two reasons. First, we concur that their specific suggestions for measuring predictor and criterion constructs are the foundation stones for practically useful predictive validities of personality measures. The implementation of an applicant response model will not compensate for poor predictor measures. We maintain that if a home is lacking a solid foundation and a good roof, no amount of interior decorating will make it habitable.

Second, the research reviewed in these summaries has historically been aimed at linking individual differences variables (i.e., personality traits) to outcome behaviors (i.e., job performance). The suggestions for improving this research reflect an increasingly strident call for incorporating situational influences in the research design. Thus, the recommendations for future directions in personality testing reflect theories, and methodologies, which examine the interactional effects of personality constructs and job settings in the prediction of job performance. We believe these are important lessons and draw on them in our development of an applicant response model which speculates on the situational characteristics that affect an applicant's choice of a response on a personality selection measure.

Hough and Oswald (2005) contend that improved predictive validities of personality tests can, and have, been accomplished by engaging in good science. We categorize their suggestions for improving the predictive validity of personality tests into three categories: (1) good measurement of the criteria (2) good measurement of the predictors and (3) consideration of the job context. Performance criteria, like predictors, are not simple constructs. Unlike general mental ability which is typically expected to be a positive predictor of all positive job behaviors, expectations for personality predictors differ across criteria. For example, we would not expect personality constructs such as emotional stability or even conscientiousness to be equally predictive of various job behaviors such as task performance, contextual performance, counterproductive behavior, creativity, leadership, and sales performance. Hough and Oswald (2005a) make this point clearly and admonish that careful explication of the criterion coupled with thoughtful, and reasonable, connections with relevant predictors is imperative for our validation studies.

Validities for personality constructs may also be hampered by the static nature of our measurement of work behaviors. As Hough and Oswald (2005) point out, personality may be a stronger predictor of an employee's change in work behavior over time than it is of his or her job performance measured immediately after being hired. This conceptualization of job performance criteria as dynamic variables represents an exciting opportunity for selection researchers. For example, Thoresen, Bradley, Bliese, and Thoresen (2004) evaluated the concurrent validities of the Big Five as predictors of sales performance measured 4 times over the period of a year. In general, they found that some variables predicted sales performance at each point in time (e.g., conscientiousness and extraversion) and some variables predicted the growth in performance over time (e.g., conscientiousness only). More importantly, in some job contexts personality variables which were only minimally related to initial performance demonstrated increased prediction over time. Bolstered by these findings, the authors suggest that "the initial gains to organizations for selection [decisions which are] based on personality traits are not epiphenomenal but ... persist (and may even increase) with increased time on the job" (p. 850). We can think of few job settings where individuals are selected solely for their performance at a single point in time. Most supervisors would be more interested in hiring individuals who would show consistently high performance and, if possible, positive growth in performance over time. We echo Hough and Oswald and Thoresen et al.'s caution that we (1) know relatively little about how personality variables relate to changes in job performance over time and (2) personality differences may become stronger predictors of distal performance contrary to the declining predictiveness that has been found for cognitive ability measures (Keil & Cortina, 2001). As personality researchers we do not have the luxury of attending only to the predictor measure; as much care needs to be given to the conceptualization and measurement of the criterion variable. More importantly, by conceptualizing and measuring job behaviors as dynamic criterion variables we could possibly increase the predictive validity of our current personality measures.

With regards to the predictors, there is strong agreement that there are many unreliable, and rather embarrassing, personality tests that are currently available to selection practitioners. Careful measurement and demonstration of validity are essential. There is also a good deal of agreement (Barrick & Mount, 2005; Borman, 2004; Hogan, 2005a; Hough & Oswald, 2005; Schmitt, 2004; Schneider, Hough, & Dunnette, 1996) that large bandwidth predictors, such as the Five Factor Model, may be inappropriate predictors in many contexts and should be substituted with narrower bandwidth predictors. When the bandwidth of the predictor measure and the criterion measure are congruent, higher validities are generated (Hogan, Hogan, & Roberts, 1996; Hogan & Holland, 2003; Hough & Oswald, 2005; Hough & Schneider, 1996; Ones, Viswesvaran, & Reiss, 1996). Even those that have typically investigated large bandwidth personality factors caution that the descriptive level of the personality trait must be commensurate with the specificity level of the criterion (Barrick & Mount, 2005)

In addition to these usual suspects of inconsistent validities, Hough and Oswald (2005) also call for additional research on more proximal predictors of job performance in order to better understand how personality impacts work behaviors. In a similar vein, Barrick and Mount (2005) suggest that personality affects work behavior through its effect on an individual's motivation. Taken at face value, these comments suggest that we should be looking for more proximal predictors of job performance than personality constructs. However, a more thoughtful interpretation is that if one is going to use personality measures as predictors of job performance, wouldn't it be nice to know *why* they worked? Research generated by most personality selection experts demonstrates a dogged focus on largely one empirical outcome: predictive validity of personality selection measures. Now, embedded in their defense of personality selection measures, they ask for research on the *process* behind this empirical relationship between personality characteristics and job behaviors. Those of us who alter the format of the personality selection items, change the instructions to applicants, and alter the frames-of-reference for a test are seeking an empirical holy grail: better validities for personality selection measures. Our applicant response model is predicated on the question, *why* do we think these types of alterations will affect an applicant's choice of a response? That is, what is the process by which instructions, frame-of-references, and item formats affect responses? If personality researchers think it is important to ask why personality predicts job performance, we don't think it is a stretch for faking researchers to ask why we think we can alter applicant responses.

Finally, summaries of the connection between personality predictor constructs and job performance criterion constructs also highlight the interactional nature of person variables and situational influences. Empirical findings of this observation have most typically centered on the search for situational moderators of the personality-performance relationship. Investigations of these moderators have generally indicated that personality traits are stronger predictors in weak situations, such as jobs with high autonomy, than in strong situations (Barrick & Mount, 1993; Beaty, Cleveland, & Murphy, 2001). In their interactionist model of job performance, Tett and Burnett (2003) further clarify that a strong work situation will exist whenever the extrinsic rewards are so large they effectively eliminate the expression of trait-relevant behaviors. That is, the characteristics of the job setting (e.g., whether I will be immediately fired for having a messy desk) will largely determine whether personality traits (e.g., my organizational skills) will have an opportunity to be predictive. Strong validities between personality constructs and job performance are therefore contingent on the characteristics of the specific job context. Hogan (2005b) adds to this conversation about situational specificity by calling for a taxonomy of situational influences on the personality-performance relationship.

Although the most common application of an interactional model of personality-performance relations can be seen in the search for various situational moderators, a separate interactional theme also emerged. In their discussion of measurement problems and the need for aggregated measures, Barrick and Mount (2005) state,

> we need to understand that although personality is fairly stable and so is the context, the response to items on a personality inventory at any one moment is determined by many traits and states and by many features of the immediate situation. (p. 366)

They further add that self-report measures generally produce lower validities than assessments of personality that are made by others and researchers should therefore explore other methods of personality assessment. Hogan (2005b) takes issue with the common conceptualization that responses to personality measures are self-reports. He suggests that a more appropriate view would be to consider responses as self-presentations. Although we are quite sure that none of these authors intended for their comments to be interpreted as supporting the development of an applicant response model, it is precisely what we have done. As a matter of fact, with the exception of Hogan's (2005a) one paragraph summary of faking research, none of the authors who most recently contributed to the defense of personality selection measures mentioned response distortion at all. But then, they were intent on providing evidence for the predictive validity of personality measures and discussions of response distortion do not typically contribute positively to this endeavor. We would like to change that. In the same way that personality researchers have embraced the importance of situational influences in understanding personality-performance relationships, we suggest that our measurement of applicant responses to personality items should also embrace situational influences.

A caveat regarding this focus on criterion-related validities is whether the validity coefficient is an appropriate indicator of the utility of the selection measure. That is, a criterion-related validity coefficient describes the group-level correspondence between predictor scores and criterion behaviors, and therefore may not reflect whether the top of the distribution which were hired are actually good job performers. Empirical investigations of this problem have demonstrated that even with acceptable validity coefficients, most individuals in an applicant-like setting who garner the highest scores on a personality selection measure are those that are faking their responses and would therefore not necessarily be good hires (Drasgow & Kang, 1984; Levin & Zickar, 2002; Rosse, Stecher, Miller, & Levin, 1998). We do not disagree with this point but would like to emphasize the context for this concern. Namely, if applicants are selected solely on the basis of their personality test scores, then we would argue that the validity coefficient may not be the appropriate evaluation of the usefulness of the predictor measure. However, we think it is a rather rare selection setting that relies on a single test score, let alone a personality test score, to make hiring decisions. Rather, it is our understanding, or assumption, that personality selection test scores are considered in conjunction with other predictors and more are typically used in an early stage of a multiple hurdle selection design. Therefore, we are concerned with the relationship between personality predictor scores and criterion measures across all individuals in the selection pool and view the criterion-related validity coefficient to be the best indicator of this.

Taken together, these reviews support the continued use of personality measures as predictors of job performance. With the indulgence of the authors cited, we would paraphrase their summaries thus: inconsistencies in predictive validities occur because all predictors are not equal, job performance is not a single and static construct, and job settings matter. In summary, substantial, and consistent, predictive validities for personality measures require: (a) good measurement of both the predictor and criterion constructs, (b) a careful match between the bandwidth of predictor and criterion, and (c) thoughtful consideration of the situational influences which may moderate the relationship between the predictor and the criterion. These researchers, as well as others, have provided convincing evidence that when we adhere to good science, personality measures generate validities that are comparable to cognitive ability indicators, explain incremental variance in performance and may even be more predictive of long term job behaviors. We concur that any evaluation of the predictive utility of a personality selection measure must meet these basic rules of good science in order to add to the extant literature. And this goes for those of us who attempt to identify intentional response distortion, calculate its effects on selection decisions, and/or attempt to reduce or eliminate it altogether.

LESSONS LEARNED FROM THE FAKING LITERATURE

In addition to attending to the tenets of good personality measurement espoused above, there is a rather large body of research which is based on the premise that one way to improve predictive validities is to reduce intentional response distortion, or faking, on personality selection measures. In this literature, diverse research designs, samples and labels abound. We organize these studies into two broad categories (1) those that address the presence and/or effects of faking and (2) those that try to fix the problem of faking. Studies that represent the first category have typically compared samples (e.g., students instructed to be honest or fake-good, incumbents, applicants) in order to get estimates of how much faking is going on in a selection setting. Additionally, many studies in this category attempt to measure the effects of faking by employing a social desirability score as a measure of faking (Ellingson, Sackett, & Hough, 1999; Ones et al., 1996; Rosse et al., 1998). These types of studies have been central to discussions about the prevalence and effect of faking but provide no direct contribution to the discussion of how to improve predictive validities of personality selection measures. Thus, we do not review them here.

Those studies that most directly address the possible improvement of predictive validities of personality selection measures fall into the second category and have two additional characteristics that we think are important to explicate. First, these studies generally attempt to demonstrate the utility of a new method for reducing faking. Simply put, they are good ideas that have been empirically tested. Second, these studies generally reflect a "magic bullet" approach in that each of them hoped that the particular method would fix the problem of faking in some noticeable and consistent way. Most of these approaches can be loosely classified into one of three categories: test construction, manipulation of the situation, and identification/correction.

Various test construction techniques have been investigated with the hope that a new method would hamper applicants' ability to fake-good on a personality measure. Changes to item grouping strategies (McFarland, Ryan, & Ellis, 2002), item formats (Christiansen, 2000; Martin, Bowen, & Hunt, 2002), and item scoring methods (Kluger, Reilly, & Russell, 1991) have been attempted but have not proven to be very popular solutions because of their limited effectiveness and the additional measurement issues they create.

A more fruitful approach has been the manipulation of the testing environment. The most obvious example of this is the research on the effects of warnings. Dwight and Donovan (2003) summarize this literature and conclude that warnings which include instructions that faking can be identified and will result in consequences, do reduce the mean level response of applicants. A more novel approach employs the use of an apparatus that participants are told can detect lying but is, in actuality, a bogus machine (Roese & Jamieson, 1993). Although this approach has a larger effect on applicant responding than warnings, it requires the use of a bulky machine and individual level testing; both are impractical for most personality selection tests. Interestingly, a similar, albeit smaller, effect has been found for some computer based testing (Feigelson & Dwight, 2000). Finally, Schmitt and Kunce (2002) produced some effects on applicant responses by requiring applicants to elaborate on their responses to personality tests.

Even though these studies were ultimately aimed at improving predictive validities of personality selection tests, many relied on a demonstration of mean differences to evaluate the effectiveness of the new approach. Although lower applicant means are generally interpreted as reflecting less applicant faking, the limited studies which have investigated the effects of these magic bullets on the criterion-related validities indicate either no change in predictive validities or very small increases (Illingworth, 2004). We interpret these results in two ways. First, the consistent effect on mean level responding indicates that warnings are, on average, altering the response choice of applicants. Second, the effects of warnings alone are not strong enough to sufficiently increase predictive validities to levels commensurate with concurrent validities.

Finally, the ultimate magic bullet has been the methods of detecting, and possibly correcting for, intentional response distortion. There has been some attempt to utilize the appropriateness index generated by the item response theory approach to identify fakers. Although this is a promising avenue of research, this technique is rather cumbersome to employ especially in light of its limited success at identifying fakers (Fernando & Chico, 2001; Zickar & Drasgow, 1996). The most ubiquitous method for mitigating response distortion has been the measurement of social desirability. In this method, applicants respond to a set of questions that are intended to measure individual differences on the tendency to engage in impression management or espouse unlikely virtues. These items are generally embedded in the personality test. The purpose of measuring this response style is to detect who is faking on the personality measure and possibly correct scores for this influence. This technique has been empirically tested (Barrick & Mount, 1996; Christiansen, Goffin, Johnston, & Rothstein, 1994; Ellingson et al., 1999; Hough, 1998; Hough, Eaton, Dunnette, Kamp, & McCloy, 1990; Ones et al., 1996) with strikingly consistent results. Use of these scales does not generate improved predictive validities for personality measures.

Based on this history, we offer two conclusions. First, no magic bullet approach has produced sufficient increases in predictive validity to fully satisfy those who continue to use personality selection tests, let alone satisfy those who have called for them to be abandoned in favor of better predictors of job performance. Second, this research does shine a little light on the process of applicant responding because some of these approaches have resulted in altered response behavior. Armed with this encouragement from the faking literature, we examine existing models of applicant responding.

MODELS OF APPLICANT FAKING

There are two types of models that are most directly linked to an applicant response model: self presentation response style or bias (Paulhus, 2002) and models of faking (McFarland & Ryan, 2000; Snell, Sydell, & Lueke, 1999). The former focus on response sets that lurk beneath self-report scores. Paulhus and his colleagues have produced an ever expanding theoretical construction for the most infamous of these response sets: social desirability. Furthermore, this model has generated an impressive set of empirical studies that have helped explicate the role, and boundaries, of social desirability across a broad array of psychological research. In this model, individuals display a consistent response bias across time and situations. Thus, dispositional attributes such as impression management, self deception, acquiescence, and malingering are the characteristics that identify the individual differences in response sets or response bias. Although useful, this purely dispositional approach is not sufficient to model applicant response behavior. Specifically, successful attempts to manipulate responses (i.e., warnings) indicate that there is also a situational influence on applicant responding. We think it is important to note that the purpose of this literature has been to understand response sets or biases and we believe research in this field has demonstrated a consistent focus on this endeavor. By way of differentiation, an applicant response model, we believe, should focus on the process by which applicants choose a response to items on a personality measure.

Models of faking attempt to identify the role that intentional response distortion plays in an applicant selection context. In the models proposed by McFarland and Ryan (2000) and Snell et al. (1999), two conditions must be operating for faking behavior to occur: individuals must be motivated and able to fake their responses to personality selection measures. In this classic person by environment approach, individual differences such as personality traits, perceptions of the selection setting, beliefs about selection tests, job experience and test taking skills are hypothesized to affect an individual's motivation to fake. In these models, characteristics of the environment such as the presence of a lie scale or warning, test format and instructions also influence an individual's faking behavior. These models were intended to provide a starting point for understanding how and why individuals engage in faking behavior on personality selection tests. However, with the exception of the limited effects for social desirability and integrity, investigations of dispositional variables which are predictive of faking behavior have generated rather disappointing results. Empirical investigations of situational effects, such as warnings, have been slightly more successful.

Taken together, investigations of the dispositional, or person level influences and the experimental manipulations of the situational influences are encouraging. However, these models were designed to illuminate, in a general way, the influences on faking behavior and provide some directions for future research. As a matter of fact, in their presentation of their model, McFarland and Ryan (2000) caution readers "the design of our current study does not allow us to isolate these factors so that we can examine their specific influences on faking behavior" (p. 818). Snell et al. (1999) clearly indicate that their "proposed model should provide a framework for understanding both individual differences … and situational differences in faking" (p. 237). Thus, both models acknowledge the importance of both situational and dispositional effects on faking behavior.

In an effort to expand these models and, hopefully, produce a theoretical structure that will aid in our endeavor to improve the predictive validities of personality selection measures, we suggest a change of focus. Instead of explicating the influences on applicant faking, we would like to understand how an applicant chooses a response to a personality selection measure. The models of faking are clearly our starting point. We have been unable to simply recast these models as applicant response models, however, for one crucial reason. These models present the dispositional influences and situation influences as static entities. Our understanding of the empirical research in this area indicates that even when dispositions such as social desirability are found to be related to faking behavior as hypothesized by the models, they account for little variance in the applicant responses (Hough, 1998). Furthermore, the clearest examples of a situational influence, such as warnings not to fake, also reveal that there is still large variance in responding (Dwight & Donovan, 2003). We do not believe the authors intended for these influences to be conceptualized as static forces; indeed, Snell et al. (1999) actually use the label "an interactional model of faking." However, the fact remains that these models do not speak to how the dispositional and situational influences would actually interact to influence applicant responding. For this reason, we searched the extant literature for models that would help us expound upon these faking models and create a more useful model of the interactional influences on applicant responding.

MODELS OF CONDITIONAL BEHAVIOR

James (1998, 2004) has pioneered a new approach to the measurement of individual attributes that employs a conditional mechanism for understanding individual response choices. James' conditional reasoning (CR) methodology conceptualizes the observed differences in attributes such as achievement motivation and aggression as the product of an implicit reasoning process. Specifically, this approach assumes that "people rely often on reasoning processes whose purpose is to enhance the logical appeal of their behavioral choices" (1998, p. 131). Justification mechanisms (JMs) act as the mechanics of this conditional approach to understanding individual differences in behavior. JMs in James' CR system are derived from empirical research which has illuminated the types, or styles, of behavior for a given personality construct (e.g., achievement motivation). For each of the identified styles (those motivated to achieve versus those who have a fear of failure), JMs are postulated for each. These JMs reflect the reasoning process underlying an individual's behavioral choice.

We believe James' conditional reasoning model holds promise because it represents a possible logical step in a long line of research of implicit bias from both the cognitive psychology literature (cf. Holland, Holyoak, Nisbett, & Thagard, 1986; Holyoak & Nisbett, 1988) and the social psychology literature (cf. Greenwald & Banaji, 1995; Nisbett & Wilson, 1977; Wilson & Brekke, 1994). Implicit biases have long been measured in very specific conditions, but, as Wright and Mischel (1987) observe, "The key task is to identify the categories of conditions in which predictable behaviors relevant to some dispositional domain are most likely to be observed" (p. 1,162). Similar conditional strategies have been proposed in additional areas as well (Dweck & Leggett, 1988; Lohman & Bosma, 2002; Mischel, 2004). These conditional models of behavior provided the framework for our applicant response model in three ways. First, they helped us conceptualize, and operationalize, the interaction of situational variables and individual differences. Second, when interactions between situations and dispositions become the focus, the research design can easily resemble the search for the proverbial "needle in the haystack." However, in each of these models, either the individual differences or the situational variables served as the primary focus to guide the development of testable research hypotheses and we gladly adopted that approach. Third, the empirical research stemming from these conditional models served as invaluable examples of how to design studies which would test these interactions. For those firmly rooted in a trait theory approach to human behavior, we suggest a perusing of the readings listed above; we all need new shoes from time to time.

We would like to present this conditional approach, which leans heavily on the conditional reasoning (CR) framework of James and his colleagues, as a possible link between existing faking models and the development of an applicant response model. Before we present this model, we must highlight the critical differences between our conditional model and the CR model. First, the CR model and development of justification mechanisms relies on the existence of types of responders. With regards to applicant responding, there is some research to suggest the possible existence of distinct classes of responses (Lueke, Snell, & Illingworth, 2002; Zickar, Gibby, & Robie, 2004). However, this literature is rather meager to support a categorical structure for applicant responding. Second, the CR model is aimed at the measurement of stable, underlying dispositions and does not, as of yet, include a substantial situational component. As we stated before, we believe an applicant response model must embrace both individual and situational sources of influence.

AN APPLICANT RESPONSE MODEL

Current faking models clearly specify that faking behavior is influenced by both individual and situational characteristics (McFarland & Ryan, 2000; Snell et al., 1999). We would like to recast their theoretical presentations, and empirical research, and assert that the process underlying an applicant's choice of responses to a personality selection test is influenced by both person-level dispositions (e.g., true level of underlying trait, personal experiences, expectations, etc.) and environment-level presses (e.g., large number of applicants, job attractiveness, presence of lie scales, etc.). This is a clear departure from the conceptualization that responses to traditional personality items reflect mere self-reports of static dispositions. The proposed applicant response model does not make this traditional assumption and we would be remiss if we were to suggest that we are the first to take the road less traveled (Hogan, 2005b; Hogan & Mills, 1978).

Conceptualizing the Reasoning Process

We conceptualize the dispositional components as the underlying biases that individuals have developed from their experiences with the world. The situational influences are characterized as environmental cues, or stimuli. Similar to existing faking models, we postulate that these personal biases and perceptual cues influence the reasoning process employed by the applicant when making a response. It is the justification mechanisms (JMs) listed in Table 7.1 that, we believe, are the proximal determinants of how, and why, an applicant chooses to provide an accurate, or inaccurate, response to a personality selection test. The situational cues and dispositional biases are the more distal influences on the choice of a response.

Modeling the Effects of the Reasoning Process

The main premise of our applicant response model is that differences in the JMs, or reasoning process, will moderate the validities of our personality selection tests. For example, if the verification JM has been activated such that individuals believe inaccurate responses will be identified, responses will be more accurate and therefore be more predictive of actual job behaviors. If the utility JM has been activated such that there is high utility for providing inaccurate responses, predictive validity will be low. However, we are not so naïve to think that criterion-related validity studies will suddenly rain from the sky to test moderator hypotheses of variables that appeared 20 minutes ago. Instead, we think it is important to first demonstrate that these JMs are at least related to applicant responses before we can reasonably justify a test of moderated validities.

With regards to responses to personality measures, we generally assume that, on average, individuals who choose to provide inaccurate responses (i.e., fake-good) will elevate their scores. By comparison, individuals who choose to provide accurate responses (i.e., honest) will, on average, generate lower scores on personality tests. Following this logic, we would therefore expect that activation of the JMs would result in mean level differences on the personality variables. From this point, one would be tempted to simply measure the JMs as individual differences variables

Table 7.1.   Justification Mechanisms for Applicant Responding

Verification JM	Response choice is determined by perceptions of whether accurate responding can be identified.
Utility JM	Response choice is determined by perceptions of whether accurate responding can be identified.
Morality JM	Response choice is determined by moralistic principles or perceptions.
Ability JM	Response choice is determined by whether the applicant believes he/she can adequately identify the "right" answer.
Framing JM	Response choice is determined by framing effects.

by asking questions such as "To what extent did you believe that providing the 'right' answer instead of an honest answer would help you get the job?" Then, this purely dispositional measure of the utility JM would be related to an applicant's response and, possibly, response accuracy or predictive validities. Relying on such explicit measures of the JMs, we believe, is akin to relying on a question such as "To what extent would you agree with a false observation just because everyone else in the room did?" in order to model the effects of social conformity. As a reasoning process, social conformity is largely an implicit process and is almost impossible to measure explicitly. However, social psychologists have been studying the antecedents and consequences of complex reasoning processes such as social conformity by manipulating the situational cues that are hypothesized to affect it and testing mean differences in behavior. With the exception of the studies of warnings to applicants, personality researchers and certainly faking researchers have been extremely reluctant to give up a correlational research paradigm. Our applicant response model seeks to break this dependency on an individual differences framework which, no doubt, has been one consequence of our alliance with a trait theory approach.

Antecedents of the Reasoning Process

We propose that the first step in testing, and further developing, the applicant response model is to experimentally manipulate the JMs by altering the cues or biases which will possibly activate them. For example, warnings should press on the verification JM which, in turn, should lead to lower average responses to a personality selection test when compared to an unwarned group. The JM is thereby serving as the reason the responses are different, that is, a mediator between the situational cue (warning) and the applicant response.

These JMs can be evoked by both individual differences (bias) and situational information (cues). For example, the JM for verification can be activated because the individual walked into the selection setting already believing that an inaccurate presentation of his/her attributes will be detected, or by the situational instructions which emphasize identification and consequences for inaccurate responding. In this way, we hypothesize that both individual dispositions and the contextual cues of the environment have the ability to press on the JM. With this rather simple framework, we would then be in search of various situational cues (e.g., warnings) and/or dispositional biases (e.g., narcissism) which affect the JMs. But, we believe most of us who have been investigating faking would feel like we've already been down this road and that we have only added a set of potential mediators (i.e., JMs) to the scenery. What is missing from the framework described above is how situational cues and dispositional biases would influence each other and thereby affect an applicant's reasoning process (i.e., JMs) for providing a response to a personality selection measure. That is, what exactly is the interactional mechanism?

Conceptualizing the Interaction of Cues and Biases

As we attempted to explicate the interactional nature of our model, we considered the various conceptualizations of an interaction. From an experimental design perspective an interaction exists if the influence of two categorical variables on a continuous third variable can only be understood in relation to each other. From a regression perspective an interaction exists if the cross product of two (typically continuous) variables explains significant variance in a continuous dependent variable above and beyond the simple effects of the two predictor variables. These statistical representations of interactions reflect the mechanics of how we actually test for interactions but bear little resemblance to our theoretical conceptualizations of interactions. In most theories, one variable is treated as the primary focus (e.g., personality measure) and the other is treated as a moderator (e.g., job autonomy). Thus, job autonomy moderates the relationship between personality predictors and job performance such that the higher the job autonomy, the stronger the relationship between an employee's personality characteristics and job performance (Barrick & Mount, 1993). In the rather brutal language of our statistical tools, degree of job autonomy and levels of personality interact in their prediction of job performance. Our statistical tests do not reflect this distinction between the primary variable of interest and the moderator. As we alluded to earlier, we surmise that we make this theoretical distinction for largely one reason: to give us focus and thereby guide our research hypotheses.

We contend that in order to understand the process by which individuals choose responses to personality selection tests, we must consider the situational cues in conjunction with dispositional biases. It is the interaction of individual differences and the context that will best explain these choices. Historically, researchers investigating faking have typically tried to understand which individuals fake their scores on personality selection tests. Thus, for many of us, the primary focus has been the dispositional influences on faking. We think this is problematic for several reasons. First, not only are the individual differences which could be at play in this scenario endless, they are not well developed theoretically. Clearly the one disposition that was heralded as *the* indicator of faking, social desirability, has not helped us generate higher predictive validities. Second, from an application perspective, even if our investigations of dispositional influences are successful and we uncover stable individual differences which are *clearly* related to the accuracy of an applicant's response, what then? We apply a correction to an applicant's score based on his or her status on this related individual difference? If this dispositional variable is strongly related to accuracy and relatively free of systematic confounds and measurement error, then this might work. But, we believe the discovery of the perfect indicator of faking is unlikely and requires traveling a rather torturous route with few signposts or resources. Finally, our reluctance to place individual differences as the primary distal influences of applicant response behavior is a practical one. We agree that understanding dispositional biases that may influence applicant response behavior through the JMs is a worthy scientific endeavor and do not want to explicitly discourage researchers from investigating them. But, in most cases, we can only measure, not control or manipulate dispositional biases. Situational cues, on the other hand, are within our control. The primary goal driving the development of our applicant response model is to increase the predictive validities of personality selection tests. Therefore, we present situational cues as the primary focus of the applicant response model and encourage research that manipulates the situational cues which are hypothesized to affect the reasoning process, or JMs, behind an applicant's response choice.

In light of these historical lessons regarding dispositional biases that affect applicant responding and our intent to *do something* that will make these responses more accurate, we offer Figure 7.1 as our applicant response model. As one can see, we speculate that the proximal causes of an applicant's response are grounded in the applicant's reasoning process. We hypothesize that the building blocks of this reasoning process are the JMs listed in Table 7.1. Thus, we contend that if you can alter the JMs, you can alter an applicant's response choice. The JMs are hypothesized to be influenced by the interaction of situational cues and dispositional

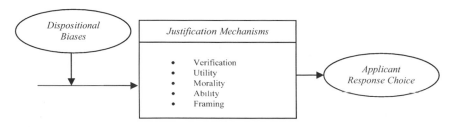

Figure 7.1.   Applicant Response Model.

biases. For the reasons explicated above, we have chosen to treat the situational cues, such as warnings, as the primary influence and dispositional biases, such as competitiveness, as the moderators of this influence on the JMs. The main research question underlying this model is: what can we do to alter the JMs so we can obtain applicant responses which, on average, look like average incumbent, or honest, responses? Moreover, what individual differences act to override or negate this situational cue and which dispositions might actually amplify these situational manipulations? It is our hope that empirical investigations of how these JMs can be altered and the effects these JMs thereby have on applicant responses would generate a clearer picture of how, and why, applicants choose to provide accurate, or inaccurate, information about themselves on personality selection tests.

Manipulating the Reasoning Process

Our rather desperate search for those "magic bullets" that would simply fix the problem of applicant faking speak to our desire as selection researchers to control, or at least influence, the way in which applicants respond to personality selection tests. In an effort to provide some empirical evidence for the existence of these JMs, we present selected research from the literature which we believe may reflect the reasoning process behind how, and why, applicants choose responses to a personality selection measure.

Verification JM

A common method for researchers to influence applicant responding is to change the salience of the consequences of intentional response distortion. For example, many studies have included warnings not to fake, hypothesizing that warnings will decrease levels of faking behavior. Indeed, Dwight and Donovan (2003) show that warnings do affect applicant responses (as operationalized by mean-level differences) on personality measures. This research has also shown that the type of warning matters: warning of both identification (we will know if you fake) and consequences (we will punish you if you fake) resulted in less extreme responding as compared with either type of warning individually. Using similar logic, Illingworth (2004) employed a within-subjects design and found that validity dropped from a baseline condition to a no-warning applicant condition and then bounced back (though not completely recovered) in a warning applicant condition. Vasilopoulos, Cucina, and McElreath (2005) have also shown that the relationship between personality test scores and cognitive ability are stronger under a warning condition. We suggest that the consistent effect of warnings on applicant responding is a due to increased salience of the verification JM.

Additional support for the verification JM can be seen in the research by Donovan, Dwight, and Hurtz (2003) which utilized the randomized response technique (RRT). The RRT allows for completely anonymous responses (lower verification cue), thereby allowing respondents to be more honest without having to worry about being identified. Results indicate that participants in the RRT group admitted to more faking on a personality test than a control group, and the perceived verifiability and severity of the items was related to the amount of faking admitted. Thus, these individual perceptions could be interpreted as moderators of the effects of the situational cues on responses. In this instance the verification JM was made salient but in an opposite direction. And, this resulted in an opposite effect on the applicant responses.

Another interesting study related to the verification JM is the use of required elaboration in which biodata items included instructions for the respondent to elaborate on why he or she provided the answer (Schmitt & Kunce, 2002). Means for respondents with required elaboration items were lower (about .6 standard deviations) than for those who did not have to explain their responses. We suggest that required elaboration served to elevate the salience of the verification bias and thereby affect an individual's response choice.

Utility JM

The basic idea behind the utility JM is that respondents will provide inflated responses or the 'right' answer when placed in situations where there is a considerable amount to gain from faking. For applied researchers and practitioners, the classic example of a highly salient situation which would cue the utility JM is the employment testing context. When responses of applicants are compared to incumbent data or control conditions, applicants tend to have higher means (Birkeland, Manson, Kisamore, Brannick, & Liu, 2002; Hough, 1998) and the personality tests used often have different factor structures between groups (Cellar, Miller, Doverspike, & Klawsky, 1996; Schmit & Ryan, 1993). These results indicate that responses to personality tests are clearly affected by changes in the context of the testing.

On the other hand, we speculate that contextual cues could also be manipulated to press on the utility JM. For instance, Pace, Xu, Penney, Borman, and Bearden (2005) developed a new set of instructions for applicants that they dubbed the "let's reason" instructions (a full discussion of their methods is described in chapter 12). In this method, they cautioned applicants that faking responses to the selection test may result in getting a job which doesn't fit them. We interpret this approach as an attempt to manipulate the applicant's perception of the utility of providing elevated responses. We believe this type of attempt to increase the predictive validity of a personality selection test has merit.

Though the job applicant context is certainly a salient or strong situation, there appears to be considerable individual differences in response behavior in these settings. In directed faking studies, personality scores are, on average, close to one standard deviation higher than honest scores (Viswesvaran & Ones, 1999). However, comparisons of applicant and incumbent samples (Birkeland et al., 2002, Hough, 1998; Stokes, Hogan, & Snell, 1993) result in closer to one half of a standard deviation difference in means. Judging from these results, it seems clear that when participants are placed in an incentive condition in which there is something to be gained from faking (such as a job, or a monetary reward in some experimental designs), at least some people will fake, and within the faking group, some will fake more than others. To some extent, this could be due to individual differences in the utility of providing inflated, rather than accurate, responses. Thus, the more the applicant wants, or needs, the job, the more likely this bias will influence his or her reasoning process in the choice of the response. These individual differences would, in our applicant response model, serve as moderators of the relationship between the situational cues and the utility JM.

Morality JM

This JM is based on the belief that response behavior is influenced by personal orientations regarding morals, values, integrity, or ethics. Theoretically, the morality JM has a strong dispositional component, as people with a low morality JM may find it hard to believe that not everyone lies on self-report surveys, while those with a high morality JM do not understand why everyone is not always honest. The belief that integrity in particular is related to faking has been put forth in the faking literature (Douglas, McDaniel, & Snell, 1996; Rosse et al., 1998). McFarland and Ryan (2000) and Griffith, English, Yoshita, and Monnot (2004) provide preliminary evidence that integrity is related to applicant response behavior.

Further evidence of the influence of morality bias on applicant responding comes from attempts to identify different patterns of applicant response behavior. Lueke et al. (2002) used a measure of individuals' perceptions of employment testing to identify three different clusters of respondents: those that believed faking happens all the time and is not equivalent to lying, those that believed faking is morally acceptable in certain circumstances, and those that equated faking on selection tests with lying and is not acceptable in any circumstance. Individuals who believed that intentional response distortion on applicant selection measures is different from lying had more elevated responses on the personality measure. Lueke et al. labeled these three clusters consistent, complex, and rare fakers, respectively. These clusters appear very similar to the item response work of Zickar et al. (2004), who produced a cluster of honest respondents, fakers, and slight fakers in an applicant sample. Though it is too early to place a great deal of faith in the reproducibility of these clusters, it is apparent that in the context of employment testing, some people choose to fake a lot, and some choose not to fake. Is this difference in applicant responding due solely to dispositional differences? Or is there a mechanism by which the context could also press on the morality JM?

We wonder if individuals' perceptions of what constitutes lying are actually consistent across situations. Contextual information such as the description of the job, the company, or the selection process, may also press on the morality JM such that individuals' perceptions of lying in a selection context may be affected. We would like to see research designs which provide situational cues that make the morality JM more salient and see if this manipulation is effective. Moreover, we believe prior research suggests that there will be strong dispositional biases (e.g., integrity) that will either enhance the effect of the manipulation (i.e., a moral person would be strongly influenced by the manipulation) or decrease or even override it completely (e.g., an amoral person would ignore the manipulation altogether). Thus, attempts to influence the morality JM would, most likely, require attention to the possible interactions between situational cues and dispositional biases.

Ability JM

Research has indicated that people who have a more complete knowledge of the skills and personality types necessary for the job can more effectively produce higher predictor scores (Frei, Snell, McDaniel, & Griffith, 1998; Mahar, Cologon, & Duck, 1995; Martin et al., 2002). These results indicate that when individuals are better able to discern which personality characteristics are important for the job, responses to personality selection tests are elevated. We maintain that this reflects a dispositional trigger of the ability JM: some applicants are equipped with the ability to figure out what the best answers are and others are not.

The more interesting consideration is whether the context could somehow influence individuals to *believe* they are ill equipped to choose the right answer. A decrease in average applicant responding has been obtained when applicants have been warned that a lie scale is embedded in the test (Dwight & Donovan, 2003). Furthermore, applicant responses that were collected via a computer were lower than responses collected with a traditional paper and pencil version of the test (Feigelson & Dwight, 2000). It could be that both of these results are due to the activation of the verification JM as we discussed earlier. However, we offer an alternative explanation: maybe the applicants believed that they would not be able to successfully determine the correct response. Given this belief, there was no incentive to look for the "right" answer and more applicants would provide less inflated scores. We have to wonder whether individual differences in ability would interact with a situational cue such as the one above to predict applicant responses. That is, are individuals with higher ability to discern the "right" answer more, or less, likely to be influenced by a situational cue that they can't fake their responses?

Framing JM

The manner in which a person frames a response to a personality item can have a major impact on the accuracy, or predictive validity, of that response. Several researchers have investigated the effects of adding the "at work" frame-of-reference to personality selection items (Bing, Whanger, Davison, & VanHook, 2004; Hunthausen, Truxillo, Bauer, & Hammer, 2003; Schmit, Ryan, Steirwalt, & Powell, 1995). These studies clearly indicate that the situational cues of the individual items (e.g., at work) produced mean level changes in applicant responses and these differences resulted in higher validities. We consider this as preliminary evidence of the framing JM and its subsequent effect on response choice and criterion-related validity. Are there any individual differences which would moderate the effect of this situational cue?

In addition to explicit modifications to a selection test which alter its frame-of-reference, we also wonder what additional framing cues and biases are in play when an applicant chooses a response to a personality selection test. Consider a person who is aggressive while driving and playing sports but not at work. Depending on which context this person is basing his or her responses while answering items about aggression, the overall score could be considered either as truthful or having been faked. We submit the notion that people often frame situations in such a way that socially desirable responding is not faking (at least in the strict explicit sense). This concept is most closely related to Paulhus' (2002) discussion of self-deceptive enhancement.

The framing JM is very similar to a concept introduced by James and colleagues known as differential framing, which basically states that people "tend to frame and interpret the world in qualitatively different ways" (LeBreton, Burgess, & James, 2000). This phenomenon is certainly not limited to faking research. Dweck and Leggett (1988) show that children's implicit orientations regarding mastery versus performance learning goals influence their perceptions of the same experience—in this case, whether struggling with a difficult task is viewed as a failure or a challenge. Robins and Beer (2001) provide evidence that positive self-illusion bias (as compared with peer ratings) can influence short-term group performance. In both studies, participants' implicit framing biases influenced their subjective interpretations of "truth" (be it regarding a novel task or evaluation of ability level).

The framing JM introduces a number of interesting implications. For example, consider a person who has frequently had problems with poor attendance; however, this person honestly believes that he or she has addressed these issues, will not have a problem missing work in the future, and responds accordingly during the application process. Where is the line between lying and self-deception? We believe that the framing JM is very dependent on the construal process, and the use of specific frames-of-reference can serve to help constrain individual differences in construal. Frame-of-reference training appears to be an area in need of more research within the personality selection literature. Future research regarding the framing JM may also focus on normative comparisons—specifically, are people more likely to frame faking behavior in a given situation as justifiable if they believe that everyone else in that context is faking as well? The personality literature and, more specifically, research on faking do not offer much insight into this justification mechanism. However, we could not create an applicant response model without the recognition that individuals perceive situations differently and that this would affect the response choice. Using other research on framing effects, such as the promotion/prevention focus in self-regulation (e.g., Higgins & Tykocinski, 1992), or the implicit person theory (Dweck, 1999) as a model, we expect that situational cues that alter the way an individual frames the selection setting will, in turn, alter the response choice. Furthermore, it is likely that differences in how individuals routinely construe situations (e.g., promotion versus prevention oriented) will moderate the effectiveness of attempts to alter the frame-of-reference via a situational manipulation.

Summary

Inspired by the conditional reasoning system of measurement, we organized a model of applicant response behavior around five justification mechanisms that we believe represent the process by which applicants choose responses to items in a personality selection measure. Our presentation of the various studies which have attempted to improve predictive validities, or just reduce intentional response distortion, barely scratches the surface of the good ideas, or magic bullets, that have been concocted and tested over the last 4 decades. We contend that the reason we have made little, if any, progress in developing a theory to explain applicant responding, or faking, lies with the fact that the magic bullets rarely work well enough to get noticed. And, since these good ideas are not tied to any theoretical structure, they just fade away. It was our goal to transform the current faking models into an applicant response model that would generate better strategies for improving applicant validities. Additionally, with the JMs we have clear focal points for developing, and organizing, new approaches for investigating the process of applicant response behavior.

Implications for Future Research

It is our hope that the applicant response model will provide personality researchers with a new tool in their quest to improve the predictive validities of personality selection measures. However, we believe the overall contribution of this new model is not just in the good ideas that we hope it will generate. Rather, by adopting a new research design when we conduct our empirical tests of the next great idea, we hope we have also created a mechanism for learning about applicant responding. With this in mind, we would like to present an example of a research design that we think will help us understand the reasoning process which underlies applicant responding.

We return to the recent attempt by Pace et al. (2005) who developed a new set of instructions and tested its effectiveness with regards to (1) lowering applicant personality scores and (2) generating higher predictive validities. As previously stated, we think it was a good idea. However, since it did not work out as planned, or was no magic bullet, what are we left with? Do we know why it did not work? Did we know why it should have worked? We contend that it should have worked because it would have pressed on the utility JM such that individuals were better able to reason that if they provided inaccurate, or faked, responses, they would potentially lose something. Namely, a good fit to a new job. Conversely, by providing accurate information, they would gain something: knowledge that this is the right, or wrong, job for them. But, it did not work. At this point we know that the distal influence (bad fit to job if you fake) did not have the expected effect on the outcome (less extreme responding and higher validity). There are three possible explanations for this result. First, the instructions did not alter an individual's reasoning process because the 'let's reason' information did not adequately press on the utility justification mechanism. Second, the utility JM (the proximal influence) does not have the hypothesized influence on the applicant response choice. And third, the manipulation did work but only for some people and the effect was therefore masked by these differences in a dispositional bias (e.g., importance of job fit for the individual). That is, there could have been an unmeasured, and critical, dispositional moderator in the relationship between the situational cue and the utility JM. We contend that if we knew why this good idea did not work, we would have a better idea of the applicant's reasoning process and would therefore be able to build a better model of applicant response behavior.

Using the proposed applicant response model as a guide, we would suggest the following steps for creating a research design that would test our good ideas for improving the predictive validities of personality selection tests. Step one: explicate the reasoning process that explains why you think your good idea is going to affect applicant responses. We have crafted the JMs with this in mind but see them only as starting points that have emerged from our review of the relevant literature. If the reasoning process that you intend to alter does not fit into one of these JMs, then add it to the model or change the model altogether. Regardless, explicate the reasoning process for all to see. Step two: hypothesize what situational cue or dispositional bias or interaction of the two will affect this reasoning process. Step three: manipulate, or measure, these cues and/or biases that you believe will affect the reasoning process. Step four: determine if these hypothesized cues and/or biases had the intended effects on the reasoning process. Step five: determine if the changes in the JM had the intended effect on the applicant responses. It is our belief that if we can demonstrate that we understand the influences on an applicant's choice of responses, we will have some great ideas that are truly deserving of step six: demonstration that your good idea actually results in improved predictive validity of personality selection tests. And, more importantly, by taking the time to test a theoretical model (steps one through five) you have helped all of us understand which cues and biases actually influence the applicant's choice of a response, and which ones do not. Thus, even when your good idea fails miserably in step six (and we are all heartbroken with you), we have learned something that will help us generate a *better* idea to improve the predictive validity of personality selection measures.

REFERENCES

Barrick, M. R., & Mount, M. K. (1993). Autonomy as a moderator of the relationships between the Big Five personality dimensions and job performance. *Journal of Applied Psychology, 78,* 111-118.

Barrick, M. R., & Mount, M. K. (1996). Effects of impression management and self-deception on the predictive validity of personality constructs. *Journal of Applied Psychology, 81,* 261-272.

Barrick, M. R., & Mount, M. K. (2005). Yes, personality matters: Moving on to more important matters. *Human Performance, 18,* 359-372.

Barrick, M. R., Mount, M. K., & Judge, T. A. (2001). Personality and performance at the beginning of the millennium: What do we know and where do we go next? *International Journal of Selection and Assessment, 9,* 9-30.

Beaty, J. C., Jr., Cleveland, J. N., & Murphy, K. R. (2001). The relation between personality and contextual performance in "strong" versus "weak" situations. *Human Performance, 14,* 125-148.

Bing, M. N., Whanger, J. C., Davison, H. K., & VanHook J. B. (2004). Incremental validity of the frame-of-reference effect in personality scale scores: A replication and extension. *Journal of Applied Psychology, 89,* 150-157.

Birkeland, S., Manson, T., Kisamore, J., Brannick, M., & Liu, Y. (2002, April). *A meta-analysis of the difference between job applicant and non-applicants on personality measures.* Paper presented at the 18th annual meeting of the Society of Industrial and Organizational Psychology, Orlando, FL.

Borman, W. C. (2004). Introduction to the special issue: Personality and the prediction of job performance: More than the Big Five. *Human Performance, 17,* 267-269.

Campion, M. A., Dipboye, R. L., Hollenbeck, J. R., Murphy, K., & Schmitt, N. W. (2004, April). *Won't get fooled again? Editors discuss faking in personality testing.* Panel discussion presented at the 19th annual meeting of the Society of Industrial and Organizational Psychology, Chicago, IL.

Cellar, D. F., Miller, M. L., Doverspike, D. D., & Klawsky, J. D. (1996). Comparison of factor structures and criterion-related validity coefficients for two measures of personality based on the five-factor model. *Journal of Applied Psychology, 81,* 694-704.

Christiansen, N. D. (2000, April). *Utilizing forced-choice item formats to enhance criterion-related validity.* Paper presented at the 15th annual meeting of the Society for Industrial and Organizational Psychology, Dallas, TX.

Christiansen, N. D., Goffin, R. D., Johnston, N. G., & Rothstein, M. G. (1994). Correcting the 16PF for faking: Effects on criterion-related validity and individual hiring decisions. *Personnel Psychology, 47,* 847-860.

Donovan, J. J., Dwight, S. A., & Hurtz, G. M. (2003). An assessment of the prevalence, severity, and verifiability of entry-level applicant faking using the randomized response technique. *Human Performance, 16,* 81-106.

Douglas, E. F., McDaniel, M. A., & Snell, A. F. (1996). The validity of non-cognitive measures decays when applicants fake. *Academy of Management Proceedings, 16,* 127-131.

Drasgow, F., & Kang, T. (1984). Statistical power of differential validity and differential prediction analyses for detecting measurement nonequivalence. *Journal of Applied Psychology, 69,* 498-508.

Dweck, C. S. (1999). *Self-theories: Their role in motivation, personality, and development.* Philadelphia: Psychology Press.

Dweck, C. S., & Leggett, E. L. (1988). A social-cognitive approach to motivation and personality. *Psychological Review, 95,* 256-273.

Dwight, S. A., & Donovan, J. J. (2003). Do warnings not to fake reduce faking? *Human Performance, 16,* 1-23.

Ellingson, J. E., Sackett, P. R., & Hough, L. M. (1999). Social desirability corrections in personality measurement: Issues of applicant comparison and construct validity. *Journal of Applied Psychology, 84,* 215-224.

Feigelson, M. E., & Dwight, S. A. (2000). Can asking questions by computer improve the candidness of responding? A meta-analytic perspective. *Consulting Psychology Journal: Practice and Research, 20,* 303-315.

Fernando, P. J., & Chico, E. (2001). Detecting dissimulation in personality test scores: A comparison between person-fit indices and detection scales. *Educational and Psychological Measurement, 61,* 997-1012.

Frei, R. L., Snell, A., McDaniel, M., & Griffith, R. L. (1998, April). *Using a within subjects design to identify the differences between social desirability and the ability to fake.* Paper presented at the 13th annual meeting of the Society for Industrial and Organizational Psychology, Dallas, TX.

Greenwald, A. G., & Banaji, M. R. (1995). Implicit social cognition: Attitudes, self-esteem, and stereotypes. *Psychological Review, 102,* 4-27.

Griffith, R.L., English, A., Yoshita, Y., & Monnot, M. J. (2004). *Differences in applicant faking behavior: One of these applicants is not like the others.* Paper presented at the 19th annual meeting of the Society for Industrial and Organizational Psychology, Chicago.

Higgins, E. T., & Tykocinski, O. (1992). Self discrepancies and biographical memory: Personality and cognition at the level of psychological situation. *Personality and Social Psychology Bulletin, 18,* 527-535.

Hogan, R. (2005a). In defense of personality measurement: New wine for old whiners. *Human Performance, 18,* 331-341.

Hogan, R. (2005b). Comments. *Human Performance, 18,* 405-407.

Hogan, R., Hogan, J., & Roberts, B. (1996). Personality measurement and employment decisions: Questions and answers. *American Psychologist, 51,* 469-477.

Hogan, J., & Holland, B. (2003). Using theory to evaluate personality and job-performance relations: A socioanalytic perspective. *Journal of Applied Psychology, 88,* 100-112.

Hogan, R., & Mills, C. (1978). A role-theoretical interpretation of personality scale item responses. *Journal of Personality, 46,* 778-785.

Holland, J. H., Holyoak, K. J., Nisbett, R. E., & Thagard, P. R. (1986). *Induction: Processes of inference, learning, and discovery.* Cambridge, MA: MIT Press.

Holyoak, K. J., & Nisbett, R. E. (1988). Induction. In R. J. Sternberg & E. E. Smith (Eds.), *The psychology of human thought* (pp. 50-91). Cambridge, England: Cambridge University Press.

Hough, L. M. (1998). Effects of intentional distortion in personality measurement and evaluation of suggested palliatives. *Human Performance, 5,* 139-155.

Hough, L. M., Eaton, N. K., Dunnette, M. D., Kamp, J. D., & McCloy, R. A. (1990). Criterion-related validities of personality constructs and the effect of response distortion on those validities. *Journal of Applied Psychology, 75,* 581-595.

Hough, L. M., & Oswald, F. L. (2005). They're right, well ... mostly right: Research evidence and an agenda to rescue personality testing from 1960s insights. *Human Performance, 18,* 373-387.

Hough, L. M., & Schneider, R. J. (1996). Personality traits, taxonomies, and applications in organizations. In K. Murphy (Ed.), *Individual differences and behavior in organizations* (pp. 31-88). San Francisco: Jossey-Bass.

Hunthausen, J. M., Truxillo, D. M., Bauer, T. N., & Hammer, L. B. (2003). A field study of frame-of-reference effects on personality test validity. *Journal of Applied Psychology, 88*, 545-551.

Illingworth, A. J. (2004). *The effect of warnings and individual differences on the criterion-related validity of selection tests.* Unpublished doctoral dissertation, The University of Akron, OH.

James, L. R. (1998). Measurement of personality via conditional reasoning. *Organizational Research Methods, 1*, 131-163.

James, L. R. (2004). The conditional reasoning measurement system for aggression: An overview. *Human Performance, 17*, 271-295.

Judge, T. A., & Ilies, R. (2002). Relationship of personality to performance motivation: A meta-analytic review. *Journal of Applied Psychology, 87*, 797-807.

Keil, C. T., & Cortina, J. M. (2001). Degradation of validity over time: A test and extension of Ackerman's model. *Psychological Bulletin, 127*, 673-697.

Kluger, A. N., Reilly, R R., & Russell, C. J. (1991). Faking biodata tests: Are option-keyed instruments more resistant? *Journal of Applied Psychology, 76*, 889-896.

LeBreton, J. M., Burgess, J. R. D., & James, L. R. (2000). *Measurement issues associated with conditional reasoning: Deception and faking.* Paper presented at the annual meeting of the American Psychological Association, Washington, DC.

Levin, R. A., & Zickar, M. J. (2002). Investigating self-presentation, lies, and bullshit: Understanding faking and its effects on selection decisions using theory, field research, and simulation. In J M. Brett & F. Drasgow (Eds.), *The psychology of work: Theoretically based empirical research* (pp. 253-276). Mahwah, NJ: Erlbaum.

Lohman, D. F., & Bosma, A. (2002). Using cognitive measurement models in the assessment of cognitive styles. In Braun, H. I., Jackson, D. N., & Wiley, D. E. (Eds.), *The role of contructs in psychological and educational measurement* (pp. 127-146). Mahwah, NJ: Erlbaum.

Lueke, S. B., Snell, A. F., & Illingworth, A. J. (2002). *The effect of different types of fakers on validity coefficients.* Paper presented at the 17th annual conference of the Society for Industrial and Organizational Psychology, Toronto, Canada.

Mahar, D., Cologon, J., & Duck, J. (1995). Response strategies when faking personality questionnaires in a vocational selection setting. *Personality and Individual Differences, 18*, 605-609.

Martin, B. A., Bowen, C. C., & Hunt, S. T. (2002). How effective are people at faking on personality questionnaires? *Personality and Individual Differences, 32*, 247-256.

McFarland, L. A., & Ryan, A. M. (2000). Variance in faking across noncognitive measures. *Journal of Applied Psychology, 85*, 812-821.

McFarland, L. A., Ryan, A. M., & Ellis, A. (2002). Item placement on a personality measure: Effects on faking behavior and test measurement properties. *Journal of Personality Assessment, 78*, 348-369.

Mischel, W. (2004). Toward an integrative science of the person. *Annual Review of Psychology, 55*, 1-22.

Murphy, K. R., & Dzieweczynski, J. L. (2005). Why don't measures of broad dimensions of personality perform better as predictors of job performance? *Performance, 18*, 343-357.

Nisbett, R. E., & Wilson, T. D. (1977). Telling more than we can know: Verbal reports on mental processes. *Psychological Review, 84,* 231-259.

Ones, D. S., Viswesvaran, C., & Reiss, A. D. (1996). Role of social desirability in personality testing for personnel selection: The red herring. *Journal of Applied Psychology, 81,* 660-679.

Pace, V. L., Xu, X., Penney, L., Borman, W. C., & Bearden, R. M. (2005, April). *Using warnings to discourage personality test faking: An empirical study.* Paper presented at the 20th annual conference of the Society for Industrial and Organizational Psychology, Los Angeles.

Paulhus, D. L. (2002). Socially desirable responding: The evolution of a construct. In H. I. Braun, D. N. Jackson, & D. E. Wiley (Eds.), *The role of constructs in psychological and educational measurement* (pp. 46-69). Mahwah, NJ: Erlbaum.

Robins, R. W., & Beer, J. S. (2001). Positive illusions about the self: Short-term benefits and long-term costs. *Journal of Personality and Social Psychology, 80,* 340-352.

Roese, N. J., & Jamieson, D. W. (1993). Twenty years of bogus pipeline research: A critical review and meta-analysis. *Psychological Bulletin, 114,* 363-375.

Rosse, J. G., Stecher, M. D., Miller, J. L., & Levin, R. A. (1998). The impact of response distortion on preemployment personality testing and hiring decisions. *Journal of Applied Psychology, 83,* 634-644.

Schmit, M. J., & Ryan, A. M. (1993). The Big Five in personnel selection: Factor structure in applicant and nonapplicant populations. *Journal of Applied Psychology, 78,* 966-974.

Schmit, M. J., Ryan, A. M., Stierwalt, S. L., & Powell, S. L. (1995). Frame-of-reference effects on personality scores and criterion-related validity. *Journal of Applied Psychology, 80,* 607-620.

Schmitt, N. (2004). Beyond the Big Five: Increases in understanding and practical utility. *Human Performance, 17,* 347-357.

Schmitt, N., & Kunce, C. (2002). The effects of required elaboration of answers to biodata questions. *Personnel Psychology, 55,* 569-587.

Schneider, R. J., Hough, L. M., & Dunnette, M. D. (1996). Broadsided by broad traits: How to since science in five dimensions or less. *Journal of Organizational Behavior, 17,* 639-655.

Snell, A. F., Sydell, E. J., & Lueke, S. B. (1999). Towards a theory of applicant faking: Integrating studies of deception. *Human Resource Management Review, 9,* 219-242.

Stokes, G. S., Hogan, J. B., & Snell, A. F. (1993). Comparability of incumbent and applicant samples for the development of biodata keys: The influence of social desirability. *Personnel Psychology, 46,* 739-762.

Tett, R. P., & Burnett, D. D. (2003). A personality trait-based interactionist model of job performance. *Journal of Applied Psychology, 88,* 500-517.

Thoresen, C. J., Bradley, J. C., Bliese, P. D., & Thoresen, J. D. (2004). The big five personality traits and individual job performance growth trajectories in maintenance and transitional job stages. *Journal of Applied Psychology, 89,* 835-853.

Vasilopoulos, N. L., Cucina, J. M., & McElreath, J. M. (2005). Do warnings of response verification moderate the relationship between personality and cognitive ability? *Journal of Applied Psychology, 90,* 306-322.

Viswesvaran, C., & Ones, D. S. (1999). Meta-analyses of fakability estimates: Implications for personality measurement. *Education and Psychological Measurement, 59*, 197-210.

Wilson, D. W., & Brekke, N. (1994). Mental contamination and mental correction: Unwanted influences on judgments and evaluations. *Psychological Bulletin, 116*, 117-142.

Wright, J. C., & Mischel, W. (1987). A conditional approach to dispositional constructs: The local predictability of social behavior. *Journal of Personality and Social Psychology, 53*, 1159-1177.

Zickar, M. J., & Drasgow, F. (1996). Detecting faking on a personality instrument using appropriateness measurement. *Applied Psychological Measurement, 20*, 71-87.

Zickar, M. J., Gibby, R. E., Robie, C. (2004). Uncovering faking samples in applicant, incumbent, and experimental data sets: An application of mixed-model item response theory. *Organizational Research Methods, 7*, 168-190.

CHAPTER 8

A SOCIOANALYTIC VIEW OF FAKING

John A. Johnson and Robert Hogan

Deception is the essence of all communication. Plants and animals constantly deceive predators and prey through mimicry, camouflage, and other duplicitous acts. Some animals, including humans, deceive members of their own species and even themselves in order to achieve status, build coalitions, attract mates, and realize other life goals. The ability to deceive is a function of natural selection and an inherent part of life. At the same time, to be deceived by others is a problem, so natural selection has also favored the ability to detect deception. The result is an arms race wherein improvements in the ability to deceive are countered by improvements in the ability to detect deception.

Faking on personality tests, and identifying that faking, can be seen as special cases of deception and the detection of deception. We believe, however, that many who study this issue take a simplistic and largely incorrect view of the test-taker-test-interpreter relationship. This (incorrect) view assumes that test-takers are detached scientific colleagues who reflect on their past thoughts, feelings, and behaviors and then use personality test items to provide an accurate account of their reflections. Those who take this view worry about factors that may distort accurate

A Closer Examination of Applicant Faking Behavior, 209–231
Copyright © 2006 by Information Age Publishing
All rights of reproduction in any form reserved.

and factual communication. One of the most feared factors is the so-called social desirability motive. Describing one's self in a socially desirable manner is generally regarded as a deceptive practice that must be detected and counteracted.

In contrast with the foregoing view, we believe that objective communication is an ideal that rarely occurs, even in science; therefore it is a mistake to assume that test-takers ever communicate with dispassionate precision. We think it is more fruitful to assume that all people, including test-takers, have agendas and that they use communication to further their agendas. Describing one's self in socially desirable terms is a part of some—but only some—people's agendas. Furthermore, even among people with these tendencies, we find (a) differences in what they consider to be socially desirable (Scott, 1963), (b) differences in behaviors designed to appear socially desirable, and (c) differences in the importance people attach to seeming respectable as compared to their other goals. Moreover, describing one's self in socially desirable terms does not necessarily involve deception, and some forms of deception have nothing to do with trying to be socially respectable. Finally, our research indicates that not all deception interferes with the valid assessment of personality and that sometimes telling the literal truth invalidates the assessment process.

We believe that the key to valid personality assessment through self-report lies neither with encouraging respondents to be detached reporters of information nor with eliminating the specter of social desirability. Rather, we think it will be more productive to allow and expect respondents to act like ordinary human beings who pursue their goals using a combination of factual communication, deceptive statements, and self-deceptive maneuvers. In our view, a comprehensive model of item responding would include all of these communication processes and would suggest methods to maximize the validity of personality measurement, regardless of the degree and type of deception that occurs.

We present our view of faking in three steps. First, we provide an analysis of what we think actually takes place in everyday social interaction. Next, we explain why we think answering personality items is a form of social interaction. Finally, we review some data that support the view that test-taking is social interaction and that provide insight into the role of faking on personality tests.

A SOCIOANALYTIC MODEL OF EVERYDAY SOCIAL INTERACTION

Consider the fact that much social interaction involves telling other people about yourself. You will mostly talk about what interests you, but your attitudes and opinions tell others a great deal more about you—far more than where you stand on certain issues. What you say and how you say it tells others whether you are judgmental or tolerant, logical or emotional, assertive or inhibited, how educated you are, and what your cultural background might be. The clothes you wear indicate whether you are bold or reserved, hip or old-fashioned, sophisticated or down-to-earth. And so forth. From our speech patterns to our clothing to our preferences in music and literature to the bumper stickers on our cars, all of these behaviors say, "This is the kind of person I am."

This line of analysis comes from the symbolic interactionists, particularly Goffman (1956). Our view, socioanalytic theory (Hogan & Holland, 2003; Hogan, 1983, 1991) begins with Goffman's observation that competent adults try to manage the impressions that others form of them. Socioanalytic theory extends Goffman's analysis in two ways. First, it explains *why* people engage in self-presentation—what motivates us to present ourselves in particular ways and what functions these self-presentations serve. Second, socioanalytic theory accounts for the *consistencies* in our self-presentational strategies. Whereas social psychologists, following Goffman, focus on the fact that we present ourselves differently to different audiences (Schlenker & Pontari, 2000), personality psychologists focus on recurring patterns of self-presentation that lead to a stable reputation.

Why Tell Others Who We Are?

Advertising who we are serves many useful functions. Seeming spontaneous and fun-loving attracts people with whom we can party. Appearing tough and crazy scares away troublemakers. Appearing tolerant, considerate, and accepting creates friends and romantic partners. And appearing intelligent and reliable may persuade an employer to hire us. These are some of the functions of interpersonal communication.

Socioanalytic theory assumes that the disposition to advertise ourselves evolved millions of years ago. Only by cooperating could our relatively small and puny ancestors survive. Coordinated activity required group members to anticipate the expectations and behavior of other members of the group. And accurately anticipating one anothers' behavior required the ability to display one's, and read others', intentions. Thus, we evolved a talent and disposition to let others know what we are up to, and this remains a part of human nature.

Although socioanalytic theory assumes that human beings are fundamentally cooperative, it also assumes the presence of a tendency toward competition, especially among males. Early American ethologist Wallace Craig (1928) noted that most social animals have evolved "rules of fair play" that constrain competition to keep it from getting out of hand and destroying the group. Males in social species normally engage in ritual-ized battles that involve a minimum of real physical violence. The winners in these competitions move up in the status hierarchy and gain access to better food, mates, and territories. Establishing relatively stable hierar-chies and territories reduces fighting and increases the stability of the group. Ritualized competition inevitably involves deceptive tactics such as bluffing, feinting, intimidation, and braggadocio. Understanding that deception more often accompanies competitive than cooperative interac-tions helps explain when faking is likely to occur on personality tests. When job applicants are focusing on "beating" other applicants or on "beating the test" in order to secure the position, they may be more likely to fake. Such a competitive mind-set is not inevitable; we suggest methods for reducing competitiveness later in this chapter.

Why Do We Present Ourselves Consistently?

Consistency of self-presentation over time is what we mean by personal-ity. If we behaved in different ways each day—cheerful and outgoing today, fearful and reserved the next, outgoing the following day—people would be unable to predict our actions and would shun us. The fact that people who are acquainted with Jane and Henry agree that she is extra-verted and he is shy signifies the existence of personality.

So, why does Jane seem extraverted and Henry seem shy? The answer lies in the evolution of the brain. According to evolutionary psychology, the brain evolved so as to handle routine activities with automatic pro-cesses, and reserve conscious control for novel problems (Gaulin & McBurney, 2004). Routines require far less energy than improvising, so our brains are designed to maximize automatic activity, and these auto-matic mental routines are what impart regularity to our behavior (Bargh & Chartrand, 1999). When we consciously learn new tasks, our perfor-mance is slow, effortful, and somewhat awkward. When we began learning how to type or drive a car, our first efforts were halting and clumsy. But with practice our performance became smooth and effortless, requiring little conscious attention. When our behavior becomes automatic, it becomes consistent.

So it is with personality traits, the behavioral consistency that others attribute to us. Personality traits are patterns of behavior rooted in tem-perament and habits; as a result, we tend to be unaware of our character-istic behavior—and how others perceive us. When individuals try to act in deceptive ways in everyday life (e.g., introverts try to act like extraverts), their natural tendencies "leak through" and observers readily detect them

(Lippa, 1976). Only good actors can make atypical performances seem convincing (Lippa, 1978).

That most of our behavior goes on routinely outside of our conscious control may surprise many of us. We normally think our behavior is voluntary, but that is an illusion. This illusion has caused some psychologists to argue that our conscious choices to behave differently in different situations (e.g., acting happy at weddings and somber at funerals) makes our behavior too inconsistent to be described in terms of personality traits such as *cheerful* or *gloomy* (Mischel, 1969). For those of us who know cheerful and gloomy people, Mischel's early critique is obviously wrong. At the same time, is it not true that people behave differently at weddings and funerals? How can we reconcile our observations that behavior changes across situations with our intuitions about personality consistency?

Mischel's (1969) error in his early work was to equate having a personality with behaving the same way in different situations (Johnson, 1999).[1] Only socially insensitive (e.g., psychotic, mentally retarded, clinically narcissistic) people behave the same way in every situation. Socially competent people (i.e., most of us) adjust their behavior to be appropriate to particular situations. This competence is not a rare and highly developed skill; it is a basic prerequisite for normal functioning. Even insects, whose behavior is completely automatized, behave differently around predators, prey, mates, and offspring. Personality is not reflected in the tendency to behave the same way across different situations; rather, it is reflected in the tendency to seek out (and avoid) particular kinds of situations and the consistent ways people present themselves in particular situations (Johnson, 1997b, 1999). Again, socioanalytic theory maintains that our behavior in most situations is guided by unselfconscious habit. People cannot help dealing with certain situations consistently, and these consistencies define their personalities.

Deception as Deviation from Consistency

Deception is a conscious, deliberate deviation from typical forms of self-presentation, a deviation that acquaintances would describe as uncharacteristic behavior. This view of deception contrasts with the view that deception involves acting in a way that is inconsistent with a single "true self" hidden inside of us. Freud may have been wrong about many things, but he was certainly correct in arguing that we are constantly prompted by conflicting impulses and ambivalent feelings, and that a single true self does not exist. Research by Gazzaniga (1998) suggests that the mind consists of independent modules rather than a single, "real" unitary self. We are not endorsing the radical, postmodernist notion that the self is only a social construction, leading to an infinite number of possible selves (Markus & Nurius, 1986). Surely personality is constrained by *some* stable structures in the brain. Yet, because neuroscience has just barely begun to identify these structures (e.g., DePue & Collins, 1999), socioanalytic theory defines the "real" person in terms of observable consistencies in social behavior and defines deception in terms of deliberate deviations from those consistencies.

Our view of interpreting item responses lies between the social constructivist position that item responses can imply virtually any personality trait (Gergen, Hepburn, & Fisher, 1986) and the "trait realist" position that valid item responses correspond veridically to behaviors or experiences guided directly by the trait being measured (McCrae & Costa, 1996). In agreement with constructivists, we see the act of item-responding as a negotiation of identity. In agreement with trait realists, we believe these negotiations are constrained by biology in nonarbitrary ways. Whereas social constructivists seem to have abandoned notions of truth and objective reality, trait realists assume that validity rests on a simple one-to-one correspondence between item content, item response, and the trait that guides the behavior or experience referenced in the item. We believe item responses are valid when respondents produce scores that correspond to their established social reputations, whether or not their responses correspond literally to actual behaviors and experiences in real life.

TEST-TAKING AS A FORM OF SELF-PRESENTATION

Personality Testing as Scientific Reporting

Tyler (1963) and others assume that the ideal form of personality assessment involves multiple observers rating a person's behavior over a period of time. Detached, disinterested, and objective, these observers record actual behaviors to derive a valid representation of personality. According to Tyler, the inconvenience of this method requires us to ask the persons we are assessing to serve as detached, disinterested, objective observers of their own behaviors. Their task, as they read each item on a personality test,[2] is to determine whether the item accurately describes their behavior and then respond according to the literal correspondence. In this view, if the item says, "I read at least 10 books a year," test-takers should think carefully about how many books they read each year and mark agreement with the item only if that number is at least 10.

This traditional view of assessment causes psychologists to worry about respondents' willingness or ability to accurately report their behavior. In particular, psychologists fear that respondents will provide socially desirable, rather than accurate, responses, especially in competitive and evaluative conditions such as preemployment screening. Consequently, much of the "self-report" personality research since the 1960s has involved efforts to work around the potential reluctance of respondents to behave the way we wanted them to. Some of these efforts appear in this volume.

Personality Testing as Social Interaction

We have a different perspective on what happens when people respond to items on a personality test. Our view is that people spontaneously use personality tests as vehicles for self-presentation (Hogan, 1991; Hogan, Carpenter, Briggs, & Hansson, 1985; Johnson, 1981; Mills & Hogan, 1978). People are naturally eager to express themselves, whether in terms of responses to personality items or bumper stickers for their cars. What are the implications of this view for understanding deception and how do they differ from the implications of the traditional view of test item responses?

Recall that the traditional view requires that people's responses correspond to their actual behaviors and worries that people will provide deceptive rather than accurate responses. In contrast, we expect that peoples' responses to items on a properly constructed test will create the same impressions that their behavior creates in real life. Furthermore, people who present themselves in negative ways (e.g., neurotics, delinquents) in real life will also provide socially *un*desirable item responses—because they are telling us who they think they are. A person's self-presentational style represents a habitual strategy designed to manipulate others. A neurotic's complaints and other signs of helplessness are designed to elicit pity, special consideration, and support from others (Watson & Andrews, 2002). A delinquent's signs of deviance and defiance are intended to create the impression of someone too dangerous to mess with (Emler, 2005).

Because people normally employ the same self-presentational gambits in response to personality test items as they use in everyday life, scores on a well-designed personality test tend to predict how the person is seen by others in everyday life. However, we do not claim that testing guarantees a personality portrait that perfectly matches a person's reputation. This is because factors other than honesty and deception affect test validity. One of the most crucial issues is whether an item's reputational significance is easily and intuitively grasped by test-takers. If the item "I read at least 10 books a year," actually conveys the impression of being smart, then people who are regarded by others as smart should endorse the item, even if they read less than 10 books and have to "lie" about the number of books they read. The trick is for the test author write items that a respondent will naturally answer in a way that accurately reflects his or her established reputation. And this accuracy must be established empirically.

The fundamental problem of valid assessment is not whether people tell the literal truth but whether they respond in ways that are consistent with their established reputations. If intellectuals fail to endorse the book-reading item, it is the test author's fault. The test author has failed to provide a clear opportunity for self-presentation. On the other hand, if most intellectuals endorse the item but a few fail to endorse it, the problem lies in the social intelligence of the few who did not get it. Presenting oneself to others is a social game, and those who are not skilled enough to play the game will provide invalid portraits of themselves on personality tests.

To recap, we assume that most social behavior consists of habitual forms of self-presentation over which we have little conscious control. The consistency of these habits creates consistent impressions on the people with whom we frequently interact, and we call these impressions *reputation*. We say that people are "authentic" when their current behavior is consistent with their reputations. Conscious attempts to act in uncharacteristic ways are usually perceived to be inauthentic. We regard personality testing as a form of social interaction that is guided by the same habits that govern everyday behavior. Socially competent people respond to well-designed test items in ways that are consistent with their reputations. Literal truth-telling is less an issue for valid testing than the social skills of the test-taker and the proper design of personality items.

Is Faking More Likely on Tests than in Real Life?

Researchers concerned with deception will probably complain that we overestimate the amount of habitual self-presentation that occurs during personality testing. Even if personality test responses are is seen as self-presentations, the process differs from everyday interactions in ways that encourage inauthentic self-presentation. For starters, in everyday life people can observe the degree to which our words and nonverbal behavior are consistent. Nonverbal behavior carries far more weight than words in communication (Mehrabian, 1972) and is a cue in detecting deception (DePaulo, Lindsay, Malone, Muhlenbruck, Charlton, & Cooper, 2003). But with personality data we have only responses to written statements without the information provided by nonverbal cues. Furthermore, in everyday life we are monitored for long periods of time so that it would be easy for acquaintances to notice an uncharacteristic self-presentation. This puts limits on who we can claim to be without contradicting ourselves (Hogan, 1987). When responding to personality test items, we are on stage for an hour at most—and surely most people can put on an act for such a short time.

Ultimately it is an empirical question as to whether people can present themselves in coherent but duplicitous ways so as to produce interpretable but duplicitous personality profiles. The next section of this chapter presents some data relevant to this question. Before looking at the evidence, consider the possibility that making a positive but inauthentic impression on a personality test is quite difficult. Falsifying a good impression requires knowing how personality is related to effective job performance. Such knowledge is available for certain jobs (e.g., successful managers are socially ascendant), but less available for others (e.g., successful architects receive low scores on measures of emotional stability). Without knowing which personality characteristics are needed for a job, applicants will not know how falsely to present themselves.

But knowing the relevant personality characteristics is still an insufficient basis for creating a desirable profile. Job applicants must also determine how to respond to individual items in order to manipulate their scores. This may be obvious for face valid items, but good inventories contain subtle and nonobvious items. Also, like the proverbial students who score poorly on multiple choice exams because they think too much about each answer, people who think too much about the implications of each item will find themselves nearly paralyzed.

To create a specific impression on a personality inventory requires intuitive knowledge and skills (Sternberg & Hedlund, 2002). Responding to a personality inventory is a kind of skilled performance; personality measures—often regarded as capturing typical performance—are more like ability tests (measures of maximal performance) than most people realize (Wallace, 1966). In the next section we present evidence indicating that the factors that facilitate skillful self-presentation in everyday life might also apply to competent self-presentation on personality tests.

EMPIRICAL RESEARCH GUIDED BY THE SOCIOANALYTIC VIEW OF TEST-TAKING

Earliest Socioanalytic Studies

If responding to personality items is a form of social interaction, then the scores for socially competent people should be more valid than scores for less socially competent people. Mills and Hogan (1978) tested this hypothesis by asking members of a suburban community service organization to complete CPI-based measures (California Psychological Inventory; Gough, 1975) of dominance, femininity, social insight, and empathy, and then asking the participants to rate one another on these characteristics. The discrepancy between standardized scores on the first three personality scales and peer ratings for these traits correlated $r = -.87$ with scores on Hogan's (1969) Empathy Scale, a well-validated measure of social competence (Johnson, Cheek, & Smither, 1983). Socially competent people were more able than less competent people to appreciate how others perceived them and their item responses more closely matched how they were described by knowledgeable acquaintances.

In a follow-up study, Johnson (1981) examined the relation between social skill and the consistency of self-presentation on personality inventories (defined as providing the same response to identical or nearly-identical items on two inventories). According to Johnson, the traditional view of personality assessment predicts that honest, earnest, and cooperative people would be more consistent, while the socioanalytic view predicts that people with good social skills would be more consistent. In three separate samples (normal adults, convicted murderers, students), all of the socioanalytic predictions, and none of the traditional predictions were verified—self-presentational skill was associated with consistency.

The Mills and Hogan (1978) and Johnson (1981) articles support the socioanalytic view of personality assessment under nonevaluative research conditions. The next two studies concern personality assessment during simulated personnel selection. Because these studies have not been previously published, the first will be described in some detail. The second study replicates the first with a different personality inventory.

Performance Under Simulated Employment Testing

Johnson (1986) asked 83 college students to complete the CPI (Gough, 1975) under standard instructions. Then participants completed four scales (Dominance, Socialization, Self-Control, and Flexibility) from Burger's (1975) short form for the CPI; these scales were chosen as markers for four of the five major factors of personality: Extraversion, Conscientiousness, Emotional Stability/Neuroticism, and Openness to Experience (McCrae, Costa, & Piedmont, 1993). Also, the validity of these scales for predicting job performance had been established by past research. As they completed the short form of the CPI, participants were asked to imagine that they were applying for six jobs representing Holland's (1985) six RIASEC occupational categories (police officer, dental technician, architect, religious counselor, business manager, and cashier/short-order cook). The order of retesting was randomized, with rest between test sessions to minimize order effects and fatigue. Participants received extra credit, and those with the best scores for each of the six jobs were promised additional extra credit as an incentive for doing well.

Full CPI scores were estimated from the short form scores using regression equations provided by Burger (1975). Two different analyses were conducted. First, changes in scores on the four CPI scales from straight-take conditions in each of the six employment testing conditions were examined, to see whether average scores increased for scales known to correlate with effective job performance. Second, a measure of success for each employment testing condition was constructed by adding scores from CPI scales known to correlate positively with job performance and subtracting scores from CPI scales known to correlate negatively with job performance. CPI scores under standard conditions were correlated with these six measures of successful employment testing to see if personality as normally measured predicted maximal performance under evaluative conditions. Also, individuals receiving the highest performance scores were interviewed to see if they had any distinguishing characteristics.

Results indicated that, across all six jobs, scores on Dominance, Socialization, and Self-Control increased, and Flexibility scores decreased. The *amount* of change differed across jobs, suggesting that participants were attempting to tailor their presentations across conditions. However, the participants failed to raise their Flexibility scores for occupations for which this scale correlates positively with job performance (e.g., dental technician) or lower their scores on Self-Control, Dominance, and Socialization for occupations where these scores are negatively correlated with job performance (e.g., manager, counselor, and architect, respectively). The participants may have reacted to the *general* social desirability of the items (low for Flexibility and high for the other three scales), but they were unable to raise or lower their scores according to the performance profiles specific to the different jobs.

Some differences in successful test performance could be accounted for by personality as measured normally (see Table 8.1). Baseline personality scores were partialled out of all predictions. Higher scores for different occupations were associated with different combinations of personality traits. Space limitations preclude a detailed discussion of the results presented in Table 8.1; in short, the ability to look good in each job category was a function of a unique set of established personality strengths.

The more informal interview data on the six best performers again illustrated that high test performance for different jobs is based on established assets. The best police officer simulator was an administration of justice major who received the Most Valuable Player honor for the college's baseball team. The best dental technician simulator was a nurse

Table 8.1.   Personality Correlates of
Employment Testing Performance

	Police Officer	Dental Technician	Architect	Religious Counselor	Business Manager	Cashier/ Cook
CPI Scale						
Dominance	13	14	12	11	16	13
Capacity for Status	08	15	04	11	28*	-05
Sociability	10	07	-08	-11	14	07
Social presence	18	25*	-13	02	14	16
Self-acceptance	14	14	-05	06	-01	18
Well-being	25*	31**	-16	03	08	08
Responsibility	14	13	09	-18	22*	08
Socialization	27*	25*	-34**	-11	-03	20
Self-control	16	18	-23*	02	-04	-08
Tolerance	19	22*	-23*	-03	16	-05
Good impression	08	15	-20	09	12	-09
Communality	10	24*	01	10	-01	14
Achievement via conformance	29**	31**	-26*	-03	-08	01
Achievement via independence	-05	05	-03	04	15	-24*
Intellectual Efficiency	17	25*	-11	-17	09	03
Psychological Mindedness	11	08	-05	10	-05	-15
Flexibility	-40**	-24*	22*	17	14	-27*
Femininity	12	06	-16	26*	-10	01
Empathy	13	-03	-03	-10	18	-02

Note: Decimal points omitted from all correlation coefficients. Criterion is the raw sum of personality scale scores related positively to effective performance minus the scale scores related negatively to effective performance. Coefficients are partial correlations corrected for scores on Dominance, Socialization, Self-Control, and Flexibility under straight-take conditions. $N = 83$.
   $*p < .05$
   $**p < .01$ (both two-tailed).

who had returned to school and was sporting a 3.9 grade-point average. The best architect simulator was an individualistic, adult returning student in the honors program and majoring in human development. The best religious counselor simulator was a soft-spoken, yet intelligent psychology major whose father is a psychiatrist. The best business manager simulator was a vivacious business major who was president of a scholastic fraternity, vice-president of the student government association, writer for the college newspaper, and recipient of four academic awards at the college's honors convocation. The best cashier/cook simulator was undistinguished academically, but had experience in the short-order food business.

Johnson (1987) investigated the degree to which these CPI-based findings would generalize to the Hogan Personality Inventory (HPI; Hogan, 1986). Seventy-eight students completed the HPI under standard instructions. They then completed the HPI at home over a 2-week period, as if they were applying for jobs as a long-haul truck driver, Navy bomb disposal technician, counselor at a private psychiatric institution, or middle-level manager. Hogan developed criterion-keyed HPI scales to predict performance in precisely these four occupations. Extra credit was given to all students, and additional extra credit was promised to the best performers in the four employment conditions.

Applicants improved their scores considerably on two criterion-keyed HPI scales for person-related occupations (counselor, manager). Scores on the HPI bomb technician scale were barely, but statistically significantly higher in the employment simulation, and scores on the HPI truck driver scale show a nonsignificant decrease. This suggests that it may be easier to enhance self-presentation for person-related occupations—perhaps because valued traits are common knowledge—but difficult to impossible for more technical occupations. Unlike the Johnson (1986) study, performance scores under simulated employment testing showed no significant relations to personality scores gathered under normal test-

Table 8.2. HPI Score Changes When Applying for Different Jobs

| Job Position | Straight-Take | | Applying | | | |
	Mean	SD	Mean	SD	Pearson r	t value
Truck Driver	134.0	19.4	131.9	32.3	.16	-0.52
Bomb Technician	38.4	6.6	40.3	5.7	.03	2.03**
Counselor	76.5	14.0	81.9	12.4	.24*	5.28
Manager	59.1	9.6	71.3	9.4	.23*	9.11**

Note: $N = 78$; significance of rs and ts based on 77 df.
*p < .05
**p < .01 (both two-tailed).

ing conditions. We need more research to understand what kind of person can present him/herself well on HPI criterion-keyed scales and whether scores gathered under normal or interview conditions are better predictors of job performance.

Valid and Invalid Lying

For decades psychologists have worried that people respond to the social desirability of personality items rather than their content. To deal with this concern, psychologists (e.g., Tellegen, 1982) developed *unlikely virtues* scales. These scales contain items describing behaviors that are so implausibly virtuous (e.g., "I have never hated anyone.") that they are unlikely actually to describe anyone. The assumption is that those who endorse unlikely virtue items are lying in order to make a good impression.

Johnson and Horner (1990) question the unlikely virtue assumption in two ways. First, they point out that, although saying something like "I have never hated anyone" may be literally untrue, in real life we often say things we do not mean to be taken literally. What people may mean by endorsing "I have never hated anyone" is that they have rarely hated anyone or that they do not carry long-term grudges. From a socioanalytic view, what matters is the image that is created by saying "I have never hated anyone." And the image is that of a forgiving, nonvindictive individual. People with reputations for being tolerant and forgiving are likely to endorse this item, even if it is literally untrue.

Johnson and Horner (1990) also question whether, when people endorse unlikely virtue items, they are responding to the social desirability of the items rather than the items' unique content. There are, after all, many different ways to be virtuous, and people may endorse only those unlikely virtues that are consistent with their reputations (Cawley, Martin, & Johnson, 2000). To test this idea, Johnson and Horner constructed six unlikely virtue (UV) scales whose contents corresponded to the content of each of the six primary scales of the Hogan Personality Inventory (HPI; Hogan, 1986): Intellectance, Adjustment, Prudence, Ambition, Sociability, and Likeability. UV Intellectance items claim unusual intellectual prowess ("In my own way, I am an intellectual giant."), UV Adjustment items claim exceptional mental health ("I have no psychological problems whatsoever.") and so forth. These UV items were embedded in the standard HPI items, and 142 students completed this augmented HPI. Participants also completed the Bipolar Adjective Rating Scales (BARS; Johnson, 1997a), designed to mea-

Table 8.3.   Correlates of Unlikely Virtue Scales

	Unlikely Virtue Scales[a]					
	UINT	*UADJ*	*UPRU*	*UAMB*	*USOC*	*ULIK*
Standard HPI Scales						
Intellectance	.56***	.19**	.11	.31**	.09	.00
Adjustment	.04	.44***	.05	.24**	.09	.08
Prudence	.01	.04	.56***	.01	-.05	.19**
Ambition	.42***	.27***	-.04	.39***	.38***	.08
Sociability	.23**	.23**	-.21**	.34***	.40***	.08
Likability	-.09*	-.03	-.01	-.10	.10	.22**
BARS Self-Ratings						
Mentality	.49***	.11	.11	.32***	.03	.04
Poise	.14*	.38***	.01	.33***	.21**	.19*
Discipline	.08	.06	.31***	.06	.01	.10
Power	.19**	.20**	.12	.30***	.10	.00
Sociality	.16*	.23**	.03	.23**	.30***	.01
Likeableness	-.07	-.07	.18*	-.06	-.07	.23*
BARS Acquaintance-Ratings						
Mentality	.16*	-.13	.08	.08	-.07	-.04
Poise	.08	.19**	.07	.16*	-.06	.04
Discipline	-.10	-.09	.24**	-.01	-.11	.16*
Power	.10	.14*	.03	.23**	.14*	.03
Sociality	.15*	.26***	.01	.24**	.28***	.06
Likeableness	-.09	-.08	.14*	-.09	-.16*	.20**

Note: Boldface coefficients indicate expected convergent validity coefficients for the unlikely virtue scales.
[a]Unlikely virtues (UV) scale labels are UINT = UV Intellectances, UADJ = UV Adjustment, UPRU = UV Prudence, UAMB = UV Ambition, USOC = UV Sociability, ULIK = UV Likability.
*$p < .05$ **$p < .01$ ***$p < .001$ (all one-tailed)

sure six HPI dimensions. In addition, they had two people who knew them well rate them with the BARS.

Table 8.3 presents the correlations between the UV scales and their HPI and BARS counterparts. A clear pattern emerges: In every case, the unlikely virtue scales have their highest correlations (on the order of .40-.50) with the HPI scale for the same dimension (e.g., unlikely Ambition's strongest correlate is with the HPI Ambition scale). Also, each unlikely virtue scale has its highest correlation with the BARS scales assessing the same dimension. The correlations with acquaintance ratings are especially noteworthy; they show that endorsing unlikely virtues items are associated with a person's reputation.

Using the same data set, Johnson (1990) tested the degree to which endorsing unlikely virtues is a generic (unitary) or selective (multidimensional) process. Four alternative models of the data were compared using LISREL. The best fitting model revealed two patterns of responding to HPI UV scales. One involved individuals selectively endorsing UV items consistent with their scores on standard HPI scales. The second was a pattern of indiscriminant exaggeration associated with elevated scores on Intellectance, Ambition, and Sociability, and low scores on Likeability. This particular profile is seen in narcissists (Raskin, Novacek, & Hogan, 1991). According to Raskin et al. and Paulhus and Reid (1991), the exaggerated self-presentational style of narcissists is an unconscious form of attention seeking, a style that is self-defeating because it leads "ultimately to rejection and interpersonal failure" (Raskin et al., p. 35).

One might conclude, then, that persons who selectively endorse items from the spectrum of virtues are engaging in genuine self-presentation, and their scores can be considered valid. Others who endorse the full range of unlikely virtue scales are genuine also—genuine narcissists. Their exhibitionistic displays and excessive need for social approval make them poor candidates for jobs that require working cooperatively with others. This study implies two things about the valid assessment of personality for employment. One is that applicants can lie (exaggerate) and still provide information that validly reflects their established reputations. The second is that different patterns of exaggeration can reflect different degrees of suitability for employment within particular jobs. We do not want narcissists in jobs that require cooperation, collaboration, and teamwork.

The Impact of Social and Cognitive Competencies on Assessment

The Mills and Hogan (1978) study was the first to suggest that the validity of personality testing varies with the respondents' capacities for appreciating how their behavior will be interpreted by others. Johnson (2002) devised a way to measure this capacity. He asked 74 students to complete the Hogan Personality Inventory (HPI; Hogan, 1986) and to have two acquaintances rate them with the Bipolar Adjective Rating Scales (BARS; Johnson, 1997a). He then asked participants to imagine, for each HPI item, how they would rate the personality of someone who had endorsed the item. The procedure for rating each item was nearly identical to the CPI item-rating procedure used by McCrae et al. (1993). Participants assigned a -2 to items that *strongly* reflected the low end of a personality factor, +2 to items that *strongly* reflected the high end, a -1 or +1 to items *somewhat* implying the low or high end of a factor, and *zero* to items that implied nothing about a factor.

Averaging these ratings across the participants established a standard for what each item implied about personality, and the agreement between an individual's ratings and this standard reflected the degree to which individuals understood how item responses would be interpreted by the community at large. The HPI scores for individuals who appreciate how item responses are interpreted predicted acquaintance ratings better than scores for those who didn't understand the implications of their responses; however, the difference was statistically significant only for sociability.

Johnson (1981) suggested that responding consistently to similar personality items indicates social competence. Johnson (2005) developed two methods for assessing consistency using an Internet sample of over 20,000 respondents to an online personality inventory. One method used the similarity of scores from odd- and even-numbered items on each scale (see Jackson, 1976). The second method identified 30 item pairs that the sample as a whole answered in different directions and then measured how closely each individual followed the normative pattern. Consistency on both measures was related to low neuroticism and high openness/intellect; this finding once again supports the competency view of self-presentational consistency. Johnson (2005) noted that in only about 1% of all cases was inconsistency severe enough to invalidate a protocol.

Implications for Constructing and Using Personality Scales in Applied Contexts

The research described in this chapter implies several recommendations for building personality tests and using them in personnel selection. Some of these recommendations are consistent with conventional wisdom, but some contradict existing practices.

First, socioanalytic research maintains that we should write personality items that encourage effortless self-presentation rather than literal self-description. We do not really want to know precisely how many books people read; we want to know how bookish people appear to others. Gordon and Holden (1998) found that, when people are asked to recall and count specific behaviors while responding to items, their scores were less valid than when they were asked to describe their general tendencies and how others see them.

What, exactly, makes an item a good vehicle for self-presentation? Traditional guides to item-writing (e.g., Wolfe, 1993) emphasize the need for simplicity and clarity of item meaning. This is not bad advice; people can easily use items such as "Others see me as shy" to communicate relative degrees of shyness. Yet, the directness and ease with which an item conveys an impression is more important than simplicity and clarity. Johnson (2004) used the term *trait indicativity* to describe how clearly an item implies a trait. He found that trait indicativity affected measurement validity more than all of the other item characteristics he measured.

Valid items with high trait indicativity may or may not be semantically ambiguous. Take the neuroticism item, "I often get headaches." The imprecision of the word "often" makes this item slightly ambiguous. Traditionally, item ambiguity is regarded as bad. Nonetheless, the very ambiguity of the item is what allows it to convey neuroticism. Neurotics exaggerate somatic complaints and are likely to endorse the item *even though the objective frequency of headaches is no greater than average*. So, we have a literally untrue response to an ambiguous item serving as a valid indicator of neuroticism.

Item ambiguity can reduce validity if the different interpretations imply different tendencies. For example, the CPI item "When I get bored I like to stir up some excitement" is ambiguous because the word "excitement" conveys very different meanings depending on the context. In some cases "excitement" means intense social stimulation. For example, McCrae et al. (1993) suggested the item implies excitement-seeking, an aspect of Big 5 Factor I (Extraversion). However, delinquents use the phrase "stir up some excitement" to refer to potentially illegal behavior. Consequently, Johnson (1997a) believes the item implies delinquency and puts it at the low end of Big 5 Factor III (Conscientiousness). On the CPI, the item is scored in the positive direction for Social Presence (a Factor I scale) and in the negative direction for Responsibility (a Factor III scale). Ultimately, it does not matter what the authors feel an item implies; what matters is how respondents understand an item. To make this determination involves asking one's target audience to evaluate the implications of items (Johnson, 2002).

Socioanalytic theory also recommends writing items that people enjoy endorsing one way or the other. People take pleasure in telling us who they are. They feel good about their bumper stickers, the letters they write to the editor, and the entries they make in their Internet blogs. Responding to personality items should tap this hedonic process. Johnson (2006) measured how much people enjoyed responding to each item on the CPI.

Items rated as most enjoyable had higher validity than less enjoyable items.

Traditional psychologists might regard these recommendations as giving away the ranch. If the constructs reflected in our items are obvious, aren't we making it easier for job applicants to fake and create the reputation that will get them the job? Earlier we presented two arguments against this possibility. First, we think that automatic habits of self-presentation usually overwhelm attempts at conscious impression management. Second, even when individuals consciously claim noncharacteristic views and preferences, their ability to fake is constrained by their knowledge, competencies and social skill. Those people who can use inventory items to make a good impression on an inventory are, in fact, socially competent; it is also likely that their competence will make them effective employees.

We offer one final suggestion for encouraging job applicants to "be themselves" (that is, respond to items in a way that is consistent with their reputations). In the beginning of this chapter, we argued that deception is more likely to occur in competitive interactions. The personnel selection process is a prototypically competitive situation. There are typically more applicants than positions, so the applicants compete for the limited slots available. Applicants also often regard the test as an enemy that must be beaten to become employed.

Although the personnel selection process is competitive, employers can make the situation better or worse by the way in which they present tests to applicants. Some employers use inventories with embedded validity scales and warn applicants that attempts to misrepresent themselves will be detected and they will be dismissed from further consideration. This tactic may prevent some conscious misrepresentation, but also creates an uncomfortable tone for those who had no intention of faking. Pace and Borman present several alternative "warning" formats in chapter 12 that employ this logic.

Instead of trying to scare applicants straight, we suggest that employers consider creating a more cooperative tone. Hiring people with the right characteristics for a job is in the best interest of employers *and* job candidates. Employers obviously want the most qualified candidates. But job applicants should also be concerned about their fit with a particular work environment, because a good fit will increase their chances of success and satisfaction (Holland, 1985). If applicants can score well on a test only by uncharacteristic impression management, they will find it hard to maintain that false impression on a daily basis in the job. We are not so Pollyannaish to think that we can prevent all faking; consequently, we advocate the use of unlikely virtue scales to help detect rare but troublesome patterns such as narcissism. We do believe, nonetheless, that treating the process of personality testing as a form of social interaction rather than scientific data gathering will in the end yield far more valid results.

NOTES

1. In fairness to Mischel, his later writings recognize that different situations require different kinds of competencies related to different kinds of personality traits, such that only a particular range of situations is even relevant to a particular trait (Shoda, Mischel, & Wright, 1993). While we continue to disagree with Mischel on a number of points (Johnson, 1999), Mischel's conceptualization of personality in terms of competencies bears a resemblance to the present view of personality as skilled self-presentation.

2. We use the term *personality test* interchangeably with terms such as personality *scale, inventory, measure,* and so forth, despite the traditional view that personality measures have no right or wrong answers while tests do. The reason for this is that we believe that responses to personality measures do in fact have correct or incorrect answers. Correct answers lead to valid scores and incorrect answers lead to invalid scores. Later in the chapter we also argue that differences in abilities underlie differences in providing correct or incorrect responses to items, thereby denying a sharp distinction between measures of typical performance and maximal performance.

REFERENCES

Bargh, J. A., & Chartrand, T. L. (1999). The unbearable automaticity of being. *American Psychologist, 54,* 462-479.

Burger, G. K. (1975). A short form of the California Psychological Inventory. *Psychological Reports, 37,* 179-182.

Cawley, M. J. III, Martin, J. E., & Johnson, J. A. (2000). A virtues approach to personality. *Personality and Individual Differences, 28,* 997-1013.

Craig, W. (1928). Why do animals fight? *International Journal of Ethics, 31,* 264–278.

DePaulo, B. M., Lindsay, J. J., Malone, B. E., Muhlenbruck, L., Charlton, K., & Cooper, H. (2003). Cues to deception. *Psychological Bulletin, 129,* 74-118.

Depue, R. A., & Collins. (1999). Neurobiology of the structure of personality: Dopamine, facilitation of incentive motivation, and extraversion. *Behavioral and Brain Sciences. 22,* 491-569.

Emler, N. P. (2005). Moral character. In V. J. Derlega, B. A. Winstead, & W. H. Jones (Eds.), *Personality: Contemporary theory and research* (3rd ed., pp. 392-419). Chicago: Nelson-Hall.

Gaulin, S. J. C., & McBurney, D. H. (2004). *Evolutionary psychology* (2nd ed.). Upper Saddle River, NJ: Pearson Prentice Hall.

Gazzaniga, M. S. (1998). *The mind's past.* Berkeley: University of California Press.

Gergen, K. J., Hepburn, A., & Fisher, D. C. (1986). Hermeneutics of personality description. *Journal of Personality and Social Psychology, 50,* 1261-1270.

Goffman, E. (1956). *The presentation of self in everyday life.* Edinburgh, Scotland: University of Edinburgh Social Sciences Research Centre.

Gordon, E. D., & Holden, R. R. (1998). Personality test item validity: Insights from "self" and "other" research and theory. *Personality and Individual Differences, 25,* 103-117.

Gough, H. G. (1975). *Manual for the California Psychological Inventory* (Rev. ed.). Palo Alto, CA: Consulting Psychologists Press.

Hogan, J., & Holland, B. (2003). Using theory to evaluate personality and job-performance relations: A socioanalytic perspective. *Journal of Applied Psychology, 88,* 100–112.

Hogan, R. (1969). Development of an empathy scale. *Journal of Consulting and Clinical Psychology, 33,* 307-316.

Hogan, R. (1983). A socioanalytic theory of personality. In M. M. Page (Ed.), *Nebraska symposium on motivation 1982: Personality-current theory and research* (pp. 55-89). Lincoln: University of Nebraska Press.

Hogan, R. (1986). *Hogan Personality Inventory: Manual.* Minneapolis, MN: National Computer Systems.

Hogan, R. (1987). Personality psychology: Back to basics. In J. Aronoff, A. I. Rabin, & R. A. Zucker (Eds.), *The emergence of personality* (pp. 79–104). New York: Springer.

Hogan, R. (1991). Personality and personality measurement. In M. D. Dunnette & L. M. Hough (Eds.), *Handbook of industrial and organizational psychology* (Vol. 2, 2nd ed., pp. 873-919). Palo Alto, CA: Consulting Psychologists Press.

Hogan, R., Carpenter, B. N., Briggs, S. R., & Hansson, R. O. (1985). Personality assessment and personnel selection. In H. J. Bernardin & D. A. Bownas, (Eds.), *Personality assessment in organizations* (pp. 21-52). New York: Praeger.

Holland, J. L. (1985). *Making vocational choices: A theory of vocational personalities and work environments.* Englewood Cliffs, NJ: Prentice-Hall.

Jackson, D. N. (1976). *The appraisal of personal reliability.* Paper presented at the meetings of the Society of Multivariate Experimental Psychology, University Park, PA.

Johnson, J. A. (1981). The "self-disclosure" and "self-presentation" views of item response dynamics and personality scale validity. *Journal of Personality and Social Psychology, 40,* 761-769.

Johnson, J. A. (1986, August). *Can job applicants dissimulate on personality tests?* Paper presented at the 94th annual convention of the American Psychological Association, Washington, DC.

Johnson, J. A. (1987, August). *Dissembling on the Hogan Personality Inventory during simulated personnel selection.* Paper presented at the 95th Annual Convention of the American Psychological Association, New York, NY. (ERIC Document Reproduction Service No. ED 290 785)

Johnson, J. A. (1990, June). *Unlikely virtues provide multivariate substantive information about personality.* Paper presented at the 2nd annual meeting of the American Psychological Society, Dallas, TX.

Johnson, J. A. (1997a). Seven social performance scales for the California Psychological Inventory. *Human Performance*, *10*, 1-30.

Johnson, J. A. (1997b). Units of analysis for description and explanation in psychology. In R. Hogan, J. A. Johnson, & S. R. Briggs (Eds.), *Handbook of personality psychology* (pp. 73-93). San Diego, CA: Academic Press.

Johnson, J. A. (1999). Persons in situations: Distinguishing new wine from old wine in new bottles. In I. Van Mechelen & B. De Raad (Eds.), Personality and situations [Special Issue]. *European Journal of Personality*, *13*, 443-453.

Johnson, J. A. (2002, July). Effect of construal communality on the congruence between self-report and personality impressions. In P. Borkenau & F. M. Spinath (Chairs), *Personality judgments: Theoretical and applied issues*. Invited symposium conducted at the meeting of the 11th European Conference on Personality, Jena, Germany.

Johnson, J. A. (2004). The impact of item characteristics on item and scale validity. *Multivariate Behavioral Research*, *39*, 273-302.

Johnson, J. A. (2005). Ascertaining the validity of web-based personality inventories. *Journal of Research in Personality*, *39*, 103-129.

Johnson, J. A. (2006). Ego syntonicity in responses to items in the California Psychological Inventory. *Journal of Research in Personality*, *40*, 73-83.

Johnson, J. A., Cheek, J. M., & Smither, R. (1983). The structure of empathy. *Journal of Personality and Social Psychology*, *45*, 1299-1312.

Johnson, J. A., & Horner, K. L. (1990, March). *Personality inventory item responses need not veridically reflect "actual behavior" to be valid*. Paper presented at the 61st Annual Meeting of the Eastern Psychological Association, Philadelphia.

Lippa, R. (1976). Expressive control and the leakage of dispositional introversion-extraversion during role-played teaching. *Journal of Personality*, *44*, 541-559.

Lippa, R. (1978). Expressive control, expressive consistency, and the correspondence between expressive behavior and personality. *Journal of Personality*, *46*, 438-461.

Markus, H., & Nurius, P. (1986). Possible selves. *American Psychologist*, *41*, 954-969.

McCrae, R. R., & Costa, P. T., Jr. (1996). Toward a new generation of personality theories: Theoretical contexts for the five-factor model. In J. S. Wiggins (Ed.), *The five-factor model of personality* (pp. 51-87). New York: Guildford.

McCrae, R. R., Costa, P. T., Jr., & Piedmont, R. L. (1993). Folk concepts, natural language, and psychological constructs: The California Psychological Inventory and the five-factor model. *Journal of Personality*, *61*, 1-26.

Mehrabian, A, (1972). *Nonverbal communication*. Chicago: Aldine-Atherton.

Mischel, W. (1969). Continuity and change in personality. *American Psychologist*, *24*, 1012-1018,

Mills, C., & Hogan, R. (1978). A role theoretical interpretation of personality scale item responses. *Journal of Personality*, *46*, 778-785.

Paulhus, D. L., & Reid, D. B. (1991). Enhancement and denial in social desirability responding. *Journal of Personality and Social Psychology*, *60*, 307-317.

Raskin, R., Novacek, J., & Hogan, R. (1991). Narcissism, self-esteem, and defensive self-enhancement. *Journal of Personality*, *59*, 19-38.

Schlenker, B. R., & Pontari, B. A. (2000). The strategic control of information: Impression management and self-presentation in daily life. In A. Tesser, R.

Felson, & J. Suls (Eds.), *Perspectives on self and identity* (pp. 199-232). Washington, DC: American Psychological Association.

Scott, W. A. (1963). Social desirability and individual conceptions of the desirable. *Journal of Abnormal and Social Psychology, 67,* 574-585.

Shoda, Y., Mischel, W., & Wright, J. C. (1993). The role of situational demands and cognitive competencies in behavior organization and personality coherence. *Journal of Personality and Social Psychology, 65,* 1023-1035.

Sternberg, R. J., & Hedlund, J. (2002). Practical intelligence, *g*, and work psychology. *Human Performance, 15,* 143-160.

Tellegen, A. (1982). *Brief manual for the Differential Personality Questionnaire.* Unpublished manuscript, University of Minnesota.

Tyler, L. E. (1963). *Tests and measurement.* Englewood Cliffs, NJ: Prentice-Hall.

Wallace, J. (1966). An abilities conception of personality: Some implications for personality measurement. *American Psychologist, 21,* 132-138.

Watson, P. J., & Andrews, P. W. (2002). Toward a revised evolutionary adaptationist analysis of depression: The social navigation hypothesis. *Journal of Affective Disorders, 72,* 1-14.

Wolfe, R. N. (1993). A commonsense approach to personality measurement. In K. H. Craik, R. Hogan, & R. N. Wolfe (Eds.), *Fifty years of personality psychology* (pp. 269-290). New York: Plenum.

CHAPTER 9

FAKING AND JOB PERFORMANCE

A Multifaceted Issue

Mitchell H. Peterson and Richard L. Griffith

One ongoing pursuit of our science has been to continuously improve the ability to select individuals who will adequately perform the tasks and duties associated with a given position, in addition to displaying a temperament that is consistent with the effective performance of these duties. The goal of prediction typically leads to the use of psychological tests that allow us to infer the presence of certain skills, abilities, or traits that are related to the successful performance of a given job. Although cognitive ability is one of the most consistent predictors of performance across a variety of jobs (Schmidt & Hunter, 1981, 1998), psychologists have searched for useful individual difference variables that may explain variance in job performance beyond that which is explained by cognitive ability. Measures of personality have been a logical next step in this quest for incremental predictive ability, and are widely supported predictors of job performance (Barrick & Mount, 1991; Tett, Jackson, & Rothstein, 1991). The accuracy and usefulness of

A Closer Examination of Applicant Faking Behavior, 233–261
Copyright © 2006 by Information Age Publishing
All rights of reproduction in any form reserved.

these measures however, may be somewhat dependent on the honesty of an applicant's responses.

Research has indicated that a substantial portion of applicants do misrepresent themselves when completing measures of personality in an employment setting (Donovan, Dwight, & Hurtz, 2003; Griffith, Chmielowski, & Yoshita, in press; McDaniel, Douglas, & Snell, 1997; Rees & Metcalf, 2003). These findings lead to perhaps the most important unanswered question in faking research: how does faking affect the inferences that are drawn from personality measures in an applicant setting? Although the current state of research in applicant faking behavior has increased our understanding of this phenomenon, we have no definitive answer to the question that is of utmost importance to practitioners who use measures of personality to select employees.

When broken down to its most basic elements, the use of any psychological measure in the selection process involves a series of inferences (Binning & Barrett, 1989). Psychological constructs that appear to be important for a given position are identified, and an attempt is made to use or design measures that adequately sample the domain of those constructs. Finally, the hope is that a statistical relationship will exist between the chosen measure of a construct and job performance. In personnel selection, the most basic inference is that using a measure (or battery of measures) will increase the likelihood of selecting individuals who will perform well on the job. Our belief is that the phenomenon of applicant faking behavior can impact this linkage in many ways, some of which may not always be evident based on an examination of the common forms of validity evidence.

The goal of this chapter is to offer a critical examination of the possible ways in which applicant faking behavior may impact hiring decisions based on personality measures. We do not attempt to offer a definitive answer to the question of whether faking is detrimental to the validity of personality measures. Instead, we will present several competing arguments examining the multi-faceted nature of this question. These arguments will likely lead to more questions than answers, with the hope that such questions will bring about continued refinement of methodologies used to study the relationship between faking and job performance.

Although this chapter will present a range of possible threats that faking poses to an organization, it seems that there are two general ways of addressing the issue. First, most faking research has addressed concerns regarding the measurement properties of personality measures and their relationship with performance criteria (Barrick & Mount, 1996; Christiansen, Goffin, Johnston, & Rothstein, 1994; Douglas, McDaniel, & Snell, 1996; Hough, Eaton, Dunnette, Kamp, & McCloy, 1990; Ones, Viswesvaran, & Reiss, 1996; Rosse, Levin, & Nowicki, 1999; Rosse,

Stecher, Miller, & Levin, 1998), or changes in the rank ordering of applicants (Donovan, Dwight, & Schneider, 2005; Griffith, Chmielowski, & Yoshita, in press) under conditions of faking. In some sense, this line of research examines the *performance of our selection procedure* under faking conditions, not the performance of those who have chosen to fake. The focus of these studies is the effectiveness of the personality instrument, and how well our current psychometric practices handle aberrant data points. However, one avenue that has received less attention from researchers is the actual *job performance of individuals who fake* during the hiring process. It is possible that as a group, individuals who fake during selection may perform differently on the job than those individuals who do not. For this line of research, the emphasis would be on the person identified as a faker and their subsequent job behavior once they are hired. While these two approaches complement each other, and use similar data, they answer conceptually distinct research questions.

This chapter will examine both of these issues in order to shed light on possible avenues for continued research on faking and job performance. We will begin with a review of the current state of research on faking and criterion-related validity. This review will highlight the various methodologies that have been implemented in examinations of validity under conditions of faking. Our focus will then shift to the discussion of possible performance differences between individuals who fake and those who do not. These differences will be discussed both in terms of their possible effects on validity, as well as in relation to postselection organizational concerns that may be associated with the hiring of individuals who elevate their responses. Finally, we will conclude this chapter by closely examining the construct of job performance to determine which of its sub-facets may be most affected by applicant faking behavior.

POSSIBLE EFFECTS OF FAKING ON THE MEASUREMENT PROPERTIES OF PERSONALITY INSTRUMENTS

Within the research on applicant faking and job performance, the primary focus has been whether faking leads to an attenuation of criterion-related validity coefficients (Barrick & Mount, 1996; Christiansen et al., 1994; Douglas et al., 1996; Hough et al., 1990; Ones et al., 1996; Rosse et al., 1998; Rosse et al., 1999). The criterion-related validity coefficient is one of the most useful and easily referenced pieces of information that exists to evaluate the quality of psychological measures used in employment settings. A demonstration of the statistical relationship between a measure and job performance can guide practitioners who are deciding whether to use a particular measure in the selection process, and is also typically an integral piece of information in any litigation surrounding claims of adverse impact. Undoubtedly, this was a logical starting point for researchers who sought to determine whether there were negative consequences associated with faking.

From a theoretical standpoint, it is easy to see how applicant faking behavior may influence criterion-related validity. In a sample of applicants, each individual brings with them a "true" level of the trait of interest for a given measure. There is likely to be some degree of variability within this sample of individuals, and the goal of the selection process is to identify those individuals with the level of that trait that has been determined to be desirable for a given job (usually a high level in the case of a monotonic relationship). As we mentioned previously, a potential concern regarding the use of personality measures in an employment setting is the fact that any score that is derived from such assessments may be subject to manipulation on the part of job applicants. Thus the observed score may not closely correspond with the true score, due to the introduction of bias.

As the prevalence of faking increases in a sample of applicants, more individuals are likely to have scores near the top of the distribution due to the score increases that are associated with faking behavior (Douglas et al., 1996; Haaland & Christiansen, 1998). The distribution of scores on the personality measure is likely to become negatively skewed as more individuals move near the top of the distribution, and a ceiling effect (at least for a subsample of applicants) is evidenced. This narrowing of scores may result in a restriction of variance on the predictor. In addition, if there are individual differences in the magnitude of applicant faking, rank ordering changes should occur, and their presence may further attenuate criterion validities due to a disruption in the monotonic relationship between the predictor and criterion.

Although at first glance this scenario seems plausible, research on the criterion-related validity of personality measures under conditions of faking has been mixed. Several studies (Barrick & Mount, 1996; Christiansen et al., 1994; Hough, 1998; Hough et al., 1990; Ones et al., 1996) have reported minimal effects on criterion-related validity. However, additional research has suggested that faking can attenuate criterion-related validity coefficients (Graham, McDaniel, Douglas, & Snell, 2002; Haaland & Christiansen, 1998; Illingworth, Snell, & Rosen, 2005; Komar, Theakston, Brown, & Robie, 2005; Mueller-Hanson, Heggestad, & Thornton, 2003; Rosse et al., 1999), and that the detrimental effects of faking are likely to be most prevalent in the upper end of the distribution of scores on measures of personality. The research examining faking and issues of criterion-related validity is anything but unified in its assertions. This lack of unity may in part be rooted in the varying operational definitions of faking behavior that exist, and the range of methodologies that have been used to measure faking and its effect on validity.

The most palpable difference between studies reporting no effects of faking and those reporting substantial reductions of criterion-related validity is their respective use of social desirability (SD) scales versus the use of methodologies that attempt to simulate, motivate, or instruct faking behavior on the part of the individuals who are completing personality measures. The usefulness of each methodology can be evaluated in terms of the questions that each can answer, along with those that each cannot.

Studies that correct for scores on measures of SD (e.g., Christiansen et al., 1994), or that attempt to partial out the effects of SD from correlations between personality and job performance (e.g., Ones et al., 1996) may offer insight into the role of SD in the selection process. Indeed, most of these studies suggest that accounting for SD has minimal effects on the relationship between personality measures and job performance. However, as Mueller-Hanson et al. (2003) suggests, SD is not likely to be synonymous with faking. If empirically supported, this notion is particularly troublesome given that the results of studies using scores on measures of SD (e.g., Ones et al., 1996) are commonly cited as evidence that faking does not affect the relationship between measures of personality and job performance. Given the findings of the authors examining SD in Chapter 5 of this text, using SD measures as a proxy for faking may not be appropriate. Measures of SD are not likely to capture true faking behavior found in an applicant setting. Instead, as Mueller-Hanson et al. (2003) and others have suggested, corrections for SD may actually lower criterion-related validity through a reduction of relevant trait variance due to the relationship between SD and other personality constructs.

Studies that have used alternative methods to investigate the impacts of faking behavior have offered insight into the question of whether faking has the *potential* to cause a reduction in criterion-related validity. Many of the studies that have used alternate conceptualizations and measures of faking behavior have reported detrimental effects on criterion-related validity. Laboratory studies have tested the effects of faking under an array of conditions by allowing for variation in selection ratios, as well as the percentage of fakers present in a sample and the severity of faking observed (Douglas et al., 1996; Komar et al., 2005; Zickar, Rosse, & Levin, 1996). These studies often involve the use of instructional sets or simulations, and have been criticized in the literature for the use of this methodology (Hough et al., 1990). Additional research has examined the effects of faking by collecting personality data under conditions that provide individuals with the requisite motivation to distort their responses to the items on the personality measure through the use of incentives (e.g., Illingworth et al., 2005; Mueller-Hanson et al., 2003). Typically, studies of this nature assume that the presence of higher scores in an incentive condition (as compared to a control condition) indicate the presence of faking behavior. The studies by Illingworth et al. (2005) and Mueller-Hanson et al. (2003) both reported lower criterion-related validity for incentive conditions. Additional studies reporting detrimental effects of faking have collected data in an organizational setting. One such study assumed that faking was more prevalent in the upper ends of the distribution of personality scores, and found that criterion related validity was lower at this point in the distribution than for scores in the bottom end of the distribution (Haaland & Christiansen, 1998).

Although the methodologies mentioned in the preceding paragraph provide insight into the effects that faking *can have* on criterion-related validity, they do not necessarily indicate the effects that faking *does have* in a true applicant setting. Studies by McFarland and Ryan (2000) and Griffith et al. (2004) have suggested that there are individual differences and variability in applicant faking behavior (See also chapter 6 of this volume). Studies using directed or simulated faking may not take such variability into account, as they tend to produce uniformly high levels of score elevation in the faking condition, which is then compared to an honest condition to assess the effects of faking on criterion-related validity. In addition, simulation studies, although useful for demonstrating the possible effects of faking under varying conditions, may not be entirely indicative of the impact of faking that occurs in a true applicant setting.

Based on the existing research examining the impact of faking on criterion-related validity, it seems a conclusion purporting that faking is not a threat to the validity of personality measures is premature. To this point, the only conclusions that can be drawn regarding this issue are: (1) Controlling for SD, a proxy for faking behavior, does not appear to affect criterion related validity, and (2) Based on simulations and studies using incentives to induce faking, it appears that faking has the potential to impact criterion-related validity. However, considerable room for improved methodology still exists. More direct, individual assessments of faking and actual performance data are still needed before firm conclusions can be drawn.

While much research has examined the impact of the applicant faking behavior on the selection process, little research has examined the performance characteristics of the faker. In general, previous faking research has highlighted the inability of our current selection paradigm to handle the faking phenomenon. While our assessment of criterion-related validity may be hampered, the critical factor for organizations is whether those who have defeated our attempts at screening are better or worse performers than their honest counterparts. If, as a group, fakers have a similar performance distribution to that of nonfakers, the inadequacies of our selection system may be less of a problem. Although attenuations of criterion-related validity coefficients due to faking should remain a primary area of concern and continued methodological refinement, the effects of faking behavior may exist beyond its role in the predictive ability of personality measures. Of particular interest is the actual job performance of individuals who have faked during the selection process.

THE JOB PERFORMANCE OF FAKERS VERSUS NONFAKERS: IMPLICATIONS FOR ORGANIZATIONS

Regardless of the information that one has regarding the criterion-related validity of an instrument, once it is put into practice in a selection setting a practitioner can only approximate the future job performance of an applicant. By comparing the criterion-related validity of two measures, one can expect that on average, decisions made based on the measure with the higher coefficient will be better than those made based on a measure with a lower degree of criterion-related validity. However, this numerical approximation of the quality of the measure tells us very little about the actual job performance of the individuals who are hired based on its results. This leads us to the issue of whether individuals who do fake are somehow different from those who do not. Research has suggested that not all individuals fake during selection (Donovan et al., 2003). This finding is not necessarily a surprise, when you consider the true variability in personality trait levels that exists between individuals. A number of individuals are likely to actually possess high levels of a trait, and therefore even if they chose to fake, they would have little room for improvement. One would expect that these are the individuals that employers would typically prefer to select. However, for the individuals who do not possess high levels of the trait, there is considerably more room for attempted score elevation. Additional studies have indicated that faking may be related to some stable individual difference variables (Griffith et al., 2004; McFarland & Ryan, 2000). The findings of these studies suggest that there are at least some individual difference variables that can influence one's propensity for engaging in faking behavior. Evidence of important individual differences in faking provides an impetus for investigation into whether there are systematic differences between fakers and nonfakers on other variables of interest, and whether these differences pose a threat to organizations using personality measures to select employees.

In keeping with this thought, we would like to turn to a consideration of the possible effects of systematic performance differences between fakers and nonfakers. An important point to note is that there is no hard and fast distinction between "fakers" and "non-fakers." Indeed, a central theme of this book is that faking is a complex phenomenon that is affected by situational, motivational, and dispositional factors. We do know that there are individual differences in faking behavior (Donovan et al., 2005; Griffith et al., in press, 2004), and that some people may be more likely to fake than others. This does not mean, however, that the same people will fake across all situations, or to the same degree. Many would be surprised to know that a substantial amount of fakers actually produce lower scores when attempting to present themselves favorably! This may be a function of inaccurate schemas about the job (Vasilopoulos, Reilly, & Leaman, 2000). Unfortunately, if we are to discuss the possible implications of hiring individuals who may have faked during the selection process, we are constrained to the use of the "faker" and "non-faker" labels. In reality, the individuals that pose the greatest threat to an organization are those that possess low levels of the trait or traits of interest to the employer, but who have faked enough to get hired or avoid being screened-out (therefore progressing further into the selection process). It is the performance of these individuals that may be of concern to the organizations that have hired them. Prior to making any comparisons, however, we need to reliably identify those who may have faked their scores.

Direct assessments of faking behavior have been very difficult to gather. Most research has demonstrated the consequences of faking, and the researcher is left to infer the cause. This has been a troublesome flaw in our research. Personality researchers to date have been on the trail of the phenomenon, but have rarely studied it directly. Instead we have studied mean level shifts, factor structure changes, and criterion-related validity. Through examination of these factors, and deductive reasoning, researchers have made a case that faking exists, and that it affects our measurement attempts (or vice versa). The faker has been elusive game, and attempts at identifying those who elevate their scores have not always hit the mark. To understand if fakers are better or worse performers we have to know who they are.

Measures of SD or frankness have traditionally been used in an attempt to approximate an individual's likelihood of engaging in response distortion during the application process. Although measures of SD are the most common form of attempting to identify fakers outside of a research laboratory, they typically have not performed as intended. At best, they should be used as a piece of convergent evidence in identifying those who may have faked.

Recent research has suggested that another index of faking may provide superior results. Some authors have advocated using the measurement properties of personality items as an indicator of applicant faking (Christiansen, Robie, & Bly, 2005). The authors constructed a covariance index (the CVI is similar to a within-person correlation) by identifying pairs of items that were not correlated in an unmotivated sample. This index was then examined in the applicant responses and if covariation existed for these item sets, the authors suggested that applicant faking might be present. This method has considerable advantages over the use of SD measures. First, no additional items are necessary. Second, the CVI index demonstrated lower correlations with measures of the big five than SD measures. These studies have demonstrated a moderate degree of overlap in the classification of fakers using the CVI and SD scores. However, the use of the CVI seems to demonstrate superior classification properties, and has been used in an examination of criterion validity with quite different results than the SD indicator. While this index may prove to be a better indicator than SD, it is nonetheless an indirect assessment.

Another method to identify fakers has employed a repeated measures design in an applicant setting (Donovan et al., 2005; Griffith et al., in press) and simulated applicant settings incorporating deception (English, Griffith, Graseck, & Steelman, 2005; Griffith et al., 2004) to measure applicant faking. Using this design, applicant responses are compared to honest responses collected at a later date under nonmotivated conditions. The discrepancies between these two scores can then be examined for faking effects. One common element of these designs has been the use of a confidence interval calculated around the honest score. If the observed applicant score falls outside the confidence interval, the individual has been categorized as a faker. To date these confidence intervals have been based on the reliability of the measure (e.g., the standard error of measurement). However, these faking cut-offs are somewhat arbitrary, and other intervals may be suggested at a later date.

While the reader does not need to settle on a definition of faking in order to consider the following scenarios, we believe these repeated measure designs are the most direct measurements of faking behavior available in the extant literature. Donovan et al. (2005) may disagree, suggesting that even this design provides only an indirect measure of faking. For any comparisons of fakers and nonfakers to be made, a reliable method of detection is necessary. These methods (as well as others such as direct reporting of faking and bogus pipeline items) in combination have the potential to put the spotlight on the faker, but new methodologies that could improve classification rates would be a welcome tool to the faking researcher.

Regardless of the operational definition adopted, once we have identified fakers we can then examine their level of job performance. In the following section we discuss these possible performance differences in a hypothetical fashion. The performance levels are presented in a somewhat generic level of specificity, and are based on literature that has either implicitly or explicitly suggested the subsequent job behavior of fakers who have been hired. These scenarios are meant to serve as an illustration of the possible performance differences between these fakers and nonfakers, and serve as a point of reference for the discussion of the implications of those differences.

These possible performance differences will first be considered in reference to their potential effects on the predictive qualities of personality instruments. This discussion will focus primarily on how systematic differences (or similarities) would be expected to impact criterion-related validity. While the criterion validity of an instrument provides valuable information to researchers and practitioners, ultimately it is the hiring decision that is of most concern to the consumers of personality based employment tests. Criterion-related validity provides information regarding the average quality of hiring decisions if we were to hire many applicants over time. For organizations that will be hiring thousands of new workers (e.g., greenfield sites, military organizations), this information may be the most salient. However for those organizations with lower hiring rates, the focus will likely be at a more individual level.

Previous simulation studies have demonstrated that if you base employment decisions on a noncognitive assessment tool, *you will* hire a faker. This phenomenon is relatively independent of the criterion validity coefficient (Haaland & Christiansen, 1998). The number of fakers who enter the organization will depend on the number of applicants who choose to fake, as well as the selection ratio.

Zickar et al. (1996) used a simulation to demonstrate the effects of faking on the rank ordering of applicants. Using a Monte Carlo design, they found that relatively few fakers are needed for the top end of the distribution to contain a high percentage of fakers. Replicating Zickar et al.'s methodology with their own data, Douglas et al. (1996) found that, as the percentage of applicants who faked increased from 0% to 25%, nine out of the top ten applicants hired would be fakers.

The Douglas et al. (1996) study used an instructional set design, and thus the level of faking was greater than that which would be expected in a true applicant setting. Griffith et al. (in press) replicated this hiring simulation in a true applicant sample and found that the selection ratio also had an effect on the percentage of fakers entering the organization. As the selection ratio decreased, more fakers were likely to be hired. When the selection ratio was varied from .50 to .10 the amount of fakers in the pool of selected applicants increased from 31% to 66%.

Previous research has suggested that 30-50% of applicants significantly elevate their scores when completing measures of personality in the application process (Donovan et al., 2005; Griffith et al., 2005). Taken together, this research suggests that in the top end of the hiring distribution, the majority of individuals will be fakers. Thus for the user of personality tests, the performance level of the newly hired faker can become much more of a concern than the validity coefficient.

To be clear, not all those who choose to fake elevate their scores. Many guess wrong when attempting to represent themselves in a positive light, and fake in the wrong direction. Thus, applicant faking behavior results in some applicants who have the desired traits being eliminated from the selection pool. In addition, many applicants who have lower levels of the desired traits are hired due to elevated scores (especially with low selection ratios). Using the signal detection framework presented by Cascio (1998) to evaluate the quality of hiring decisions, faking increases false positives and false negatives.

The impact of these detection errors can ultimately be linked to the performance level of the individuals selected. While criterion-related validity is a useful index for evaluating the overall effectiveness of the measure, the employer wants to know how much he or she can count on the new employee, and cares little about the efficiency of the measure.

If many of the new employees entering the organization may have faked their way into the position, the performance level of "fakers" gains incremental importance above that of the criterion-related validity coefficient. In the following section, we review the literature that has examined the faking/performance relationship, and present hypothetical scenarios in a criterion-related validity framework. In each of these scenarios, the performance level of the faker is likely to impact criterion-related validity. However these performance levels have a much more direct effect on the selection decisions that managers make based on noncognitive measures. If 9 of 10 new employees are good performers, all is well. However, other scenarios leave more to chance.

Keep in mind that these scenarios are hypothetical in nature. Although empirical evidence is obviously necessary to substantiate these hypotheses, the fact is that relatively little research has offered direct measurements of faking (beyond approximations via SD scales) and job performance in the same sample of individuals. The goal of discussing possible scenarios is designed to be concordant with the general tone of this chapter in that such a discussion raises issues that warrant consideration, but have not yet been directly tested.

In the section following our discussion of the performance scenarios, we will focus more directly on the issue of job performance, rather than prediction. More specifically, we will argue that considerations of job performance and faking should move beyond core task performance and into the realm of contextual performance and counterproductive behavior. Once again, due to a dearth of research examining this issue, much of this section will be devoted to building an argument for continued investigation on the part of personality researchers.

Since much of the faking research has started with the assumption (at least implicitly) that fakers may be poor performers; we will start our discussion of the performance scenarios there.

POSSIBLE SCENARIOS HIGHLIGHTING PERFORMANCE DIFFERENCES BETWEEN FAKERS AND NONFAKERS

Fakers as Low Performers

Much of the impetus for faking research comes from the concerns voiced by practitioners. This concern, in part, is centered on fears that applicants who have been dishonest will be hired, and that this dishonesty may transfer to the workplace. Our culture places a strong negative emphasis on dishonesty, and it is generally associated with a lack of integrity. Those who see faking as dishonesty may be more likely to suggest that fakers would be dishonest at work (e.g., lie about being sick to get a day off) or engage in more severe violations of integrity (such as steal money or merchandise).

If fakers are indeed worse performers, then the relationship between the personality measure and the measure of job performance would be negative such that high test scores would equate to low performance (Figure 9.1.). As mentioned by previous researchers, this situation places a strain on our assumption of linearity (Burns & Christiansen, 2004; Haaland & Christiansen, 1998). While the linear relationship would hold for the portion of the sample answering honestly, a nonlinear (or linear in a reverse fashion!) relationship would hold for the fakers. Without being able to distinguish between fakers and nonfakers, this nonlinearity would likely register as outliers or increased residuals.

While there is little research that examines the actual performance of fakers, some empirical support has been offered for the fakers-as-worse-performers scenario. Rosse et al. (1999) examined the performance of customer service agents who had been assessed using a noncognitive selection measure. Faking was approximated through the use of an unlikely virtue scale, which functions in much the same way as measures of SD. Therefore, the results of this study should be interpreted with some caution given the criticisms of such scales. The data suggested that those employees that had faked the selection device scored lower on the positive elements in the performance appraisal, and higher on counterproductive behaviors such as lying, exaggerating, and stealing sales from

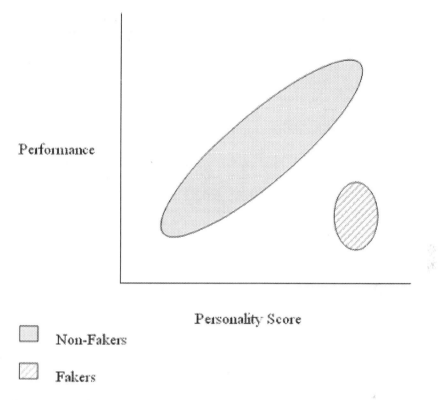

Performance

Personality Score

☐ Non-Fakers

▨ Fakers

Figure 9.1.   Fakers as low performers.

other agents. Faking scores were negatively correlated with positive behaviors (r = -.21), and positively related to counterproductive behaviors (r = .22). However mean levels of performance data were not compared across fakers and nonfakers. This is an essential analysis to complete if statements are to be made about fakers being worse performers. While a negative correlation may imply that the fakers were worse performers, much of the direction of the correlation may be artifactually due to the nonlinearity issue. If scores on the personality test are extremely high, even moderate levels of performance may result in a negative correlation. In addition, no results were presented for the nonfakers, so direct comparisons were not possible from this study.

Donovan et al., (2005) also examined the job performance of fakers versus nonfakers using a repeated measures design. The authors conducted a field study that utilized a self-report personality measure of goal orientation. Initial data were obtained from job applicants, and then again as incumbents once they were hired. Fakers were identified by the discrepancy between their applicant and incumbent scores. In an exploratory analysis, the authors examined performance data for both groups. While the differences in performance were not significant, there was over a half standard deviation difference in performance scores between fakers and nonfakers (with fakers being lower). Closer examination revealed that the lack of statistical significance of this difference was likely impacted by a low sample size.

Of the three scenarios we present, the situation in which fakers are worse performers than nonfakers would likely be the most detrimental to an organization. In this case, it would be likely that individuals who scored high on the personality measure would perform as poorly as, or worse than many of the nonfakers who scored in the lower end of the distribution. If this were the case, the use of a personality measure may harm the quality of the selection decision substantially, regardless of the overall level of criterion-related validity. Previous research has demonstrated that even with a relatively low percentage of fakers, the upper end of the hiring distribution becomes saturated, and decision quality is negatively impacted (Douglas et al., 1996; Zickar et al., 1996).

With regard to criterion-related validity, the magnitude of the effects on the criterion coefficient would be somewhat related to the ratio of fakers to nonfakers. If the number of fakers is relatively low, the criterion coefficient for the overall sample may be only slightly attenuated. However, recent research has demonstrated that the number of fakers may be as high as 50% of the applicant pool (Donovan et al., 2005; English et al., 2005; Griffith et al., in press). This scenario would lead to a severe decrement to the overall validity coefficient. The fakers' high scores on the personality measure, paired with lower job performance, would lead to the possibility of an inverse relationship between the personality measure and job performance for this sub-sample of applicants, leading to a substantial reduction in criterion-related validity for the overall sample.

Fakers as Top Performers

Research has posited that (at least for sales jobs) individuals who score high on measures of SD may actually be better performers than those who score lower (Hogan, Hogan, & Roberts, 1996) (Figure 9.2.). The rationale behind this argument is that some sales and service jobs require a form of self-presentation that is very similar to faking (or SD). Again, although this research has considered assessments of SD to be synonymous with assessments of faking behavior, it is our contention that this may be an incorrect assumption. According to the rationale regarding the role of self-presentation in sales jobs, individuals who misrepresent themselves

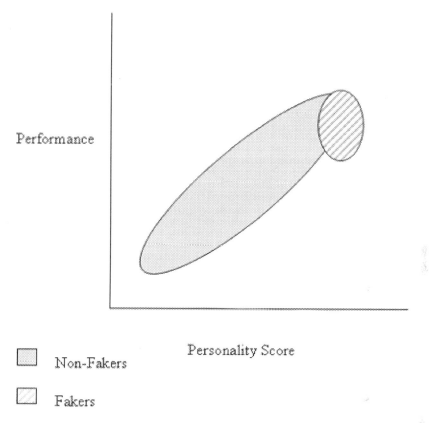

Figure 9.2.   Fakers as top performers.

during the application process are thought to be more effective in jobs that require some degree of interpersonal flexibility. The issue of cross-situational specificity would be particularly relevant for this argument. More recent views of faking suggest that it is a complex process where the applicants' responses may require an understanding of the job, and an ability to formulate a schema of a "desirable" candidate (Vasilopoulos et al., 2000). To the extent that this situation is duplicated in the employment environment, the skill of faking may transfer and manifest itself in improved performance.

So for certain jobs, or certain job activities, this argument is plausible. There is a growing body of research in the emotional labor (highly relevant to customer service or sales jobs) literature that suggests that effective employees do not just paste on a smile, they "deep act" and actually change their cognitions to fit the requirements of the situation (Grandey,

2003). For the basis of the fakers-as-good-performers argument to hold, applicant faking behavior would need to be similar to other forms of deception and impression management. Unfortunately the research has provided us with a limited understanding of the dynamics of faking behavior, so research linking these two phenomena is sparse (perhaps nonexistent?). If this linkage were empirically established, the meta-analytic research on influence tactics (Higgins, Judge, & Ferris, 2003) suggests a mechanism that would support the notion that fakers would receive higher performance ratings. This study demonstrated that several impression management tactics do result in higher performance evaluations. So following this argument, fakers might not actually be better performers, but their higher performance ratings may be due to effective use of influence tactics. Thus an observed positive correlation between the personality measure and job performance would be due to an upwards bias for both the predictor and criterion.

One flaw in this potential argument does emerge from the Higgins et al. study. The influence tactic of self-promotion (conceptually most similar to faking) does not show a positive relationship with performance assessments. The authors note: "It appears that such tactics backfire in obtaining favorable assessments from supervisors" (p. 100). We suspect that there may be some overlap between the two forms of deception, but that impression management on paper and in person might be two different animals. The ability to fake on paper may be a necessary but not sufficient condition for influence tactics used in person.

If fakers are actually better performers (at least in some jobs), then we have a situation where the criterion-related validity of the measure is not likely to be affected at all by faking. In fact, a slight increase in validity could occur under these conditions. In effect, if fakers were rising to the top of the distribution of applicants, many of them would be hired, and these would be ostensibly good selection decisions given the expectation that fakers should tend to perform better on the job. Those applicants with high scores would also have high performance ratings and the predictor criterion relationship would remain unchanged. This effect would remain relatively stable if there were a low to moderate percentage of fakers overall. As the percentage of fakers increased past a "tipping point," criterion-related validity would begin to attenuate due to a restriction of variance on the predictor.

Fakers as Equivalent Performers

Consistent with the theme of much of this book, we believe that faking is a complex phenomenon and that fakers are a varied group. While some antecedents of faking (such as low levels of integrity) may be associated with poor performance, other antecedents may covary with performance in a different fashion. Many of our beliefs about those who fake are based on untested assumptions, and a good scientist should stick with the null hypothesis until the data suggest otherwise! In the absence of data, sound theoretical reasoning can be applied to the issue in hopes of making those tests of the null refutable. An inspection of the theoretical model of faking offered by McFarland and Ryan (2000) gives some insight into the factors influencing faking which might also be associated with poor performance. Of the 11 factors presented in the model, perhaps two, morals and values, could be argued to be directly associated with poor performance. Even if this were the case, these factors would need to be more clearly defined before that linkage could be established. An important fact to note is that these factors alone would not likely predict faking. Recent thinking on the issue implies an interaction model between the antecedents of faking behavior.

If research concluded that fakers were no different in terms of their job performance, we would still be faced with a problem in our predictive model. If fakers, as a group, had relatively equal levels of performance to that of nonfakers (Figure 9.3.), we could expect a very different predictive relationship between a measure of personality and job performance for the two groups of applicants. For the group of nonfakers, we could expect that this measure would have a criterion-related validity coefficient that is somewhat higher than that which has been observed in the literature for a full group of applicants (roughly .20 - .25). In general, individuals who score higher on the measure would be expected to exhibit higher levels of job performance. Keep in mind that because this group of individuals is responding relatively honestly, we would expect that there would be a fair degree of variability in their scores on the measure (and in their respective levels of performance).

When we consider the group of fakers separately, however, the picture changes dramatically. If the group of fakers has relatively similar levels of job performance to that of the nonfakers, there is likely to be a similar degree of variability in job performance for the two groups as well. The fakers are likely to have high scores on the personality measure, and we would expect reduced variability between applicants because many of them may approach the highest possible score on the measure. A reduction in criterion-related validity would occur for the group of fakers because there would be varying levels of job performance for a group of people that is likely to have high scores on the measure of personality. The prediction would essentially be random. The implication associated with this situation is that for individuals who are faking, we may have absolutely no ability to predict their job perfor-

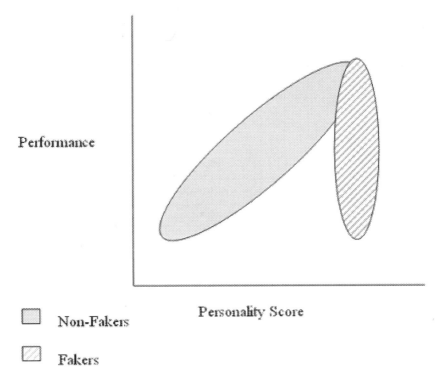

Figure 9.3.   Fakers as equivalent performers.

mance based on a measure of personality. Although our predictive ability is less than perfect in general, it may be akin to a coin flip for a group of individuals who are not providing accurate representations of themselves. When considering the two groups in combination, the occurrence of faking may also lower the overall criterion-related validity coefficient as well.

The plausibility of this scenario may be evidenced by previous research examining differences in validity coefficients obtained in concurrent, as opposed to predictive validation strategies. Hough (1998) reported that predictive studies of personality measures yielded validity coefficients that were .07 lower than those obtained in concurrent validation efforts. In a similar finding from a study examining the validity of integrity measures, Ones, Viswesvaran, and Schmidt (1993) reported a validity coefficient of r = .31 for predictive studies, and r = .38 for concurrent strategies. Given that applicants are more likely to be motivated to fake than incumbents, it is possible that the lower validity coefficients obtained in predictive studies could be a result of the prominence of faking in the applicant sample.

In addition to being evidenced by comparisons of predictive and concurrent validation studies, this statistical pattern has received preliminary support in other recent research (Griffith, Yoshita, Peterson, & Malm, 2006). When the conscientiousness factor of the NEO-FFI was correlated with performance in a combined group of applicants the correlation ($r = .22, p < .05$) was similar to those reported in Barrick and Mount (1991). When we split out the fakers and examined only the nonfaking group the correlation rose to $r = .33, p < .02$. The correlation for the faking group was nonsignificant ($r = -.15, p = .57$). We did not find mean level differences in performance across the two groups. Caution must be taken with any interpretation of these preliminary results due to a relatively low sample size ($n = 63$). Perhaps the most important implication of this finding is that having a sample of nonfaking applicants leads to a level of criterion-related validity rarely reached by measures of personality.

Taking all three scenarios into consideration, it seems that the latter is likely to be the most plausible. Given the lack of empirical studies in the area, no conclusions can be drawn, but an examination of applicant/incumbent differences in validity support the fakers-as-equivalent-performers model. As we explained earlier, if fakers had better performance than their nonfaking counterparts we would expect no drop in criterion-related validity. However the empirical data does suggest otherwise, with predictive designs consistently yielding lower validities (Hough, 1998; Ones et al., 1993). The drop in validities are generally small (in Cohen's effect size terms) with the difference between validation designs being less than .10. These findings would be inconsistent with the fakers-as-poor-performers-model in which we would expect a considerable drop in the criterion-related validity of the measure if the prevalence of faking is anywhere near what has been reported (Donovan et al., 2003, 2005; English et al., 2005; Griffith et al., in press). If the fakers-as-equivalent-performers model were supported, validities would drop somewhat due to the random noise and nonlinearity of the personality/performance relationship of the fakers. We would not expect a drastic drop, because even with the randomness of the prediction, we would be right by default perhaps a third of the time. Examining the figure, it is apparent that some of the fakers have high scores and high performance, falling in line with our assumptions (at least of criterion-related validity; construct validity is a completely different issue!). If this fakers-as-equivalent-performers scenario were actually the case, it would offer some explanation as to why personality measures have typically exhibited only moderate predictive ability (Barrick & Mount, 1991).

Taking Donovan et al.'s (2003) finding that at least 30% (and perhaps as many as 50%) of applicants may elevate their scores on personality measures during the application process into account, the lower validity coefficient for this substantial portion of the sample would cause a reduction in the coefficient for the overall sample of applicants. This reduction in criterion-related validity may be substantial in some cases and relatively small in other cases. This variation would likely be due to the percentage of the sample that is faking, as well as the distribution of job performance levels across the sample. Understanding this phenomenon better may allow us to develop correction formulas that account for faking to give us a better indication of our true level of predictive power.

It is important to note that both the fakers-as-worse-performers and fakers-as-better-performers models are subsumed in the equivalent model (That should stop some squabbling among faking researchers!). It is very likely that at least some of the variance in faking is attributed to a lack of integrity (Griffith et al., 2004). There are individuals who will lie on the application, the resume, and the personality test to achieve a desired goal of employment. These people will likely lie in similar self-serving situations at work, and may engage in worse transgressions of integrity. Conversely, some applicants figure out "the game" of employment testing and elevate their score. These individuals may be likely to figure out "the game" as it relates to job performance as well, and be top performers. In our emphasis on the faking phenomenon we may have forgotten one simple fact. The same holds true for the applicants who did not fake, and while on average our tests help with prediction, they fail us routinely. It is also important to note that these graphical representations are a simplification of applicant faking behavior. Not every applicant who fakes tops out the selection instrument. There seems to be a great deal of variability in how much people actually fake in the employment setting, so the actual distributions would not be so clean. The fakers who choose to elevate their score slightly to moderately don't disturb the linearity much, and do not deviate from their true scores enough to change hiring decisions except at the smallest selection ratios. Those who have a large magnitude of faking, however, do affect our estimations of criterion-related validity.

MOVING BEYOND THE TASK PERFORMANCE DOMAIN IN RESEARCH ON FAKING AND PERFORMANCE

When we have examined the faking performance relationship, the emphasis has been on the predictor side and the deviation from normal response patterns. However, little attention (as usual) has been paid to issues surrounding performance criteria. One avenue that has yet to be pursued by researchers studying applicant faking has been an examination of the relationship between faking and various forms of job performance. The apparent distinction between task performance and contextual performance (Motowidlo & Van Scotter, 1994) raises another question for researchers who are interested in applicant faking. Could faking have different effects on the relationship between personality and these two dimensions of job performance?

Contextual Performance

Although this chapter does not permit an in-depth discussion of fine distinctions between task and contextual performance (see Motowidlo & Van Scotter, 1994 for a more detailed discussion of these differences), a brief definition of these constructs is necessary. As defined by Borman and Motowidlo (1997), task performance involves successfully carrying-out essential job functions that contribute directly to the technical core of an organization. Contextual performance tends to encompass more altruistic behaviors that involve employees going out of their way to improve organizational functioning or assist coworkers. There is a possibility that performance differences between fakers and nonfakers may exist outside of the traditional task performance domain, and thus be masked in a comparison that did not specifically examine these nontask dimensions.

Consider for a moment the nature of task performance. Every job has attached to it a set of essential tasks and duties. These tasks and duties vary widely from job to job, and the requisite knowledge, skills, and abilities that facilitate effective performance are also likely to vary. Measures of job knowledge, task-relevant skill, and cognitive or physical ability are typically instituted into a selection system to tap these desirable qualities that can tell us a great deal about what an applicant will be able to do (or what they will be able to do with some additional training) once they are on the job. Focusing purely on variables related to task performance, however, may not reveal a great deal about the day-to-day activities of employees. As Borman and Motowidlo (1993) suggest, these actions that occur outside the realm of task performance,

> Can either help or hinder efforts to accomplish organizational goals by doing many things that are not directly related to their main task functions, but are important because they shape the organizational, social, and psychological context that serves as the critical catalyst for task activities and processes. (p. 71)

Borman and Motowidlo go on to note that these contextual activities have been frequently ignored in considerations of performance criteria. Additionally, Borman and Motowidlo (1997) note that task related activities are more likely than contextual activities to exist as dimensions on an employee's performance appraisal form.

The characterization of tasks as being essential is an important point to consider with respect to the role of personality in predicting job performance. Because employees must perform these essential tasks to a satisfactory level in order to remain employed, personality is unlikely to be the primary influence on their performance. Some research has indicated that personality variables have varying predictive ability for contextual as compared to task performance (Borman & Motowidlo, 1993; Borman & Motowidlo, 1997; Bott, Svyantek, Goodman, & Bernal, 2003; Motowidlo & Van Scotter, 1994), and are better predictors of typical performance as opposed to maximal (DuBois, Sackett, Zedeck, & Fogli, 1993). As these studies suggest, personality tends to be a better predictor of behaviors associated with contextual performance, while measures of ability, experience, or knowledge are more effective predictors of task performance. While ability and experience may have a substantial bearing on what an employee will be able to do, measures of personality may paint a better picture of an individual's typical work behavior.

We believe that investigations of faking and contextual performance could represent a potentially fruitful avenue of research within the realm of applicant faking. If measures of personality tend to predict contextual performance better than task performance, it seems that the most prominent effects of faking may be present in the prediction of contextual performance or nontask performance. As Borman and Motowidlo (1993) note however, prediction of contextual performance is not typically a primary consideration in the personnel selection process; or for that matter, in any procedure aimed at establishing evidence of criterion-related validity. Taking this into consideration, it is unlikely that any effects of faking on the prediction of nontask related performance would be evident unless a study was specifically designed to test this assumption. If applicants who fake are selected in a traditional selection process however, it is possible that their contextual or extra-role behaviors may differ from those of applicants who do not fake, once they are actually on the job.

Motowidlo, Borman, and Schmit (1997) make an interesting distinction among variables that contribute to contextual performance, and it seems that this distinction may be relevant to research on applicant faking. The authors argue that there are three variables that contribute to contextual performance: Contextual knowledge, contextual skill, and contextual habits. Contextual knowledge involves an understanding of the ways in which one can help fellow coworkers and the organization in ways that reach beyond the boundaries of task performance. Contextual skill involves actually being able to engage in behavior that accomplishes these goals. Contextual habits, on the other hand, may be the most relevant to considerations of applicant faking. As defined by Motowidlo et al., "Contextual work habits are patterns of responses that either facilitate or

interfere with effective performance in contextual work situations"
(p. 82). These habits could be thought of as tendencies to engage in
behaviors that characterize contextual performance. According to Motowidlo et al., personality variables such as extraversion and agreeableness
are likely to be predictors of these habits.

So what will we know about applicants who only have high scores on
measures of extraversion and agreeableness because they elevated their
responses on a measure of personality used during selection? Are they
going to be just as likely to exhibit positive contextual work habits as those
individuals who had higher true levels of extraversion and agreeableness?
The truth is that we do not know. Individuals who were truly agreeable
and extraverted should be more likely to exhibit positive contextual work
habits than their fellow coworkers who may have tried to convey a more
positive image of themselves during the selection process. Individuals
who did fake may also be expected to exhibit what Motowidlo et al. (1997)
refer to as more negative or politically motivated contextual work habits
such as Machiavellianism. Once again, using a situational specificity
framework, we can look for situational cues that might elicit similar
behavior. The application setting is a competitive environment in which
the "winners" are rewarded with desirable outcomes. So in similar situations such as competition for promotion or limited organizational
resources, we could expect behavior similar to faking. Although there is
no literature that has tested this hypothesis, it seems that it is worth
exploring. If individuals are willing to alter their responses to measures
used in the selection process in order to further their chances of obtaining
employment, it is possible that they could be willing to engage in other
similar self-serving behaviors on the job.

Counterproductive Work Behaviors

A final area that we would like to consider technically falls outside of
the task and contextual performance domain. Counterproductive work
behaviors do not refer to aspects of job performance per se, but research
has indicated that these behaviors are related to contextual performance
(Sackett, 2002). Sackett lists several categories into which counterproductive behaviors can fall. Theft, destruction of property, misuse of time, misuse of information and resources, unsafe behavior, poor attendance, poor
quality work, alcohol use, drug use, inappropriate verbal actions, and
inappropriate physical actions are all possible forms of counterproductive
work behavior. Sackett reported a correlation of approximately $r = -.60$
for studies that had examined the relationship between elements of coun-

terproductive behavior and citizenship behaviors (a similar construct to contextual performance).

The study by Rosse et al. (1999) may provide some evidence for a link between faking and counterproductive work behaviors. Rosse et al. correlated scores on a frankness scale with supervisor ratings of positive and negative work behaviors. The results of this study indicated that scores on the frankness scale (which was reverse coded such that higher scores were indicative of higher expected levels of faking) were positively related to supervisor ratings of negative behaviors such as misleading and making impossible promises to customers. An additional point worth noting is that scores on the frankness scale were negatively related to supervisor ratings of positive work behaviors such as listening carefully to customers, and suggesting products that fit the customer's needs. This study suggests that faking (as approximated by a frankness scale) may be predictive of negative work behaviors that involve deceptive behavior on the part of employees.

By suggesting that individuals who fake on measures of personality may engage in more counterproductive work behaviors than those who do not fake, we are by no means categorizing fakers as a group of delinquent individuals who will wreak havoc in organizations. We do believe however, that *some* individuals who use deceptive tactics to further their chances of gaining employment could use deception in other work-related situations. Although faking may not directly predict counterproductive behaviors, the individual difference variables that predict both of these tendencies could be similar. Obviously individuals who engage in more severe forms of counterproductive behavior such as destroying property, engaging in sexual harassment or physical assault of other employees are likely to have a set of problems that range well beyond their propensity for elevating their responses on a personality measure. However, we do believe that this is another possible path for researchers to pursue in an attempt to understand all of the possible implications of applicant faking behavior.

CONCLUDING REMARKS

The overarching theme of this chapter has been that the true nature of the relationship between faking and job performance may be more complex than the current state of research would suggest. Given the fact that the majority of research examining the faking and job performance relationship has focused primarily on issues surrounding criterion-related validity, we have argued that researchers have only scratched the surface of this issue. While research targeted at considerations of faking and criterion-related validity is undoubtedly one important piece of the faking and job performance puzzle, it should by no means be a stopping point.

Regardless of this assertion, there are undoubtedly those in the field of personality research and industrial/organizational psychology who will contend that the lack of a clear-cut finding regarding the effects of faking (as operationalized by scores on SD scales) on the predictive ability of personality measures is evidence of a dead end, rather than an opportunity for further investigation. We believe that this chapter has provided logical arguments contrary to this sentiment, and that it has shed light on possible paths that could be pursued in an investigation of the relationship between faking and job performance. Although psychometric properties of psychological tests offer invaluable information regarding the quality of a measure, concerns regarding the influence of faking on the utility of these measures may reach beyond psychometrics, into the day-to-day functioning of organizations. Obviously no measure is going to tell us everything about how an employee is likely to perform or behave once they are on the job. The prediction of human behavior may be one of the most daunting tasks facing any science. This is no excuse however, for not making an attempt to know everything that we can about how a phenomenon like applicant faking behavior can impact hiring decisions and subsequently, the day-to-day activities of organizations. Any assumptions left untested could pose problems for researchers and practitioners alike.

An additional point that may be raised is that because personality measures are typically part of a battery of measures used in the selection process, research may eventually find that the overall effects of faking on the selection process may be minimal. Personality measures are not typically (and should not be) used as the primary basis of hiring decisions. However, any threat to the validity of a selection system, even if the threat only affects one component of that system, should be of utmost concern to researchers and practitioners alike. If we are to use personality in the process of making hiring decisions, we should continuously strive for precise measurement and understanding of personality constructs, and any phenomena that can affect the measurement of these constructs should be of concern. As this chapter has attempted to point out, the effects of faking on the utility of personality variables in personnel selection may reach beyond an attenuation of criterion-related validity coefficients. This, however, is no reason to hastily conclude that faking is not a concern to those who use personality measures in the selection process.

Perhaps the most important point to be made is that if we can agree that faking is a phenomenon that does indeed occur during the selection process, it is worthy of continued investigation because the full effects of faking are certainly anything but well-understood. Our goal as researchers and practitioners should be to strive for the best prediction of individual behavior at work. We agree with the sentiment offered by Hough (1998) that regardless of how faking affects criterion-related validity, the fact that it exists as a phenomenon in the selection setting should lead researchers to wonder exactly what it is that we are measuring. It seems hasty to conclude that faking does not matter when its very presence has the potential to call into question many of the inferences that are made during the selection process.

REFERENCES

Barrick, M. R., & Mount, M. K. (1991). The Big Five personality dimensions and job performance: A meta-analysis. *Personnel Psychology, 44,* 1-26.
Barrick, M. R., & Mount, M. K. (1996). Effects of impression management and self-deception on the predictive validity of personality constructs. *Journal of Applied Psychology, 81,* 261-272.
Binning, J. F., & Barrett, G. V. (1989). Validity of personnel decisions: A conceptual analysis of the inferential and evidential bases. *Journal of Applied Psychology, 74,* 478-494.
Borman, W. C., & Motowidlo, S. J. (1993). Expanding the criterion domain to include elements of contextual performance. In N. Schmitt & W. C. Borman (Eds.), *Personnel selection in organizations* (pp. 71-98). San Francisco: Jossey-Bass.
Borman, W. C., & Motowidlo, S. J. (1997). Task performance and contextual performance: The meaning for personnel selection research. *Human Performance, 10,* 99-109.
Bott, J. P., Svyantek, D. J., Goodman, S. A., & Bernal, D. S. (2003). Expanding the performance domain: Who says nice guys finish last? *International Journal of Organizational Analysis, 11,* 137-152.
Burns, G. N., & Christiansen, N. D. (2004). *Effects of faking on the linear construct relationships of personality test scores.* Paper presented at the 19th annual conference of the Society for Industrial and Organizational Psychology, Chicago.
Cascio, W. F. (1998). *Applied psychology in human resource management* (5th ed.). Upper Saddle River, NJ: Pearson Prentice Hall.
Christiansen, N. D., Goffin, R. D., Johnston, N. G., & Rothstein, M. G. (1994). Correcting the 16PF for faking: Effects on criterion-related validity and individual hiring decisions. *Personnel Psychology, 47,* 847-860.
Christiansen, N. D., Robie, C., & Bly, P. R. (2005, April). Using covariance to detect applicant response distortion of personality measures. In M. J. Zickar (Chair), *Faking research: New methods, new samples, and new questions.* Symposium conducted at the 20th annual conference for the Society of Industrial and Organizational Psychology, Los Angeles.
Donovan, J. J., Dwight, S. A., & Hurtz, G. M. (2003). An assessment of the prevalence, severity, and verifiability of entry-level applicant faking using the randomized response technique. *Human Performance, 16,* 81-106.

Donovan, J. J., Dwight, S. A., & Schneider, D. (2005, April). Prevalence and impact of faking in an organizational setting. In S. A. Dwight (Chair), *Faking it: Insights and remedies for applicant faking.* Symposium conducted at the 20th annual conference for the Society of Industrial and Organizational Psychology, Los Angeles.

Douglas, E. F., McDaniel, M. A., & Snell, A. F. (1996, August). *The validity of non-cognitive measures decays when applicants fake.* Paper presented at the annual meeting of the Academy of Management, Cincinnati, OH.

DuBois, C. L., Sackett, P. R., Zedeck, S., & Fogli, L. (1993). Further exploration of typical and maximum performance criteria: Definitional issues, prediction, and White-Black differences. *Journal of Applied Psychology, 78,* 205-211.

English, A., Griffith, R. L., Graseck, M., & Steelman, L. A. (2005). *Frame of reference, applicant faking and the predictive validity of non-cognitive measures: A matter of context.* Manuscript submitted for publication.

Graham, K. E., McDaniel, M. A., Douglas, E. F., & Snell, A. F. (2002). Biodata validity decay and score inflation with faking: Do item attributes explain variance across items? *Journal of Business and Psychology, 16,* 573-591.

Grandey, A. A. (2003). When "the show must go on": Surface acting and deep acting as determinants of emotional exhaustion and peer-rated service delivery. *Academy of Management Journal, 46,* 86-96.

Griffith, R. L., Chmielowski, T. S., & Yoshita, Y. (in press). Do applicants fake? An examination of the frequency of applicant faking behavior. *Personnel Review.*

Griffith, R. L., English, A., Yoshita, Y., Gujar, A., Monnot, M., Malm, T., et al. (2004, April). Individual differences and applicant faking behavior: One of these applicants is not like the other. In N. D. Christiansen (Chair), *Beyond social desirability in research on applicant response distortion.* Symposium conducted at the 19th annual conference for the Society for Industrial and Organizational Psychology, Chicago.

Griffith, R. L., Yoshita, Y., Peterson, M. H., & Malm, T. (2006). Addressing elusive questions: Investigating the faking-performance relationship. In R. G. Griffith & Y. Yoshita (Chairs), *Deceptively simple: Applicant faking behavior and prediction of job performance.* Symposium conducted at the 21st annual conference for the Society for Industrial and Organizational Psychology, Dallas, TX.

Haaland, D., & Christiansen, N. D. (1998). *Departures from linearity in the relationship between applicant personality test scores and performance as evidence of response distortion.* Paper presented at the 22nd annual International Personnel Management Association Assessment Council Conference, Chicago.

Higgins, C. A., Judge, T. A., & Ferris, G. R. (2003). Influence tactics and work outcomes: A meta-analysis. *Journal of Organizational Behavior, 24,* 89-106.

Hogan, R., Hogan, J., & Roberts, B. W. (1996). Personality measurement and employment decisions: Questions and answers. *American Psychologist, 51,* 469-477.

Hough, L. M. (1998). Effects of intentional distortion in personality measurement and evaluation of suggested palliatives. *Human Performance, 11,* 209-244.

Hough, L. M., Eaton, N. K., Dunnette, M. D., Kamp, J. D., & McCloy, R. A. (1990). Criterion-related validities of personality constructs and the effect of response distortion on those validities. *Journal of Applied Psychology, 75*, 581-595.

Illingworth, A. J., Snell, A. F., & Rosen, C. C. (2005, April). Effects of warnings and individual differences on the criterion-related validity of non-cognitive tests. In J. P. Bott & C. C. Rosen (Chairs), *Moving from laboratory to field: Investigating situation in faking research*. Symposium conducted at the 20th annual meeting of the Society for Industrial and Organizational Psychology, Los Angeles.

Komar, S., Theakston, J., Brown, D. J., & Robie, C. (2005, April). *Faking and the validity of personality: A monte-carlo investigation*. Paper presented at the 20th annual meeting for the Society for Industrial and Organizational Psychology, Los Angeles.

McDaniel, M. A., Douglas, E. F., & Snell, A. F. (1997, April). *A survey of deception among job seekers*. Paper presented at the 12th annual conference of the Society for Industrial and Organizational Psychology, St. Louis, MO.

McFarland, L. A., & Ryan, A. M. (2000). Variance in faking across noncognitive measures. *Journal of Applied Psychology, 85*, 812-821.

Motowidlo, S. J., Borman, W. C., & Schmit, M. J. (1997). A theory of individual differences in task and contextual performance. *Human Performance, 10*, 71-83.

Motowidlo, S. J., & Van Scotter, J. R. (1994). Evidence that task performance should be distinguished from contextual performance. *Journal of Applied Psychology, 79*, 475-480.

Mueller-Hanson, R., Heggestad, E. D., & Thornton III, G. C. (2003). Faking and selection: Considering the use of personality from select-in and select-out perspectives. *Journal of Applied Psychology, 89*, 348-355.

Ones, D. S., Viswesvaran, C., & Reiss, A. D. (1996). Role of social desirability in personality testing for personnel selection: The red herring. *Journal of Applied Psychology, 81*, 660-679.

Ones, D. S., Viswesvaran, C., & Schmidt, F. L. (1993). Comprehensive meta-analysis of integrity test validities: Findings and implications for personnel selection and theories of job performance. *Journal of Applied Psychology, 78*, 679-703.

Rees, C. J., & Metcalf, B. (2003). The faking of personality questionnaire results: Who's kidding whom? *Journal of Managerial Psychology, 18*, 156-165.

Rosse, J. G., Levin, R. A., & Nowicki, M. D. (1999, April). *Assessing the impact of faking on job performance and counter-productive job behaviors*. Paper presented in Paul Sackett (Chair) New empirical research on social desirability in personality measurement, symposium for the 14th annual meeting of the Society for Industrial and Organizational Psychology, Atlanta, GA.

Rosse, J. G., Stecher, M. D., Miller, J. L., & Levin, R. A. (1998). The impact of response distortion on preemployment personality testing and hiring decisions. *Journal of Applied Psychology, 83*, 634-644.

Sackett, P. R. (2002). The structure of counterproductive work behaviors: Dimensionality and relationships with facets of job performance. *International Journal of Selection and Assessment, 10*, 5-11.

Schmidt, F. L., & Hunter, J. E. (1981). Employment testing: Old theories and new research findings. *American Psycholgist, 36*, 1128-1137.

Schmidt, F. L., & Hunter, J. E. (1998). The validity and utility of selection methods in personnel psychology: Practical and theoretical implications of 85 years of research findings. *Psychological Bulletin, 124,* 262-274.

Tett, R. P., Jackson, D. N., & Rothstein, M. (1991). Meta-analysis of personality-job performance relationships. *Personnel Psychology, 47,* 157-172.

Vasilopoulos, N. L., Reilly, R. R., & Leaman, J. A., (2000). The influence of job familiarity and impression management on self-report measure response latencies and scale scores. *Journal of Applied Psychology, 85,* 50-64.

Zickar, M., Rosse, J., & Levin, R. (1996). *Modeling the effects of faking on personality scales.* Paper presented at the 11th annual conference of the Society for Industrial and Organizational Psychology, San Diego, CA.

CHAPTER 10

FORCING CHOICES IN PERSONALITY MEASUREMENT

Benefits and Limitations

Patrick D. Converse, Frederick L. Oswald, Anna Imus, Cynthia Hedricks, Radha Roy, and Hilary Butera

Although the use of personality measures in personnel selection settings has had a mixed history in the past several decades (Guion & Gottier, 1965; Hough & Oswald, 2005; Hough & Schneider, 1996), it is now commonplace in selection practice, and the practice is supplemented by research evidence indicating that some measured aspects of personality, such as conscientiousness, exhibit modest but consistent criterion related validity across most occupations (e.g., Barrick & Mount, 1991; Ones, Viswesvaran, & Schmidt, 1993; Tett, Jackson, & Rothstein, 1991). Despite today's Zeitgeist of optimism (see also Hough & Oswald, 2000), some organizations remain reluctant to include personality assessments in their selection systems because of a reasonable concern that some job applicants may intentionally misrepresent themselves on personality tests, resulting in an increase in hiring applicants that are not qualified (i.e., false positives).

A Closer Examination of Applicant Faking Behavior, 263–282
Copyright © 2006 by Information Age Publishing
All rights of reproduction in any form reserved.

Some organizational researchers have argued that applicant faking on personality measures is not a major issue in selection contexts (e.g., Barrick & Mount, 1996; Hogan, 1991; Ones, Viswesvaran, & Reiss, 1996), but several other lines of research suggest this type of response distortion represents a legitimate concern. For example, evidence indicates that individuals can (e.g., Viswesvaran & Ones, 1999) and often do (e.g., Donovan, Dwight, & Hurtz, 2003) fake on noncognitive assessments in selection contexts, and that this misrepresentation can affect the construct validity of personality measures (e.g., Schmit & Ryan, 1993) and the quality of personnel selection decisions (e.g., Mueller-Hanson, Heggestad, & Thornton, 2003; Rosse, Stecher, Miller, & Levin, 1998).

Researchers and practitioners have dedicated a lot of time and energy in attempting to address this concern, such as by making it more difficult for job applicants to "fake good" on personality measures, by using subtly worded items, or by correcting for faking by partialling out the covariance statistically between the personality scores and scores on measures of social desirability or impression management (see Hough, Eaton, Dunnette, Kamp, & McCloy, 1990). However, for a variety of reasons, these attempts have not been particularly successful (e.g., see Christiansen, Goffin, Johnston, Rothstein, 1994; Ellingson, Sackett, & Hough, 1999; Ones et al., 1996). This chapter focuses instead on one method of reducing faking that has received support in recent organizational research (e.g., Christiansen, Burns, & Montgomery, 2005; Jackson, Wroblewski, & Ashton, 2000), namely using a forced-choice (FC) item format to measure personality, where for each item, test-takers have to select from a set of statements. For some FC items, then, the test-taker may not be able to avoid endorsing an undesirable statement from this set; by contrast, in a traditional Likert-scale format, each statement is presented independent of one another, and the test-taker can more easily "fake good" by not rating the undesirable statements highly. Recent studies have produced promising findings related to the reduced fakability and comparable validity of FC scales relative to Likert scales, but many key questions remain regarding the actual implementation of these measures in selection settings. Thus, this chapter reviews practical and conceptual issues associated with the use of FC personality measures in selection contexts, focusing on five main areas: item format, psychometric properties, applicant faking, criterion-related validity, and applicant reactions.

FORMAT OF FC ITEMS

By developing FC personality measures, researchers have attempted to reduce the potential for faking a priori, versus attempting to make faking corrections post hoc. Although some have criticized these types of measures on psychometric grounds (e.g., because of the negative correlations between scales that generally result; Hicks, 1970), recent research has produced promising findings regarding the fakability and validity associated with FC measures (e.g., Christiansen et al., 2005; Jackson et al., 2000).

FC personality measures generally consist of sets of two or more statements, with each set comprising one "item." Here is an example of an FC item:

Choose the one option below that is "most" like you, and the one that is "least" like you:

(a) I tend to be calm even in the most chaotic situations.
(b) Sometimes I get angry at the smallest things.
(c) When in a group, I prefer having a discussion before a decision is made.
(d) I do not tend to trust other individuals at first.

This item is FC because individuals must choose one statement as "most like me" and one statement as "least like me" out of the four statements given. Statements selected as most consistent with one's personality are typically scored +1, those selected as least consistent are scored –1, and all others are scored 0. Scores for those statements measuring the same personality dimensions are then combined to form scale scores. This sample FC item is multidimensional, where each statement reflects a different construct. FC items can also be unidimensional, where each statement reflects a different level of a single construct, although a set of statements arranged along one dimension may be harder to write (e.g., trying to write an item that reflects "average" conscientiousness), and the item may be easier to fake to the extent that the test-taker can discern the trait level each statement reflects.

Statements within an FC item are typically balanced in terms of social desirability so that, for example, respondents choose between two desirable options as "most like me" and two undesirable options as "least like me," or they rank order four options of approximately equal desirability (see Norman, 1963, for a discussion of approaches to matching options for FC item development). Given this balancing of item attractiveness, test-takers cannot respond solely on the basis of desirability. It is therefore assumed that, between items that are equally desirable (or undesirable), they will then select statements that most accurately reflect their personalities. More than 50 years ago, Gordon (1951) explained the rationale as follows:

A major assumption underlying this technique in personality measurement is that if two items are equally derogatory from the point of view of the social group, individuals to whom one of the items is more applicable will tend to perceive that item as being the less derogatory. Thus, if an individual, who is motivated to make socially acceptable responses, is forced to select one of the items as being *least* like himself, he will select the item that he perceives to be most derogatory, which will tend to be the item that *is* less like himself. It is assumed that the converse holds when he is forced to select one of a pair of complimentary items as being *most* like himself. (p. 407)

Thus, it is assumed that in situations in which test-takers are motivated to respond in a socially desirable manner, FC items that attempt to equalize statements on their social desirability will tend to result in more accurate assessments of personality than Likert items.

PSYCHOMETRIC PROPERTIES OF FC MEASURES

Over the years, several disadvantages of FC methodology have been identified. These limitations primarily involve the psychometric properties of FC scales that result from forcing choices among item options. Hicks (1970) provides a simple illustration of the effects of FC methodology, where two factors (A and B) are assessed by two normative measures that are uncorrelated, and so $r_{AB} = 0$. If an FC measure of these factors was developed in which test-takers were forced to choose an item option representing factor A (scoring one point for scale A and therefore zero points for scale B) or an option representing factor B (scoring one point for scale B and therefore zero points for scale A), scores between factors A and B would correlate $r_{AB} = -1.00$ instead of 0. This huge difference in correlations tends to be reduced when more factors are involved, but this general approach of forcing choices nevertheless results in primarily negative correlations between scales.

Despite some evidence to the contrary (Saville & Willson, 1991), this dependence among scales generally produces psychometric difficulties, such as when conducting a factor analysis (e.g., see Closs, 1996; Cornwell & Dunlap, 1994; Hicks, 1970; Johnson, Wood, & Blinkhorn, 1988; Meade, 2004). Even if the normative scores for a set of measures are orthogonal, an FC scale measuring the same characteristics will tend to produce negative observed correlations that reflect a bias of the FC format, not a substantive relationship between personality constructs (Cornwell & Dunlap, 2004).

This scale dependence also has implications for reliability. As Tenopyr (1988) demonstrated, the internal consistencies of scales from multidimensional FC measures are interdependent, and therefore it is possible for one highly reliable scale to inflate the reliability of other scales artificially (see also Johnson et al., 1988). However, Tenopyr also noted that in some circumstances, internal consistencies can be lowered as a result of scale dependence. Thus, internal consistency reliability estimates for FC measures may be artifactually high or low due to item format. Note also that test-retest reliability may be influenced by scale interdependence, and therefore this alternative reliability index does not necessarily provide a solution to this issue (Tenopyr, 1998). In addition, Baron (1996) concludes that reliability estimates for FC scores tend to be comparatively lower than those for normative scores, particularly when few scales are involved or when the underlying dimensions are strongly positively correlated (see also Bartram, 1996).

Meade (2004) has also outlined additional psychometric complexities associated with the FC format. He argues that FC items produce not only scale-level interdependence, but also item-level interdependence as a result of grouping statements and forcing choices among them. Specifically, in responding to an FC item, test-takers must engage in a cognitive decision-making process to choose among the given statements. This evaluation process is assumed to be affected by several factors including not only the test-taker's true trait levels, but also the item thresholds (i.e., the trait level necessary for endorsing a statement), and person-specific properties (i.e., unique characteristics that influence selecting a statement over others in the set). As a result, scores obtained from FC measures for a given trait are influenced by the true and error scores for *all* of the traits represented in the statement sets. Thus, from a classical test theory perspective, the item-level psychometric properties of FC measures are much more complex than those of more traditional measure formats, because observed scores are affected by more than just the true score and error associated with the single scale of interest. Meade thus provides detailed reasoning behind the psychometric difficulties associated with FC formats.

Perhaps the most important implication of the measurement limitations of FC measures for selection contexts relates to the inferences that can be appropriately drawn from scores on these measures. Specifically, FC items involve comparing item options, rather than directly rating each item option individually on a Likert scale, and therefore FC data indicate each individual's relative level on the constructs measured, rather than his or her absolute level. Therefore, these types of data are argued to measure intraindividual differences (e.g., whether a given individual is higher on conscientiousness than extraversion), but not the interindividual differences necessary in personnel selection (e.g., whether one individual is higher on conscientiousness than another individual; e.g., see Closs, 1996; Cornwell & Dunlap, 1994; Hicks, 1970; Johnson et al.,

1988). Although this is true in principle, for practical purposes these types of scores may still provide useful information on interindividual differences in some contexts. The following sections discuss this further.

Purely Ipsative and Partially Ipsative Measures

FC measures where test-takers are only given two options and they must choose one are called *purely ipsative* measures, and they tend to be highly problematic on both conceptual and statistical grounds. From a conceptual standpoint, consider the following item:

Choose the one option below that is "most" like you:

(a) Once I give priority to a project, I follow it through. (conscientiousness)
(b) I'm usually the first person to strike up a conversation with strangers. (extraversion)

One individual may indicate that she is more conscientious than extraverted, and another individual may indicate the opposite, that he is more extraverted than conscientious. However, if the first person is low on both traits, and the second person is high on both, then of course it is incorrect to assume from ipsative scores on the item that the first person is higher than the second on conscientiousness. Statistically, as mentioned previously, purely ipsative items force negative correlations onto the data: Choosing conscientiousness in the previous example necessarily means not choosing extraversion. For reasons such as these, scores obtained from purely ipsative FC measures should not be used to make the sort of interindividual comparisons that are required in selection contexts.

However, all FC measures do not necessarily result in purely ipsative measurement (see Hicks, 1970). Some FC assessments are *partially ipsative*, where test items still require test-takers to make choices, but there is more variability built into the scores than in purely ipsative FC measures, which allows one to draw more appropriate psychological and psychometric conclusions from the data. Hicks outlined several circumstances that produce partially ipsative measures: (a) test-takers only partially order item alternatives, rather than ordering them completely (e.g., choosing "most" and "least" from four alternatives, rather than rank ordering all four); (b) scales have differing numbers of items; (c) not all ranked alternatives are scored; (d) scales are scored differently for test-takers with different characteristics; (e) the scored alternatives are differentially weighted; (f) one or more of the scales are deleted when the data are analyzed; and (g) the test contains items that are normative in nature. Thus, purely ipsative and partially ipsative measures can be very similar or even identical in appearance, differing only in how test-takers are asked to respond or in how responses are scored. Under any of the circumstances outlined, the psychometric limitations of resulting scores are not as severe as with purely ipsative measurement. Scores on each individual scale of a partially ipsative measure are still based on relative responses, however, and therefore relaxing the measurement limitations allows for—but in no way guarantees—the use of partially ipsative scores to provide appropriate information on interindividual differences (McCloy, Heggestad, & Reeve, 2005).

FC Scales and Trait Levels

As Christiansen et al. (2005) noted, the appropriateness of making interindividual comparisons from FC data is to some degree an empirical question, and previous research suggests such comparisons may be supported. Several studies have found fairly strong correlations between FC and non-FC measures of the same constructs (e.g., Bowen, Martin, & Hunt, 2002; Christiansen et al., 2005). For example, Christiansen et al. (Study 1) found correlations ranging from .52 to .68 between FC and corresponding Likert-scale measures of conscientiousness and extraversion. Correlations of this magnitude do not indicate complete convergence, but they suggest that FC measures do in fact provide information about absolute trait levels. Heggestad, Morrison, Reeve, and McCloy (2006) found even stronger relationships between Likert and FC versions of a Big Five inventory consisting of the same statements, with uncorrected convergent validity coefficients ranging from .75 to .87 and correlations corrected for unreliability ranging from .92 to 1.00. Furthermore, these researchers compared percentile rank differences between these two measures (i.e., the average difference in test-takers' ranks between the two versions), and the consistency between versions of hypothetical pass-fail decisions at several selection ratios. Results from these additional methods of assessing similarity in scores also indicated substantial consistency between the two measure formats, further supporting the idea that FC measures can provide useful information on absolute trait levels. Similarly, Meade (2004) used scores from a sample of job applicants that completed a personality inventory using both normative (Likert-scale) and FC ("most like me" and "least like me") response formats to directly compare hypothetical hiring decisions stemming from these two sets of scores. Using a 25th percentile cut-score, the percentage of agreement between the hiring decisions based on normative scores and those based on FC scores ranged from 71% to 78%; for a 50th percentile cut-score agreement ranged from 57% to 67%; and for a 75th percentile cut-score agreement ranged from 62% to 70%. As expected, there is far from perfect agreement (the formats are in fact different, and neither are perfectly reliable), but there is substantial overlap in the selection decisions that would be made based on scores from these two measures.

This previous research suggests that traditional approaches to FC measurement potentially result in information useful for interindividual comparisons needed for selection decisions. Perhaps even more promising, however, is recent work suggesting that it may be possible to go a step further and extract normative information from FC items (McCloy et al., 2005; Stark, Chernyshenko, & Drasgow, 2005). For example, McCloy et al. have outlined a technique for this extraction based on Coombs's unfolding model of the psychological processes involved in responding to multidimensional FC items (Coombs, 1964). This multidimensional unfolding model is similar to item response theory (IRT), in suggesting that both test-takers and test items can be represented as points in a joint k-dimensional construct space. According to the model, test-takers evaluate each statement that measures a particular construct by first determining the distance between their location on the construct continuum and the location implied by the statement. Then, given a set of such statements to compare (as is the case for multidimensional FC items), they select as "most like me" that statement corresponding to the smallest distance and as "least like me" that statement corresponding to the largest distance.

Based on this model, McCloy et al. (2005) have shown how it may be possible to obtain normative information from FC items. Their method first administers a set of items to a group of individuals motivated to answer honestly (e.g., a research group or group of job incumbents). These are traditional Likert-scale items, and IRT is applied to the data to determine where items are located on the continua of the constructs they measure. Next, the same items are given in FC format to a group motivated to fake (e.g., job applicants). The item locations established through IRT in the "honest" sample now serve as boundaries for the data in the "faking" sample, where an individual's FC trait level response to an item leads to an estimate of whether the individual is above or below the boundary on a given construct continuum. Assuming Coombs's unfolding model holds, one can apply similar inferences to all of an individual's responses to a set of FC items, which ultimately leads to a point estimate and confidence interval regarding the individual's location on each construct being measured. McCloy et al. present simulation results indicating that the technique potentially provides fairly accurate estimates of trait levels. Thus, this approach may provide a means of capitalizing on the potential advantages of FC items, while avoiding the measurement limitations typically associated with this format.

However, more work is clearly required. For instance, further research should examine the extent to which the assumptions of the multidimensional unfolding model actually align with the cognitive processes involved in responding to FC items. This model assumes that test-takers respond to each item in a very systematic, rational manner, which may not be true in many cases. Another potential limitation relates to using a Likert-scale measure to establish item parameters and then combining these items into groups to form an FC measure. Specifically, this grouping may affect the item parameters, leading to less accurate measurement. Empirical work should also determine whether having Likert/normative scoring influenced by faking affects personnel selection decisions any worse than having FC/ipsative scoring, which may serve to reduce the effects of faking but carries the tradeoff of greater variability in measuring individuals' normative trait levels.

Summary

FC items require test-takers to choose from among statements that have been balanced in terms of social desirability. Although this is intended to reduce the potential for response distortion, it also results in several measurement limitations related to the interdependencies among items and scales that are imposed by this item format. An important implication of these limitations for selection is that in many cases, interindividual comparisons cannot be adequately supported. However, the severity of these limitations can vary depending on a number of important factors, and empirical evidence suggests that some FC measures can in fact provide useful information about absolute trait levels. Furthermore, recent work suggests the possibility of extracting normative information from FC items. Combined, these considerations suggest that some FC measures may be useful in selection contexts, but users of these measures should be especially aware of important limitations when analyzing and interpreting results on the reliability and validity of such measures.

APPLICANT FAKING OF FC MEASURES

Arguments in favor of the FC format generally focus on its potential to reduce faking. The Likert-response format appears to have the advantage of simplicity in placing less of a cognitive burden on the test-taker, but the downside to this simplicity is that the desired answer to a statement may be relatively transparent for these formats (e.g., it is obviously good to endorse a positive statement "strongly agree"), where job applicants may distort their responses in the direction of an ideal job candidate and undermine the criterion-related validity of the measure. In contrast, by forcing job applicants to choose among statements that are equally desirable (or undesirable) in nature, the FC format would appear to make this type of distortion more difficult, and thus to the extent that faking is irrelevant to job performance, the FC format has the potential for higher criterion-related validity.

Recent research supports this line of reasoning, if faking resistance is operationalized as an overall reduction in mean differences between faking and nonfaking conditions (e.g., job applicant vs. incumbent or "honest" samples). For example, Jackson et al. (2000) found that, when test-takers were instructed to complete a Likert-type integrity test as though they were attempting to make a good impression to gain a desirable job, they scored 0.95 SD higher on average than when instructed to respond honestly. By contrast, respondents completing an FC version of the measure were only able to increase their scores 0.32 SD on average. Bowen et al. (2002) and Christiansen et al. (2005) obtained similar results. Comparing scores from an applicant instructions condition with those from an honest instructions condition, Bowen et al. reported an average d of 0.41 for selected scales using a normative format, but only 0.24 for corresponding scales using an FC format. Christiansen et al. reported similar findings across three separate studies.

Using somewhat different procedures, Martin, Bowen, and Hunt (2002) also obtained evidence suggesting FC measures are less susceptible to faking; however, rather than examining score reductions, faking was measured as the similarity between test-takers' personality scores across 30 scales and their ratings of how an ideal junior manager would score on these same scales. Results indicated no significant difference in profile similarity for an FC measure between honest instructions and applicant instructions conditions (where test-takers were told to try to look good to compete for a junior manager position), but for a normative measure, there was significantly greater similarity between the actual personality profile and the ideal profile in the applicant condition than there was in the honest condition (see also Bowen et al., 2002). This again suggests FC measures are generally more resistant to response distortion. However, this particular finding should be viewed with some caution. As Meade (2004) pointed out, ideal profiles were obtained with a normative ten-point scale, and therefore these profiles were not subject to the same measurement constraints as the actual profiles obtained with the FC measure. This difference may be at least partially responsible for the obtained results, and thus results are not directly comparable. These recent studies taken as a whole, however, are suggestive of the fact that FC measures may be more resistant to intentional distortion than measures using more traditional item formats.

Despite some promising research findings of the resistance of FC measures to faking, we wish to note explicitly that the FC format is no panacea. Most of the research discussed above, as well as earlier investigations of FC methodology (e.g., Bass, 1957; Borislow, 1958; Dunnette, McCartney, Carlson, & Kirchner, 1962; Longstaff & Jurgensen, 1953; Maher, 1959; Mais, 1951) demonstrated that, when motivated or instructed to distort responses, test-takers are able to do so on FC measures (although the mean shifts in scores tend to be much smaller than those obtained with non-FC measures). Furthermore, no personnel selection research has identified the nature of faking with any great amount of conceptual precision, and therefore no definitive indicator of actual faking exists. Thus, although the FC format may reduce score inflation compared with Likert scales, it may introduce other forms of faking, particularly if FC tests become more commonplace, and test-takers become more accustomed to the format and more strategic about how to take these types of tests.

In addition, despite some positive findings at the group level (e.g., smaller mean differences between faking and honest conditions for FC measures than for non-FC measures), recent evidence focusing on the individual level suggests that FC formats may not have similar desired effects at this level. Heggestad et al. (2006) found larger mean differences between scores obtained under honest and faking instructions for a Likert measure than for an FC measure, consistent with other research findings. However, despite a reduction in mean differences between honest and faking, examination of same-scale correlations, percentile rank differences, and hypothetical selection decisions for honest versus faked scores indicated that the FC measure did not result in more accurate reflections of "true" (i.e., honest) scores than the Likert measure.

These recent studies investigating the fakability of FC personality measure formats in comparison to non-FC formats have tended to rely on one methodology: comparing scores obtained under honest instruction conditions with those obtained under applicant instruction conditions. Future research might profitably extend these studies by comparing faking on FC and Likert-response personality measures in situations where test-takers are job applicants or are otherwise naturally motivated to distort responses, rather than those in which they are directed to fake.

CRITERION-RELATED VALIDITY OF FC MEASURES

As mentioned previously, there is some promising evidence regarding the criterion-related validity of FC measures as well. Compared with other measures, studies have demonstrated that FC measures have comparable validities (e.g., Villanova, Bernardin, Johnson, & Dahmus, 1994) or even higher validities (e.g., Bartram, 2005; Gordon, 1951). Furthermore, some evidence suggests that FC scales maintain their validity better than non-FC measures in situations involving faking. For instance, Jackson et al. (2000) found that criterion-related validity for a Likert-response integrity test predicting self-reported workplace delinquency dropped from .48 to .18 when participants were directed to make a good impression as a job applicant. On the other hand, criterion-related validity for an FC integrity test predicting the same criterion only dropped from .41 to .36 when respondents were asked to respond like a job applicant. Even increases in criterion-related validity associated with FC measures have been reported. Using supervisor ratings of job performance for a group of currently and recently employed students, Christiansen et al. (2005) found that criterion-related validity increased for an FC measure of conscientiousness under job applicant instructions ($r = .17$ for honest instructions versus .46 for instructions to complete the measure as if they were applying for a customer service job), whereas validity decreased for a non-FC measure ($r = .21$ for honest instructions versus .08 for applicant instructions).

However, evidence from older studies indicates that intentional distortion maintains a strong negative effect on the criterion-related validities associated with FC measures. For instance, Dunnette et al. (1962) had salesmen complete an FC measure of personality with honest instructions and with faking instructions (answering as a successful salesman would). Relationships between both sets of scores and manager ratings of sales effectiveness were then examined. Faking had a clear negative effect on these correlations (e.g., the correlation for assertiveness decreased from .38 for the honest condition to -.14 for the faking condition). Maher (1959) obtained similar results. In this study, criterion-related validity for an FC study activity questionnaire predicting grade point average (GPA) was .45 and .56 for two honest conditions; these relationships were .10 and .07 for two faking conditions.

Thus, the evidence for the effect of intentional distortion on the criterion-related validity of FC scales is quite mixed: it can decrease, increase, or have little effect on criterion-related validity. Unfortunately, it is unclear at this point how the conditions responsible for these different outcomes vary. As Christiansen et al. (2005) observed, one relevant factor may be the extent to which the criterion is cognitively loaded. For example, these researchers found that faking increased the correlation between an FC measure of conscientiousness and a measure of job performance (the latter being cognitively loaded; Schmidt & Hunter, 1998), but faking decreased the correlation between this FC conscientiousness measure and another Likert-scale measure of conscientiousness (which is unlikely to be cognitively loaded). This suggests that cognitive ability may play a role in faking on FC measures, resulting in stronger correlations for criteria that are more related to cognitive ability, but weaker relationships for criteria that are less related to cognitive ability. In other words, ability may be one of the person-specific properties discussed by Meade (2004) that influence responses to FC items in situations involving faking. This may then have implications for FC measures predicting task versus contextual performance (e.g., Borman & Motowidlo, 1993, 1997; Motowidlo, Borman, & Schmit, 1997; Motowidlo & Van Scotter, 1994). If ability variables are better predictors of task performance and personality variables are better predictors of contextual performance (e.g., see Motowidlo et al., 1997), then faking on FC measures may increase correlations with task performance, and at the same time decrease correlations with contextual performance.

Other factors also appear to be involved, as both Dunnette et al. (1962) and Maher (1959) actually found decreases in validity with cognitively loaded criteria (job performance and GPA, respectively). It is likely that other issues related to the FC measure itself (e.g., the degree to which statements within the FC items have been successfully balanced in terms of social desirability) and the sample involved (e.g., level of knowledge regarding the dimensions measured by the FC scale) are key influences on the effect of faking on criterion-related validity. Future research that systematically examines factors such as these might allow for clearer conclusions regarding the conditions that affect the criterion-related validity of FC measures.

The incremental validity of FC measures is another area ripe for future research. Previous research has examined the criterion-related validity of FC personality measures in isolation, leaving relatively unexplored the incremental validity associated with these tests when additional predictors are involved, relatively unexplored. Villanova et al. (1994) examined the incremental validity associated with an FC measure of job compatibility, finding that this measure (really measuring person-job fit, rather than personality) explained additional variance in turnover and a turnover/performance composite over measures of verbal and numerical ability. Thus, in the area of person-job fit, where FC measures may be particularly useful because they involve within-person comparisons, there is evidence for incremental validity. However, additional research would be required to determine whether these results generalize to FC measures of personality.

Incremental validity is a particularly key issue for FC personality measures because previous research has suggested that cognitive ability—a well known and generally strong predictor of job performance (see Schmidt & Hunter, 1998)—is an individual-differences factor that predicts responses to these measures, and therefore at least some of the criterion-related validity observed for this format may be attributable to this association with cognitive ability. In particular, Christiansen et al. (2005) found that cognitive ability was related to scores on both an FC and non-FC measure of personality in an applicant instructions condition, but the correlation was stronger for the FC measure ($r = .25$ for FC versus $.15$ for non-FC). In addition, results suggested that individuals high in cognitive ability were more able to increase their scores on the FC measure in part because they had a better understanding of the requirements of the job described in the applicant instructions. As Christiansen et al. (2005) note, these results indicate that faking on FC measures is more cognitively demanding than faking on Likert-format measures and also requires that the test-taker have a good understanding of the personality factors important for the job involved. This suggests that in actual or simulated applicant settings, in which respondents are motivated to fake, scores on FC personality measures may partially reflect cognitive ability, and therefore it is possible that these personality measures have limited incremental validity over ability measures. Thus, future research could test this empirically by examining both cognitive ability and FC personality scales together as predictors of relevant criteria to determine the extent to which personality measures using the FC format explain additional criterion variance above and beyond a traditional measure of cognitive ability.

On a final note for this section, it appears that results pertaining to criterion-related validity are often assumed to support the notion that FC measures provide information about absolute trait levels. However, as Heggestad et al. (2006) have pointed out, demonstration of criterion-related validity for a purely ipsative measure does not necessarily indicate that the ipsative measure has captured information about absolute trait levels; rather, it indicates that individuals who are relatively higher on the focal trait (e.g., conscientiousness) compared with the other traits measured (e.g., the rest of the Big Five factors) also tend to score higher on the criterion. Interpretation of criterion-related validity associated with partially ipsative measures is somewhat less clear-cut, but again these validity coefficients do not necessarily imply that predictor scores represent absolute levels. However, the research reviewed previously suggests reasonable consistency between scores from FC and non-FC measures, and thus it appears that—despite involving relative responses—scores from FC measures to some extent reflect absolute levels and may be useful in making interindividual comparisons in selection contexts. It is an empirical question to ask whether FC profiles, or patterns of relative highs and lows across several personality dimensions, provide incremental validity in selection contexts.

APPLICANT REACTIONS TO FC MEASURES

One other potentially important, but relatively neglected, practical issue related to the use of FC measures involves test-taker reactions. Research has linked perceptions of selection procedures to several important outcomes such as organizational attractiveness and behavioral intentions related to job offer acceptance, recommendation of the organization to others, and purchasing behavior (for reviews see Hausknecht, Day, & Thomas, 2004; Ryan & Ployhart, 2000). In addition, there is reason to believe that test-takers may react more negatively to personality measures using an FC format than to those using more traditional formats. First, researchers have suggested that applicant perceptions should be related to the length of the selection procedures (e.g., Ryan & Ployhart, 2000) and to perceived test ease (e.g., Hausknecht et al., 2004). Although there may be exceptions, FC versions of personality measures can often take longer to complete than non-FC versions (e.g., Bowen et al., 2002), because responding requires not only reading each statement and deciding how characteristic it is of oneself, but also making relative judgments. Related to this, and consistent with Christiansen et al.'s (2005) results suggesting that making a good impression on FC measures is cognitively demanding, test-takers may view this format as more difficult to respond to than more traditional formats. Second, applicant perceptions have also been linked to the "opportunity to perform," which involves the test providing sufficient opportunity to allow the applicant to demonstrate knowledge, skills, and abilities (see Gilliland, 1993), and such opportunities might be extended to include the opportunity to convey one's job-related personality characteristics. This opportunity may be perceived to be more limited with FC measures than with Likert-scale measures, because with FC measures test-takers must make choices among statements, whereas with Likert-scale measures they are free to rate themselves high (or low) on each statement.

Finally, consistent with these points, Bowen et al. (2002) reported preliminary evidence suggesting test-takers react more favorably to personality measures using a Likert-response format than those using an FC format. In this study, participants indicated the Likert-response format was generally easier to answer, less confusing, more interesting, and more conducive to presenting oneself as desired. Note, however, that Bowen et al.'s assessment of reactions was fairly limited, for one, because it involved just four items. In addition, the format of the measure (Likert versus FC) was a within-subjects factor, and therefore test-takers were able to compare their experiences with the first format when providing reactions to the second. It is unclear whether similar results would hold when individuals are exposed to either one format or the other, but not both (i.e., in a between-subjects design), a situation that is more likely to be encountered in selection settings where the test applicants take (or the specific test content at least) would be in a single format. Thus, future research should include a more thorough and practice-oriented exploration of test-taker reactions to different test formats, focusing on several relevant perceptions identified in previous conceptual and empirical work, and using between-subjects designs.

CONCLUSIONS

Faking on personality measures remains an important concern for personnel selection researchers and practitioners, because of the potentially serious consequences of faking on effectively selecting individuals into organizations (e.g., Mueller-Hanson et al., 2003). Given research evidence related to the prevalence (e.g., Donovan et al., 2003) and potential effects (e.g., Rosse et al., 1998) of response distortion in applicant settings, continued research focusing on methods of dealing with this distortion appears warranted. This should be done through substantive means: No amount of linear correction for faking based on social desirability scores (or any other quantitative index of faking) will improve the mean performance of those selected by any practical amount (Schmitt & Oswald, in press). For instance, despite receiving substantial criticism (e.g., Johnson et al., 1988), recent empirical research (e.g., Christiansen et al., 2005; Jackson et al., 2000) suggests that FC personality measure formats represent a promising means of reducing the potential for faking.

This review explored both conceptual and practical issues associated with the use of FC personality measures, focusing on item format, psychometric properties, applicant faking, criterion-related validity, and applicant reactions. Traditional approaches to FC measurement clearly lead to serious psychometric limitations, but several findings suggest that for some FC measures, it may nonetheless be possible and practical to make the interindividual comparisons that are necessary for making selection decisions. Furthermore, recent work (e.g., McCloy et al., 2005; Stark et al., 2005) indicates that it may be possible to retrieve normative information from multidimensional FC items. The available evidence regarding faking suggests that FC measures are not immune to intentional response distortion; however, they appear to be more resistant to this distortion than measures using more traditional item formats, at least at the group level. Results concerning criterion-related validity appear to be mixed: Some research suggests that FC measures have equal or even higher validity compared with non-FC measures in situations involving faking, but other studies have revealed decrements in validity. Further research is required to uncover the factors related to these different effects. Finally, there is little evidence comparing test-taker reactions to FC formats versus traditional item formats. Previous work on test-taker perceptions suggests such a comparison may favor traditional formats, but these are in within-subjects designs, when in practice both measures would not be given to the same applicant.

In all, these findings suggest that researchers and practitioners should continue to explore FC formats as a promising approach to dealing with faking on personality measures. Clearly, there are tradeoffs associated with the use of FC measures and the evidence regarding expected benefits is not universally positive. Furthermore, development of a well designed FC measure can be a difficult undertaking, involving several issues and considerations beyond those involved in developing measures using more traditional formats. However, the available evidence suggests that in some circumstances, FC measures may provide practical benefits over non-FC measures. Thus, continued research on this format is clearly needed, and the use of FC measures in applied settings requires careful weighing of the potential benefits and limitations associated with these measures.

REFERENCES

Baron, H. (1996). Strengths and limitations of ipsative measurement. *Journal of Occupational and Organizational Psychology, 69*, 49-56.

Barrick, M. R., & Mount, M. K. (1991). The big five personality dimensions and job performance: A meta-analysis. *Personnel Psychology, 44*, 1-26.

Barrick, M. R., & Mount, M. K. (1996). Effects of impression management and self-deception on the predictive validity of personality constructs. *Journal of Applied Psychology, 81*, 261-272.

Bartram, D. (1996). The relationship between ipsatized and normative measures of personality. *Journal of Occupational and Organizational Psychology, 69*, 25-39.

Bartram, D. (2005). *Increasing validity with forced-choice item formats*. Unpublished manuscript.

Bass, B. M. (1957). Faking by sales applicants of a forced-choice personality inventory. *Personnel Psychology, 41*, 403-404.

Borislow, B. (1958). The Edwards Personal Preference Schedule and fakability. *Journal of Applied Psychology, 42*, 22-27.

Borman, W. C., & Motowidlo, S. J. (1993). Expanding the criterion domain to include elements of contextual performance. In N. Schmitt & W. C. Borman (Eds.), *Personnel selection in organizations* (pp. 71-98). San Francisco: Jossey-Bass.

Borman, W. C., & Motowidlo, S. J. (1997). Task performance and contextual performance: The meaning for personnel selection research. *Human Performance, 10*, 99-109.

Bowen, C. C., Martin, B. A., & Hunt, S. T. (2002). A comparison of ipsative and normative approaches for ability to control faking in personality questionnaires. *The International Journal of Organizational Analysis, 10*, 240-259.

Christiansen, N. D., Burns, G. N., & Montgomery, G. E. (2005). Reconsidering forced-choice item formats for applicant personality assessment. *Human Performance, 18*, 267-307.

Christiansen, N. D., Goffin, R. D., Johnston, N. G., & Rothstein, M. G. (1994). Correcting the 16PF for faking: Effects on criterion-related validity and individual hiring decisions. *Personnel Psychology, 47*, 847-860.

Closs, S. J. (1996). On the factoring and interpretation of ipsative data. *Journal of Occupational and Organizational Psychology, 69*, 41-47.

Coombs, C. H. (1964). *A theory of data*. New York: Wiley.

Cornwell, J. M., & Dunlap, W. P. (1994). On the questionable soundness of factoring ipsative data: A response to Saville & Willson (1991). *Journal of Occupational and Organizational Psychology, 67*, 89-100.

Donovan, J. J., Dwight, S. A., & Hurtz, G. M. (2003). An assessment of the prevalence, severity, and verifiability of entry-level applicant faking using the randomized response technique. *Human Performance, 16*, 81-106.

Dunnette, M. D., McCartney, J., Carlson, H. C., & Kirchner, W. K. (1962). A study of faking behavior on a forced-choice self-description checklist. *Personnel Psychology, 15*, 13-24.

Ellingson, J. E., Sackett, P. R., & Hough, L. M. (1999). Social desirability corrections in personality measurement: Issues of applicant comparison and construct validity. *Journal of Applied Psychology, 84*, 155-186.

Gilliland, S. W. (1993). The perceived fairness of selection systems: An organizational justice perspective. *Academy of Management Review, 18*, 694-734.

Gordon, L. V. (1951). Validities of the forced-choice and questionnaire methods of personality measurement. *Journal of Applied Psychology, 35*, 407-412.

Guion, R. M., & Gottier, R. F. (1965). Validity of personality measures in personnel selection. *Personnel Psychology, 18*, 135-164.

Hausknecht, J. P., Day, D. V., & Thomas, S. C. (2004). Applicant reactions to selection procedures: An updated model and meta-analysis. *Personnel Psychology, 57*, 639-683.

Heggestad, E. D., Morrison, M., Reeve, C. L., & McCloy, R. A. (2006). Forced-choice assessments of personality for selection: Evaluating issues of normative assessment and faking resistance. *Journal of Applied Psychology, 91*, 9-24.

Hicks, L. E. (1970). Some properties of ipsative, normative, and forced-choice normative measures. *Psychological Bulletin, 74*, 167-184.

Hogan, R. (1991). Personality and personality measurement. In M. D. Dunnette & L. M. Hough (Eds.), *Handbook of industrial and organizational psychology* (Vol. 2, 2nd ed., pp. 55-89). Palo Alto, CA: Consulting Psychologists.

Hough, L. M., Eaton, N. K., Dunnette, M. D., Kamp, J. D., & McCloy, R. A. (1990). Criterion-related validities of personality constructs and the effect of response distortion on those validities. *Journal of Applied Psychology, 75*, 581-595.

Hough, L. M., & Oswald, F. L. (2000). Personnel selection: Looking toward the future—Remembering the past. *Annual Review of Psychology, 51*, 631-664.

Hough, L. M., & Oswald, F. L. (2005). They're right, well ... mostly right: Research evidence and an agenda to rescue personality testing from 1960s insights. *Human Performance, 18*, 373-387.

Hough, L. M., & Schneider, R. J. (1996). Personality traits, taxonomies, and applications in organizations. In K. R. Murphy (Ed.), *Individual differences and behavior in organizations* (pp. 31-88). San Francisco: Jossey-Bass.

Jackson, D. N., Wroblewski, V. R., & Ashton, M. C. (2000). The impact of faking on employment tests: Does forced choice offer a solution? *Human Performance, 13*, 371-388.

Johnson, C. E., Wood, R., & Blinkhorn, S. F. (1988). Spuriouser and spuriouser: The use of ipsative personality tests. *Journal of Occupational Psychology, 61*, 153-162.

Longstaff, H. P., & Jurgensen, C. E. (1953). Fakability of the Jurgensen Classification Inventory. *Journal of Applied Psychology, 37*, 86-89.

Maher, H. (1959). Studies of transparency in forced-choice scales: I. Evidence of transparency. *Journal of Applied Psychology, 43*, 275-278.

Mais, R. D. (1951). Fakability of the classification inventory scored for self confidence. *Journal of Applied Psychology, 35*, 172-174.

Martin, B. A., Bowen, C. C., & Hunt, S. T. (2002). How effective are people at faking on personality questionnaires? *Personality and Individual Differences, 32*, 247-256.

McCloy, R. A., Heggestad, E. D., & Reeve, C. L. (2005). A silk purse from the sow's ear: Retrieving normative information from multidimensional forced-choice items. *Organizational Research Methods, 8*, 222-248.

Meade, A. W. (2004). Psychometric problems and issues involved with creating and using ipsative measures for selection. *Journal of Occupational and Organizational Psychology, 77*, 531-552.

Motowidlo, S. J., Borman, W. C., & Schmit, M. J. (1997). A theory of individual differences in task and contextual performance. *Human Performance, 10*, 71-83.

Motowidlo, S. J., & Van Scotter, J. R. (1994). Evidence that task performance should be distinguished from contextual performance. *Journal of Applied Psychology, 79*, 475-480.

Mueller-Hanson, R., Heggestad, E. D., & Thornton, G. C. (2003). Faking and selection: Considering the use of personality from select-in and select-out perspectives. *Journal of Applied Psychology, 88*, 348-355.

Norman, W. T. (1963). Personality measurement, faking, and detection: An assessment method for use in personnel selection. *Journal of Applied Psychology, 47*, 225-241.

Ones, D. S., Viswesvaran, C., & Reiss, A. D. (1996). Role of social desirability in personality testing for personnel selection: The red herring. *Journal of Applied Psychology, 81*, 660-679.

Ones, D. S., Viswesvaran, C., & Schmidt, F. L. (1993). Comprehensive meta-analysis of integrity test validities: Findings and implications for personnel selection and theories of job performance. *Journal of Applied Psychology, 78*, 679-703.

Rosse, J. G., Stecher, M. D., Miller, J. L., & Levin, R. A. (1998). The impact of response distortion on preemployment personality testing and hiring decisions. *Journal of Applied Psychology, 83*, 634-644.

Ryan, A. M., & Ployhart, R. E. (2000). Applicants' perceptions of selection procedures and decisions: A critical review and agenda for the future. *Journal of Management, 26*, 565-606.

Saville, P., & Willson, E. (1991). The reliability and validity of normative and ipsative approaches in the measurement of personality. *Journal of Occupational Psychology, 64*, 219-238.

Schmidt, F. L., & Hunter, J. E. (1998). The validity and utility of selection methods in personnel psychology: Practical and theoretical implications of 85 years of research findings. *Psychological Bulletin, 124*, 262-274.

Schmit, M. J., & Ryan, A. M. (1993). The Big Five in personnel selection: Factor structure in applicant and nonapplicant populations. *Journal of Applied Psychology, 78*, 966-974.

Schmitt, N., & Oswald, F. L. (in press). The impact of corrections for faking on the validity of noncognitive measures in selection settings. *Journal of Applied Psychology*.

Stark, S., Chernyshenko, O. S., & Drasgow, F. (2005). An IRT approach to constructing and scoring pairwise preference items involving stimuli on different dimensions: The multi-unidimensional pairwise-preference model. *Applied Psychological Measurement, 29*, 184-203.

Tenopyr, M. L. (1988). Artifactual reliability of forced-choice scales. *Journal of Applied Psychology, 73*, 749-751.

Tett, R. P., Jackson, D. N., & Rothstein, M. G. (1991). Personality measures as predictors of job performance: A meta-analytic review. *Personnel Psychology, 44*, 703-742.

Villanova, P. V., Bernardin, J. H., Johnson, D. L., & Dahmus, S. A. (1994). The validity of a measure of job compatibility in the prediction of job performance and turnover of motion picture theater personnel. *Personnel Psychology, 47*, 73-90.

Viswesvaran, C., & Ones, D. S. (1999). Meta-analysis of fakability estimates: Implications for personality measurement. *Educational and Psychological Measurement, 59*, 197-210.

CHAPTER 11

THE USE OF WARNINGS TO DISCOURAGE FAKING ON NONCOGNITIVE INVENTORIES

Victoria L. Pace and Walter C. Borman

Research findings on the validity of personality measures in predicting job performance (Barrick & Mount, 1991; Barrick, Mount, & Judge, 2001; Tett, Jackson, & Rothstein, 1991) have resulted in their increasing use in organizational settings. Of course, the use of personality measures for prediction of job-related criteria has not always received wide support. Challenges to the use of these tests, such as those voiced by Guion and Gottier (1965) and Mischel (1968), led to the unpopularity of these measures among researchers for an extended period of time.

One of the most controversial uses of noncognitive measures in the workplace today is in the selection of employees. There are many who question the utility and even ethics of using personality in a selection context (Campion, Dipboye, Hollenbeck, Murphy, Ryan, & Schmitt, 2004). As we will discuss, there are clearly concerns that must be considered in the use of such tests. However, it is our view that the use of measures that have been shown to have incremental validity for prediction of performance is justified when one considers the alternative. In other words,

A Closer Examination of Applicant Faking Behavior, 283–304
Copyright © 2006 by Information Age Publishing
All rights of reproduction in any form reserved.

inclusion of a valid though less-than-perfect predictor is preferable to the limitation inherent in a smaller set of predictors. For example, the use of fewer predictors may have the drawback of reduced coverage of the job performance criterion space. We and others (e.g., Borman, Penner, Allen, & Motowidlo, 2001; Hurtz & Donovan, 2000; Motowidlo & Van Scotter, 1994) have demonstrated that personality variables are better predictors of contextual or citizenship performance than task performance. On the other hand, general cognitive ability appears to be a better predictor of task performance compared to citizenship performance (Motowidlo, Borman, & Schmit, 1997). Thus, if we are interested in predicting citizenship performance as well as task performance, considering personality in the mix of predictors seems warranted.

In addition, the reduction of adverse impact realized through the use of personality measures as compared to cognitive ability tests (Hogan & Roberts, 2001) is an important practical benefit of these measures. Few would advise the use of personality measures alone for important selection decisions. However, as one component of a comprehensive selection battery, such tests can be useful for gaining a well-rounded view of job applicants and enhancing selection decision-making processes.

In sum, personality assessment is clearly useful for organizational as well as individual purposes. Improving the assessment process, as well as the assessment tools themselves, should be our next focus as we strive to optimize validity for prediction of important criteria.

CONCERNS ABOUT PERSONALITY MEASURES IN THE WORKPLACE

In order to make improvements to noncognitive assessment, we must seriously consider the arguments put forth by those who discount the usefulness of these assessments. Because of the important consequences for individual lives and careers as well as organizational interests that are sometimes greatly affected by test results, many concerns have been raised about the use of tests in the workplace, including those that measure personality.

One concern mentioned by some, is the lower level of validity typically observed for personality measures than for cognitive measures. Although the .30 ceiling for validity of noncognitive measures has been extensively discussed (e.g., Mischel, 1968), it appears that this ceiling can be exceeded when corrections for methodological artifacts are used or if we improve measurement procedures directly. One promising direction for improvement is through increased work specificity of test items. Studies have shown that personality items related to the work or school context have higher validity for prediction of performance in those contexts than do items of a more general nature (Pace, 2005; Schmit, Ryan, Stierwalt, & Powell, 1995). Even if noncognitive measures do not reach the validity of cognitive measures, as mentioned, personality measures can be expected to predict some nonoverlapping criteria (i.e., citizenship performance). Furthermore, it can be shown that even with relatively low levels of validity, improvements in prediction can be substantial. As an illustration, Hogan and Roberts (2001) point out that given a base rate of .50 (i.e., 50% of the applicant pool has an acceptable level of the trait or ability of interest), a validity coefficient of .20 increases the rate of correct decisions from 50% to 60%. A validity coefficient of .40 increases the probability of correct decisions to 70%.

Others have objected to the use of personality tests in selection on the grounds that the amount of variance in criteria such as job performance that is attributable to situation and person x situation effects is of such magnitude that examining only individual differences without looking at the situation is inappropriate (e.g., Mischel, 1968). A small number of studies have taken situation into account. Beaty, Cleveland, and Murphy (2001) found that personality predictors were more valid in a setting with relatively weak situational demands compared to a setting with stronger situational demands. In addition, Hogan, Rybicki, Motowidlo, and Borman (1998) found that agreeableness was a good predictor of job performance in an organization where "getting along" was valued, and ambition was a superior predictor of performance when "getting ahead" was more valued. Thus, the organizational context can have an effect on the validity of personality for predicting job performance, but it can also *increase* our understanding of personality-job performance linkages.

Most relevant to this book, concerns about the use of personality measures exist due to the possible influence of response distortion or faking (Christiansen, Goffin, Johnston, & Rothstein, 1994). Although test-taker attempts to appear either more or less favorable than justified can each lead to poor selection and other employment decisions, distortion in the favorable direction is more likely in most employment contexts. Faking in a socially desirable direction has been referred to as impression management, response distortion, social desirability, self-enhancement, intentional distortion and dissimulation (McFarland & Ryan, 2000).

FAKING AND ITS CONSEQUENCES

Research on Faking

Faking or response distortion on noncognitive measures has been extensively explored. Not only has research indicated that people have the ability to fake when instructed to do so (Furnham, 1986; Hough, Eaton, Dunnette, Kamp, & McCloy, 1990; Rees & Metcalfe, 2003), it appears that applicants often do distort their responses in employment settings (e.g., Donovan, Dwight, & Hurtz, 2003). According to Viswesvaran and Ones (1999), the mean of applicant samples is generally 0.48 to 0.65 standard deviations higher than that of incumbent samples.

There is a lack of consensus, however, on the effect of faking. Whereas faking is sometimes found not to seriously affect criterion-related validities of personality measures in work settings (Hough, 1998; Ones, Viswesvaran, & Reiss, 1996), it has also been shown to decrease variance and test validity (Topping & O'Gorman, 1997). In addition, applicant faking has been shown to introduce more error for those at the high end of the predictor distribution (Arthur, Woehr, & Graziano, 2001; Mueller-Hanson, Heggestad, & Thornton, 2003) and may cause changes in the rank order, especially at the top of the test score distribution. This outcome can substantially impact which persons are selected, most seriously when the selection ratio is small (Rosse, Stecher, Miller, & Levin, 1998). Therefore, most researchers and practitioners concerned with personality assessment and selection recommend taking precautions to prevent or reduce faking on these measures (Hough & Ones, 2002; McFarland, 2003). However, depending on the testing situation and use of test results, faking is more problematic in some scenarios than in others.

Concerns About Faking in Specific Testing Situations and for Specific Test Uses

Certain situations exist in which faking is not of such great import. Specifically, faking is not a very important threat when motivation to fake is likely to be low. This will be the case when assessment results are not tied to important outcomes. For example, when results are used to foster understanding and communication among team members and to enhance team effectiveness, there should be little motivation to slant responses.

Individual motivation to fake might also be low when applicants feel they possess the necessary characteristics for job attainment. When applicants are confident of a positive outcome if they provide honest responses, they are unlikely to distort responses. Normally, this would be expected to apply to some applicants and not to others. However, this might be the case for entire groups of applicants when selection ratios are high or when the job market is favorable for workers. Of course, selection testing is not as useful in this situation, as there is low utility for the use of tests in a high selection ratio setting.

Another scenario where faking is not so great a threat is when noncognitive measures are used for de-selection rather than top-down selection (Mueller-Hanson et al. 2003), and are followed by additional selection procedure components. Even if applicants are motivated to "fake good," Mueller-Hanson et al. have shown that the criterion-related validity of the predictor measure may be relatively high for the lower third of the predictor distribution ($r = .45$, $p < .05$, in their study). This contrasts with the nonsignificant validity found for the upper third of the predictor distribution in Mueller-Hanson et al.'s "fake good" group ($r = .07$, $n.s.$). Thus, lower scoring applicants can be more confidently categorized as lacking necessary levels of the characteristics that are important for job performance. This is not to say that all individuals with low true scores will be successfully screened out this way. Additional selection components may be helpful in eliminating less desirable applicants who have managed to make their way into further stages of the selection process.

Unlike the examples just discussed, faking is very consequential in certain other situations. Of greatest concern is the occurrence of faking when selection ratios are low and a top-down selection procedure is being followed. Assuming that there are individual differences in the extent of faking behavior (Robie, Born, & Schmit, 2001; Rosse et al. 1998), the rank ordering of applicant scores is likely to change when applicants are motivated to fake. This will impact who is selected using a top-down procedure, to the point that fakers are likely to predominate in the group of selected applicants, particularly if the selection ratio is low (Rosse et al.)

Faking of noncognitive tests for classification or assignment after selection is slightly less worrisome. In classification, faking is in a sense more complex and difficult for the test-taker. Faking requires test-takers to slant their responses in the direction of a particular job's personality requirements rather than simply trying to "look good" in general.

Although not all situations call for intensive efforts to deter or detect faking of personality assessments, there are many circumstances in which faking is a major consideration, justifying a call for more extensive research into methods to discourage its occurrence and to address its possible effects. It is these circumstances, particularly the selection context, which we primarily consider.

Model of Faking

To guide systematic study on how and why faking occurs, McFarland and Ryan (2000) proposed a model (see Figure 11.1). The model not only links beliefs about faking, intention to fake, and faking behavior, but also incorporates situational influences, ability to fake, and opportunity to

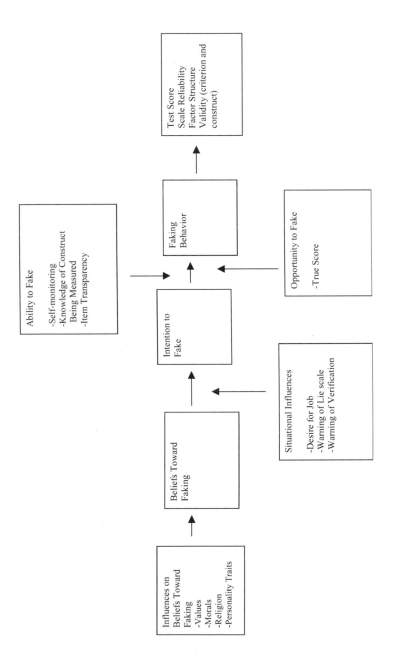

Source: McFarland and Ryan (2000). Reprinted with permission.

Figure 11.1.   A model of faking. From "Variance in Faking across Noncognitive Measures."

fake. Specific influences on beliefs toward faking listed by McFarland and Ryan include personal values, morals, religion, and personality traits. Situational influences include desire for the job, warning of a lie scale, and warning that responses might be checked and verified. Factors influencing the ability to fake include high self-monitoring, knowledge of the construct being measured, and high item transparency. Individuals' opportunity to fake is expected to be dependent on their true scores (McFarland & Ryan). In other words, individuals with high true scores cannot improve their scores much by faking. Therefore, these individuals do not have as great an opportunity to "fake good" as individuals whose true scores are lower.

The McFarland and Ryan model provides a convenient structure for the discussion of theory and research findings relevant to faking. In a correlational study, Seiler and Kuncel (2005) examined relationships between a number of variables and intention to fake. Moral convictions were the best predictor of intention to fake (correlation in the negative direction.) This result is consistent with the influence of morals on beliefs toward faking that is hypothesized by McFarland and Ryan. Using measures adapted from McFarland (as cited in Seiler & Kuncel), Seiler and Kuncel also found that attitudes toward faking, subjective norms, and perceived control were related to intention to fake. Belief in detection was found to correlate negatively with intention to fake. This result is quite relevant to our discussion of warnings later in this chapter.

From a theoretical point of view, understanding antecedents and consequences of faking on noncognitive measures, and the mechanisms involved, is of particular interest. From a practical view, we have made the point that faking is of concern to organizations.

Addressing the Faking Problem

Previous attempts to reduce faking have employed a variety of strategies that can be seen as targeting different aspects of the McFarland and Ryan (2000) model. First, statistical correction can be performed. This procedure attempts to fix the effects of faking after faking behavior has occurred, and is commonly done using lie or social desirability scales (Ellingson, Sackett, & Hough, 1999). The considerable use of such faking corrections was indicated in Goffin and Christiansen's (2003) survey, as the majority of researchers who participated in their study favored using such scales. However, as Goffin and Christiansen cautioned, these scales may not adequately measure response distortion. In fact, personality scores corrected for faking have not shown significantly improved validity (Ellingson et al., 1999; Ones et al. 1996; Schmitt & Oswald, in press). Second, relying on other predictors such as structured interviews in addition to personality inventories can reduce "ability to fake," which influences the link from intentions to actual test-taking behaviors. Fortunately, including multiple predictors in a context as critical as employee selection is not only sound advice, but also a generally accepted practice.

Using nontransparent items has also been suggested as a way to reduce ability to fake (e.g., McFarland & Ryan, 2000). However, some research has shown lower validities for nontransparent items as compared to more face-valid items (e.g., Burkhart, Gynther, & Fromuth, 1980). Additionally, nontransparent items are sometimes perceived by test-takers as lacking job relevance, thereby possibly impacting fairness perceptions and cooperation. Some researchers have found the use of forced-choice item formats to be an acceptable method (e.g., Christiansen, Burns, & Montgomery, 2005). Finally, warnings against faking have been employed in survey instructions to decrease the intention to fake. Although a fair amount of research has examined the earlier mentioned methods and their effectiveness in dealing with faking, it is beyond the scope of this chapter to cover this research in depth. Instead, it is this last strategy, the use of warnings, which we will address in greater depth.

RESEARCH ON THE USE OF WARNINGS TO DETER FAKING

Although warnings are most clearly associated with situational influences in that they affect the testing situation directly, warnings can also be seen as having effects on earlier model components, because they may influence the test-taker's cognitions about the testing process and their underlying beliefs about whether faking will produce positive results. These cognitions are likely to influence the test-taker's intention to fake. Past research has clearly shown that intention is a good predictor of behavior (Fishbein & Ajzen, 1975). Therefore, faking behavior should decrease if warnings lead to negative beliefs about faking and decreased intention to fake.

Empirical research has generally shown that warnings do deter faking (e.g., Dwight & Donovan, 2003; McFarland, 2003). In most research, effectiveness was evidenced by a significant difference in group-level mean scores between warned and unwarned conditions, though some studies have revealed interesting findings about individual behavior using within-subjects designs (e.g., Griffith, Yoshita, Gujar, Malm, & Socin, 2005). Griffith et al. found that the effects of warnings on individual faking may be partially masked by the existence of individuals whose scores are actually lower in the warned applicant condition than in the honest condition. These lower scores would tend to counterbalance scores from persistent fakers who exhibit higher scores in the applicant than the honest condition regardless of the presence of a warning. For this reason, these researchers suggest that reducing the percentage of applicants who fake may be more effective than reducing mean scores at the group level if the primary goal is to improve selection decisions.

Dwight and Donovan (2003) computed a meta-analytic sample size-weighted average of effect sizes from 10 studies that looked at the effectiveness of warnings. This computation yielded a weak to modest effect size ($d = .23$), indicating lower predictor scores for warned test-takers as compared to unwarned test-takers. However, the included studies used a variety of warning types. This fact may help to explain some of the variation in individual study effect sizes as well as the fairly small average effect size.

Types of Warnings

Most warnings used in testing situations to this point include a detection warning, a consequence warning, or both. In a detection-type warning, the test-taker is informed of the existence of lie detection methods such as inclusion of a social desirability scale or verification of responses with other sources. "We will catch you" is the main point. In the McFarland and Ryan (2000) model, detection warnings can be expected to lower test-taker confidence in their ability to fake successfully, thereby leading to a likely decrease in intention to fake. This warning is sometimes, but not always, combined with explicit consequences of being "caught."

In a consequence warning, specific consequences of dishonest responding are outlined. "You will be punished" is the focus here. Usually these consequences involve invalidation of individual test results and can include elimination from the applicant pool. Warnings of negative consequences might decrease intention to fake as warnings of detection do, but by a somewhat different process. The warning must elicit doubt on the part of the test-taker about their ability to fake undetected, and then induce some level of avoidance of the accompanying consequences.

A combination type warning, including both detection and consequence elements, has also been used. Research exploring the effectiveness of warnings has found that different types of warnings have different effects on faking. Dwight and Donovan (2003) found that a warning indicating faking could be detected, and that there were consequences for faking, produced significantly lower scores and a reduced number of "fakers" as compared to an unwarned condition. However, scores in conditions that included only part of this warning, either the detection part or the consequence part, did not differ significantly from scores produced by the unwarned condition.

We recently developed warning instructions that focused more on test-taker interests than organizationally delivered consequences (Pace, Xu, Penney, Borman, & Bearden, 2005). Rather than indicating distrust of applicants by warning them of detection of faking and corresponding punishment, a "let's reason together" approach included in this study used a more positive tone that encouraged applicants to consider their best interests by responding honestly. The instructions point out to the applicant that slanting responses might be detrimental to their long-term goals because faking might result in getting into a job the applicant is not very good at or may not enjoy. This warning content may be most effectively used in a setting in which a variety of job openings exist. The results of our study found that mean personality test scores under this instruction set were about the same as mean scores with "for-research-only" instructions. Future research in this direction needs to compare means for the "let's reason together" instructions with those obtained in an operational setting and, more importantly, the criterion-related validity against job performance for each of these conditions. The "let's reason together" type of warning probably addresses beliefs about faking such that faking is viewed as not helpful for achieving personal goals. This negative belief toward faking can then be expected to decrease or eliminate intention to fake.

Similar to reasoning with the test-taker, another approach might clearly explain the purposes for testing and how this is related to the applicant's best interests. We might call this an educational instruction set because it attempts to educate the test-taker about the positive aspects of testing. Perhaps fairness perceptions of the testing process can be enhanced and reductions in faking achieved by further explanation of how testing contributes to better selection decisions. Based on the norm of reciprocity (Gouldner, 1960), we might hypothesize that applicants will form intentions to respond with honesty in return for fair, open, and honest treatment by the organization. However, too much information about the personality constructs being targeted may be detrimental, as some research has shown that individuals more knowledgeable about what is being measured by the test are better able to fake responses successfully (Cunningham, Wong, & Barbee, 1994; Dwight & Alliger, 1997). This kind of test-related information might increase faking by influencing the ability-to-fake component of the McFarland and Ryan (2000) model. Accordingly, the strategy of providing more information about the goals and purpose of the selection process should probably stop short of identifying and describing the constructs being measured.

Another approach might include the use of warnings that make moral conviction more salient. Based on the Seiler and Kuncel (2005) result of a relationship between moral conviction and intention to fake, we might expect such warnings to be effective in deterring faking. It is not clear exactly how instructions should be developed to influence applicants' moral attitudes toward the test-taking process. Table 11.1 summarizes the different types of warnings.

It may be that warnings that address more aspects of the faking model antecedents are more effective in reducing faking behavior, particularly at the group level. This could be one reason why Dwight and Donovan (2003) found greater effectiveness for warnings that include both detection and consequences over those that included only one of these components. It might be helpful to incorporate warning components that influence individuals' intrinsic beliefs about the usefulness of faking, as well as components that outline extrinsic factors that bear on perceived ability to fake.

Table 11.1.   Some Current and Potential Instruction Set/Warning Types

Instruction Set/Warning Type	Description	Affected Model Component (and Possible Motivational Aspect)
Detection	Alerts test takers to the presence of methods such as lie scales or checking with other sources to detect inaccurate responses.	Intention to fake (Reduced confidence in ability to fake successfully)
Consequence	Provides information about what will happen if response distortion is suspected.	Intention to fake (Avoidance of negative consequences)
Reasoning	Uses a friendly tone to encourage test-takers to consider that honest responses will lead to a job that fits them best.	Beliefs toward faking—not useful Intention to fake (Seeking most positive outcome over time)
Educational	Explains the reasons for testing and how accurate test responses aid the organization in enhancing the fairness of selection decisions.	Beliefs toward faking—not fair Intention to fake (Desire to reciprocate with fairness and honesty toward organization)
Moral conviction	Increases the salience of test-takers' associations of honesty with appropriate behavior and positive self-image.	Beliefs toward faking—not moral Intention to fake (Maintenance of positive self-image)

Perhaps the wording of the warning used can activate additional, more distal antecedents, such as values. Considering such distal inputs may increase warning effectiveness because these model components may affect all subsequent ones. However, warnings aimed at influencing proximal antecedents to faking behavior, such as intention to fake, may yield stronger effects on behavior than those that seek to influence only distal antecedents. Snell and Fluckinger provide insight to the potential interaction of proximal and distal mechanisms of faking in chapter 7 of this book. Future research using a greater variety of warning approaches should help to clarify this question.

With the exception of the Dwight and Donovan (2003) study and the one by Pace et al. (2005), mentioned above, there has been little research about exactly how warnings should be delivered or what content they should include for maximal effectiveness. For the most part, research is still exploratory in that we seem to be at the stage of "let's try this intervention and see what happens." Further studies are now needed using a greater variety of warning components and formats in order to better understand how these treatments affect test-taker propensity toward truthful or distorted responses. As we continue exploration into new types of warnings, we should also examine faking behavior more closely, develop interventions aimed at addressing specific antecedents of this behavior, and then test the effectiveness of these interventions in realistic selection contexts.

The mechanisms through which warnings work and theoretical guidance regarding how we might improve these warnings must be further examined. Although the McFarland and Ryan (2000) model outlines key components of the process around faking behavior, additional theoretical work is needed to delineate the cognitions leading to faking behavior and how these might be altered to reduce the likelihood of faking. Focus groups or think-aloud procedures might provide useful information about applicant beliefs and intentions to fake. Specific formulations of warning components and formats might effectively impact certain links between test-taker beliefs and cognitions about faking and their intentions to fake. We may gain a clearer view of test-taker cognitions and how interventions such as warnings interact with faking model components by considering the role of motivation.

Motivation and Warnings

Regardless of how effectively warnings address certain antecedents of faking behavior, applicants may still decide that other factors favorable to faking are more compelling. For example, applicants who desperately need a job may be willing to take their chances at being caught and may feel that they have too little to lose to be concerned with consequences that are outlined in the warning. These applicants may not be overly concerned with their long-term goals even if they are convinced that qualifying fraudulently for a job is not in their best interest over the long run. As support for this point, Schmit and Ryan (1992) found reduced validity of noncognitive measures among those with greater motivation to do well than among those with less at stake. There will probably always be some individuals who are sufficiently motivated to fake despite our best efforts. A good next step is to seek greater understanding of the likely motivators of faking behavior and how factors that motivate individuals toward honest responding can be made more salient for applicants and given more weight than factors leading to faking.

Most warnings implicitly encourage test-takers to project into the future the likely consequences of faking. Usually the consequences suggested are negative. In the case of warnings that merely state that faking can be detected, it is somewhat unclear what the consequence of faking might be after detection. Warnings may be most effective when possible consequences of faking are made clear and are very meaningful to those who take the test. In reality, there is often a complex mix of positive and negative consequences for the test-taker. Consequences are attached to both faking and not faking. In a selection context, appearing to be the perfect job candidate in order to obtain the position is the positive consequence that test-takers probably have in mind. The test-taker must estimate the likelihood that faking will assist this goal and then weigh the value of this option against the possible negative consequences of faking, often outlined in warnings. As with other decisions, the expected utility of each course of action and the value of each possible outcome influence the choice that is made. Effective warnings should lower the expected utility of faking or raise the value of honest response.

Related to positive and negative consequences, two questions worth considering are: (1) do certain types of warnings induce a temporary promotion focus or prevention focus in test-takers (Higgins, 1987; Shah, Higgins, & Friedman, 1998); and (2) what is the implication for the effectiveness of these warnings? In other words, a warning may encourage a test-taker to form positive goals and focus on achievement of positive outcomes, or a warning may induce formation of goals that focus on avoidance of punishing outcomes. Determining which type of motivation is more effective in deterring faking has important implications for the framing of warning sets.

Detection and consequence warnings are likely to focus individuals' thoughts on preventive strategies. Motivation to avoid the negative consequences implied or explicitly stated by the warnings is meant to lead to a decrease in faking. A reasoning type warning may be made more promotive in nature by focusing the test-taker on obtaining valued outcomes such as good job fit through honest responding.

Further exploration of the effectiveness of positive approaches in which clearly outlined benefits of honest responding are the focus rather than threats of detection or external punishment could be informative. Griffith et al. (2005) thought a less punitive warning might reduce the effect of a conservative response bias that they found many individuals exhibited when warned of negative consequences. In other words, some individuals appeared to become overly modest about their own good qualities, resulting in lower scores in the applicant warning condition than in a later honest condition. It is possible that they feared the consequences to such a degree that they presented themselves as lower on the tested traits than they actually were in an effort to avoid elimination from the applicant pool. Although warnings generally have not addressed positive consequences of being honest in the past, it is probably possible to make such positive consequences more salient through the framing of the warning.

Two related areas deserving more attention are self-regulation (Bandura, 2001) and goal setting (Locke & Latham, 2002). Perhaps we should look further into applicant goals and attempt to predict their cognitions concerning faking. If we can counter test-taker strategies that are based on positive beliefs about faking and intentions to fake, we may have a better chance of, essentially, talking applicants out of faking. We might also want to know how applicants evaluate their performance as they are progressing through a personality selection instrument. Would it help to incorporate warning reminders into the test rather than only at the beginning? Although it is probably unnecessary to remind test-takers of any deeply held personal beliefs about honesty in test-taking, occasional "boosts" into salience may be required to maintain newer, subtler, or more temporary perceptions such as those involving mutual trust and fairness between the organization and the applicant, or a positive focus on achieving the best person-job fit. Such "reminders" might be placed at the bottom of some pages and could take a positive tone as in "Help us find your perfect job—please answer accurately" or one that is less positive such as "We reserve the right to verify responses."

Using a different approach, warnings could make the virtue of honesty more salient, perhaps bringing to mind ideal self-standards (Higgins, 1987). Bandura (2002, p. 101) states, "The self-regulatory mechanisms governing moral conduct do not come into play unless they are activated." Making self-standards regarding honesty more salient might cause the test-taker to avoid behavior that is discrepant with these standards in an effort to maintain a positive self-image.

Exploring how self-efficacy may directly influence tendencies to fake or may moderate the effectiveness of warnings is another area in need of further research. Seiler and Kuncel (2005) found that applicant perceptions of their own ability correlated negatively with intention to fake. These perceptions involved the individual's beliefs about how his or her honest responses might be perceived by the employer. If these perceptions were favorable, it is not surprising that applicants would feel little need to fake.

Still another way to view the effects of warnings on faking behavior is to attempt through test instructions to increase applicants' levels of personal responsibility toward test responding. When individuals feel that they control processes and outcomes important to their goals, a sense of personal responsibility may be more likely to result. Arguably, an individual-centered reasoning approach may be more successful in evoking this sense of responsibility than organization-centered detection or consequence approaches. How an increased sense of personal responsibility will affect tendencies to respond truthfully or with distortion remains to be researched. But by fostering a sense of mutual respect and caring between the individual and the organization, we might be more successful in our persuasion efforts. Finally, individuals with an internal locus of control may be more responsive to a reasoning approach than those whose locus of control is external.

Among those for whom a sense of autonomy and control is important, the reasoning approach may generate less threat and elicit greater satisfaction with the selection process. This approach is likely to do so by fostering a sense of trust and consideration in the applicant's relationship with the organization. Even those who experience the selection process, but are not ultimately selected, may exit with a better feeling about the respect given to individuals in their dealings with the organization. These perceptions can continue to impact recruiting, marketing, and other relations with the public for much longer than the selection period. For such far-reaching effects, it is important that we examine the perceptions of applicants associated with warnings and noncognitive measures.

Warnings and Participant Perceptions

Some researchers and practitioners have expressed concerns about applicant perceptions of the testing process, specifically related to the use of warnings. An interesting research topic is the effect of warnings on applicant views of the selection process and more general views of the organization. McFarland (2003) explored the effect of a combination (detection and consequences) warning on applicant responses. She found that warning test-takers that a social desirability scale was included and that those with high scores on this scale would not be eligible for a monetary incentive had no negative effect on applicant perceptions. Although this is a reassuring result, we may want to consider possible differential effects on applicant perceptions as we generate a greater diversity of warning components and formats.

It is entirely possible that instead of the testing situation influencing test-taker perceptions of the selection process, rather earlier test-taker perceptions of the organization and the fairness of previous portions of the selection procedure might influence their intentions to fake and subsequent faking behavior. In the Seiler and Kuncel (2005) study mentioned earlier, fairness perceptions correlated negatively with intention to fake. It is important for the deterrence of faking that we continue to consider how applicant perceptions of fairness relate to faking. Does a warning of detection enhance or detract from applicant perceptions of fairness compared to an unwarned condition? Applicants may view warnings about verification as an invasion of privacy, or they may fully accept attempts to check with other sources as part of improving the fairness of the application process. If the warning mentions or implies the existence of a "lie scale," applicants may consider this underhanded, or a favorable indication of technical sophistication. How does inclusion of negative consequences within the warning affect fairness perceptions? Consequences might differ quite a lot in severity. For example, warning about reexamination if test response irregularities are observed is less threatening than warning of removal from the applicant pool. These different levels of severity may affect fairness perceptions as well as other applicant responses. Could a promotive warning that aims to convince applicants that honest responding is in their best interest work as well or better than a preventive (negative consequence-oriented) warning in influencing applicant perceptions? It seems likely that a more positively framed warning could induce a more positive outlook toward the organization.

The Effect of Warnings on Validity

Although it appears clear that at least some types of warnings successfully deter faking, the primary reason so many are concerned about faking is its possible impact on the criterion-related validity of noncognitive measures. Essentially, faking that leads to hiring poor performers is the real threat (see Peterson & Griffith in this volume). Unfortunately, there is a lack of published research comparing criterion-related validity of applicant noncognitive scores in warning and no warning conditions. For this reason, it is imperative that future studies of faking gather criterion-related validity data. Even a small, perhaps undetected, amount of faking by a minority of applicants can produce meaningful differences in rank order, thus impacting who is hired. Assessing and comparing the validity of noncognitive measures in honest, warned applicant, and unwarned applicant conditions is a logical next step for research. One way to determine whether we have met our ultimate goals related to deterrence of faking is to assess whether the use of a warning produces a validity coefficient comparable to that of the honest condition and superior to that of the unwarned applicant condition. It would be particularly helpful to ascertain whether a warning exists or can be developed that consistently improves the criterion-related validity for the upper end of the predictor distribution.

In contrast to criterion-related validity, there have been a few studies that have examined the effect of warnings on construct validity. For example, Vasilopoulos, Cucina, and McElreath (2005) found that the use of a detection (verification) warning increased the correlation of job-relevant personality scores with cognitive ability. These researchers found further evidence (differing response latencies) in support of their hypothesis that warnings increase the cognitive difficulty of the response decision by causing test-takers to develop more complex strategies in order to present themselves positively without being detected. The higher correlations of cognitive ability with personality scores indicate reduced construct validity and a possible undermining of one of the very desirable properties of noncognitive tests, lack of adverse impact.

On the other hand, McFarland (2003) found improved construct validity as evidenced by reduced multicollinearity among noncognitive and social desirability scores when a detection (inclusion of social desirability scale) and consequence (withholding of monetary incentive) combination warning was used.

SUMMARY AND FUTURE RESEARCH SUGGESTIONS

Although concerns about noncognitive measures should be carefully considered, it is important to recognize that noncognitive testing enables us to improve the accuracy of job performance predictions. Although there remains a great deal of work to be done to improve the validity of personality tests, such tests are tools that enhance the quality of selection decisions and should be included in a battery of selection procedures to provide as comprehensive a view of the applicant as possible.

In a number of testing contexts, concerns about faking are not so critical as to seriously reduce the usefulness of noncognitive test results. These situations include the use of test results for individual or interpersonal understanding, training applications, and de-selection. The degree of faking in the classification or job assignment context is dependent on the extent of inequity in the desirability of jobs being assigned, but with classification we would expect less faking in general. When a top-down selection procedure depends on results of a noncognitive measure and the selection ratio is small, faking is of great concern. For this reason, several methods for reducing the impact of applicant faking on critical selection decisions have been implemented in the past. These methods include development of fake-resistant items and situational interventions prior to testing, as well as identification of individuals suspected of response distortion and subsequent correction or invalidation of scores after testing. The use of warnings to deter faking on noncognitive measures has been the primary focus of this chapter.

Several different types of warnings have been studied and used in the past, with differing levels of success. The most researched types are warnings about detection, consequences, or a combination of the two. A relatively new type involves reasoning with the test-taker that faking is not in their best interest. Although current warnings generally show promise for deterrence of faking, they should continue to be studied in both laboratory and applied settings, particularly through studies aimed at assessing and improving their effectiveness. Although we commend the work done with warnings, we believe that other possible approaches to warnings remain to be developed and researched.

We recommend that the role of theory in faking be expanded, as it may contribute valuable insight into potential practical remedies. Although a very useful model of faking has been put forth by McFarland and Ryan (2000), additional ideas borrowed from motivation and other psychological theories such as decision theory should be brought to bear on attempts to address faking. Specifically mentioned were concepts related to goal setting, self-regulation, promotion versus prevention focus, and utility and value perceptions of possible behaviors and outcomes. Also discussed were the impact of positive perceptions of job-related ability on faking behavior, locus of control and personal responsibility, as well as moral conduct standards and maintenance of self-image. With such a wealth of theory on which to draw, a great deal of research using a variety of approaches is possible.

Initially, think aloud procedures during testing or questioning immediately after testing might be used in laboratory studies to investigate cognitions and conscious strategies underlying applicant decisions to fake or respond honestly. A comparison of the cognitions voiced by individuals in various warning conditions could be made with those in a no warning condition. Systematic isolation and combination of warning components, instruction framing, and format are more easily accomplished within a laboratory setting. An additional benefit of laboratory studies might be a better understanding of how clearly and consciously conceptualized applicant strategies are. Based on typical cognitions expressed in these studies, a confidential survey could be developed that would allow researchers to question applicants in a field setting about the strategies they employ when completing noncognitive measures. This type of survey could help to determine the prevalence of certain cognitions and accompanying responses in applicant groups who are given different instruction sets. Within-subjects designs appear likely to produce additional information about individual faking behaviors and how they may change in response to warnings. Once we understand the cognitions and behaviors that exist within applicant settings, we can more effectively develop interventions. Perhaps the ideal warning would be one that counters the most common cognitions favorable to faking and does so by addressing several antecedents to faking behavior, from distal to proximal, as outlined by the McFarland and Ryan (2000) model.

The effect of warnings on applicant perceptions considered by McFarland (2003) is not yet fully examined. Although McFarland found no negative effects on perceptions, more research with different kinds of warnings should be conducted. Any negative perceptions that are triggered could be long lasting and may ultimately influence subsequent job performance, organization-directed citizenship behavior, and general public relations.

Effects of warning on validity, especially criterion-related validity, must also be more thoroughly examined. In fact, this is probably the most critical next step for research on the effectiveness of warnings. Now that we have seen that warnings can lower mean scores to levels closer to honest conditions, it is time to determine whether this is sufficient. In other words, does this effect on mean scores really address our concern about ensuring the quality of selection decisions and accurately predicting future job performance?

In sum, although response distortion on noncognitive measures is of concern to organizations; based on previous literature, several strategies to address faking are currently being examined. Using warnings at the time of test administration appears to be a potentially effective strategy for discouraging faking, but development of additional warning approaches grounded in theory and continued research efforts are recommended.

REFERENCES

Arthur, W. Jr., Woehr, D. J., & Graziano, W. G. (2001). Personality testing in employment settings: Problems and issues in the application of typical selection practices. *Personnel Review, 30,* 657-676.

Bandura, A. (2001). Social cognitive theory: An agentic perspective. *Annual Review of Psychology, 52,* 1-26.

Bandura, A. (2002). Selective moral disengagement in the exercise of moral agency. *Journal of Moral Education, 31,* 101-119.

Barrick, M. R., & Mount, M. K. (1991). The big five personality dimensions and job performance: A meta-analysis. *Personnel Psychology, 44,* 1-26.

Barrick, M. R., Mount, M. K., & Judge, T. A. (2001). Personality and performance at the beginning of the new millennium: What do we know and where do we go next? *International Journal of Selection and Assessment, 9,* 9-30.

Beaty, J. C., Cleveland, J. N., & Murphy, K. R. (2001). The relation between personality and contextual performance in "strong" versus "weak" situations. *Human Performance, 14,* 125-148.

Borman, W. C., Penner, L. A., Allen, T. D., & Motowidlo, S. J. (2001). Personality predictors of citizenship performance. *International Journal of Selection and Assessment, 9,* 52-69.

Burkhart, B. R., Gynther, M. D., & Fromuth, M. E. (1980). The relative predictive validity of subtle vs. obvious items on the MMPI Depression scale. *Journal of Clinical Psychology, 36,* 748-751.

Campion, M. A., Dipboye, R. L., Hollenbeck, J. R., Murphy, K. R., Ryan, A. M., & Schmitt, N. W. (2004, April). In F. P. Morgeson (Chair) *Won't get fooled again? Editors discuss faking in personality testing.* Panel discussion conducted at the annual meeting of the Society for Industrial and Organizational Psychology, Chicago.

Christiansen, N. D., Burns, G., & Montgomery. G. E. (2005). Reconsidering the use of forced-choice formats for applicant personality assessment. *Human Performance, 18,* 267-307.

Christiansen, N. D., Goffin, R. D., Johnston, N. G., & Rothstein, M. G. (1994). Correcting the 16PF for faking: Effects on criterion-related validity and individual hiring decisions. *Personnel Psychology, 47,* 847-860.

Cunningham, M. R., Wong, D. T., & Barbee, A. P. (1994). Self-presentation dynamics on overt integrity tests: Experimental studies of the Reid Report. *Journal of Applied Psychology, 79,* 643-658.

Donovan, J. J., Dwight, S. A., & Hurtz, G. M. (2003). An assessment of the prevalence, severity, and verifiability of entry-level applicant faking using the randomized response technique. *Human Performance, 16,* 81-106.

Dwight, S. A., & Alliger, G. M. (1997). Reactions to overt integrity test items. *Educational and Psychological Measurement, 57,* 937-948.

Dwight, S. A., & Donovan, J. J. (2003). Do warnings not to fake reduce faking? *Human Performance, 16,* 1-23.

Ellingson, J. E., Sackett, P. R., & Hough, L. M. (1999). Social desirability corrections in personality measurement: Issues of applicant comparison and construct validity. *Journal of Applied Psychology, 84,* 155-166.

Fishbein, M., & Ajzen, I. (1975). *Belief, attitude, intention, and behavior: An introduction to theory and research.* Reading, MA: Addison-Wesley.

Furnham, A. (1986). Response bias, social desirability and dissimulation. *Personality and Individual Differences, 7, 385-400.*

Goffin, R. D., & Christiansen, N. D. (2003). Correcting personality tests for faking: A review of popular personality tests and an initial survey of researchers. *International Journal of Selection and Assessment, 11,* 340-344.

Gouldner, A. W. (1960). The norm of reciprocity: A preliminary statement. *American Sociological Review, 25,* 161-178.

Griffith, R., Yoshita, Y., Gujar, A., Malm, T., & Socin, R. (2005). *The impact of warnings on applicant faking behavior.* Manuscript submitted for publication.

Guion, R. M., & Gottier, R. F. (1965). Validity of personality measures in personnel selection. *Personnel Psychology, 18,* 135-164.

Higgins, E. T. (1987). Self-discrepancy: A theory relating self and affect. *Psychological Review, 94,* 319-340.

Hogan, J., Rybicki, S. L., Motowidlo, S. J., & Borman, W. C. (1998). Relations between contextual performance, personality, and occupational advancement. *Human Performance, 11,* 189-207.

Hogan, R., & Roberts, B. W. (2001). Introduction: Personality and industrial and organizational psychology. In B. W. Roberts & R. Hogan (Eds.), *Personality psychology in the workplace* (pp. 3-16). Washington, DC: American Psychological Association.

Hough, L. M., Eaton, N. K., Dunnette, M. D., Kamp, J. D., & McCloy, R. A. (1990). Criterion-related validities of personality constructs and the effect of response distortion on those validities. *Journal of Applied Psychology, 75,* 581-595.

Hough, L. M. (1998). The millennium for personality psychology: New horizons or good ole daze. *Applied Psychology: An International Review, 47,* 233-261.

Hough, L. M., & Ones, D. S. (2002). The structure, measurement, validity and use of personality variables in industrial, work and organizational psychology. In N. Anderson, D. S. Ones, H. K. Sinangil, & C. Viswesvaran (Eds.), *International handbook of work and organizational psychology: Personnel psychology* (Vol. 1, pp. 233-277). London: Sage.

Hurtz, G. M., & Donovan, J. J. (2000). Personality and job performance: The big five revisited. *Journal of Applied Psychology, 85,* 869-879.

Locke, E. A., & Latham, G. P. (2002). Building a practically useful theory of goal setting and task motivation: A 35-year odyssey. *American Psychologist, 57,* 705-717.

McFarland, L. A., & Ryan, A. M. (2000). Variance in faking across noncognitive measures. *Journal of Applied Psychology, 85,* 812-821.

McFarland, L. A. (2003). Warning against faking on a personality test: Effects on applicant reactions and personality test scores. *International Journal of Selection and Assessment, 11,* 265-276.

Mischel, W. (1968). *Personality and assessment.* New York: Wiley.

Motowidlo, S. J., Borman, W. C., & Schmit, M. J. (1997). A theory of individual differences in task and contextual performance. *Human Performance, 10,* 71-83.

Motowidlo, S. J., & Van Scotter, J. R. (1994). Evidence that task performance should be distinguished from contextual performance. *Journal of Applied Psychology, 79,* 475-480.

Mueller-Hanson, R., Heggestad, E. D., & Thornton, G. C. III. (2003). Faking and selection: Considering the use of personality from select-in and select-out perspectives. *Journal of Applied Psychology, 88,* 348-355.

Ones, D. S., Viswesvaran, C., & Reiss, A. D. (1996). Role of social desirability in personality testing for personnel selection: the red herring. *Journal of Applied Psychology, 81,* 660- 670.

Pace, V. L. (2005). *Creative performance on the job: Does openness to experience matter?* Unpublished master's thesis, University of South Florida, Tampa.

Pace, V. L, Xu, X., Penney, L. M., Borman, W. C., & Bearden, R. M. (2005, April). Using warnings to discourage personality test faking: An empirical study. In J. P. Bott & C. C. Rosen (Chairs), *Moving from laboratory to field: Investigating situation in faking research.* Symposium conducted at the annual meeting of the Society for Industrial and Organizational Psychology, Los Angeles.

Rees, C. J., & Metcalfe, B. (2003). The faking of personality questionnaire results: who's kidding whom? *Journal of Managerial Psychology, 18,* 156-165.

Robie, C., Born, M. P., & Schmit, M. J. (2001). Personal and situational determinants of personality responses: A partial reanalysis and reinterpretation of the Schmit et al. (1995) data. *Journal of Business and Psychology, 16,* 101-117.

Rosse, J. G., Stecher, M. D., Miller, J. L., & Levin, R. A. (1998). The impact of response distortion on preemployment personality testing and hiring decisions. *Journal of Applied Psychology, 83,* 634-644.

Schmit, M. J., & Ryan, A. M. (1992). Test-taking dispositions: A missing link? *Journal of Applied Psychology, 77,* 629-637.

Schmit, M. J., Ryan, A. M., Stierwalt, S. L., & Powell, A. B. (1995). Frame-of-reference effects on personality scale scores and criterion-related validity. *Journal of Applied Psychology, 80,* 607-620.

Schmitt, N., & Oswald, F. L. (in press). The impact of corrections for faking on the validity of noncognitive measures in selection settings. *Journal of Applied Psychology.*

Seiler, S. N., & Kuncel, N. R. (2005, April). *Modeling the individual difference determinants of faking: Integration and extension.* Paper presented at the 20th annual conference of the Society for Industrial and Organizational Psychology, Los Angeles.

Shah, J., Higgins, T., & Friedman, R. S. (1998). Performance incentives and means: How regulatory focus influences goal attainment. *Journal of Personality & Social Psychology, 74,* 285-293.

Tett, R. P., Jackson, D. N., & Rothstein, M. (1991). Personality measures as predictors of job performance: A meta-analytic review. *Personnel Psychology, 44,* 703-742.

Topping, G. D., & O'Gorman, J. G. (1997). Effects of faking set on validity of the NEO FFI. *Personality and Individual Differences, 23,* 117-124.

Vasilopoulos, N. L., Cucina, J. M., & McElreath, J. M. (2005). Do warnings of response verification moderate the relationship between personality and cognitive ability? *Journal of Applied Psychology, 90,* 306-322.

Viswesvaran, C., & Ones, D. S. (1999). Meta-analyses of fakability estimates: Implications for personality measurement. *Educational and Psychological Measurement, 59,* 197-210

CHAPTER 12

FAKING ON NONCOGNITIVE MEASURES

The Interaction of Cognitive Ability and Test Characteristics

Nicholas L. Vasilopoulos and Jeffrey M. Cucina

Over the past 15 years, researchers interested in personality testing have debated the impact of applicant faking on validity. Much of the faking research has focused on the effectiveness of social desirability scales used to correct scores. On the whole, research suggests that correcting scores for social desirability does little to improve criterion-related validity (Barrick & Mount, 1996; Ones & Viswesvaran, 1998; Ones, Viswesvaran, & Reiss, 1996). Some have viewed these findings as evidence that applicant faking doesn't matter (Ones et al., 1996). Others have viewed the findings as misleading, arguing that social desirability scales fail to accurately capture applicant faking (e.g., Goffin & Christiansen, 2003; Snell & McDaniel, 1998; Snell, Sydell, & Lueke, 1999).

Much of the recent research on applicant faking has focused on what Vasilopoulos, Cucina, and McElreath (2005) referred to as *proactive approaches*. In proactive approaches the test developer takes steps to deal

A Closer Examination of Applicant Faking Behavior, 305–331
Copyright © 2006 by Information Age Publishing
All rights of reproduction in any form reserved.

with faking before the test is administered to the applicant. Examples of proactive approaches to faking include the use of a warning of response verification in the instructions, forced-choice item format, subtle item content, second hand item content, and the dispersion of items across an inventory either randomly or in an alternating sequence. These approaches seem to have some promise in reducing score inflation due to faking,[1] however little consideration is given to whether the use of proactive approaches changes the psychological constructs being assessed.

In this chapter, we discuss the possibility that many of these proactive approaches to faking increase the extent to which cognitive ability is measured. The rationale behind using a proactive approach to faking is that applicants find it more difficult to inflate their scores. If this is correct, personality scores should capture individual differences in cognitive abilities such as reasoning and problem solving in addition to the personality trait of interest. Thus, in our attempt to reduce faking, we may have inadvertently introduced a construct irrelevant cognitive load into personality measures.

TEST INSTRUCTIONS AND THE PERSONALITY-COGNITIVE ABILITY RELATIONSHIP

Attempts to deal with faking by altering test instructions typically involve including a warning of response verification. The most common types of warnings used by researchers interested in faking on noncognitive measures are *warnings of response verification by others* and *warnings of a lie scale included the inventory*. Pace and Borman (see chapter 11 of this book) provide a discussion of the different types of warnings used in personnel selection. Examples of both types of warnings are provided in Table 12.1. The premise behind warnings is that they induce applicants who were planning to fake to instead answer honestly. An alternative possibility is that the warning increases the complexity of the item response process, making it harder for applicants to figure out which items to fake.

Warning of Verification by Others

In both a field and laboratory study, Vasilopoulos et al. (2005) demonstrated the effects that warnings of response verification by others can have on the relationship between personality and cognitive ability test scores. In forming their hypothesis, they drew on Tetlock's (1985) social contingency model, which suggests response complexity is inversely

Table 12.1. Examples of the Personality Item Formats, Instructions, and Keys Discussed in This Chapter

Item Type	Item	Instructions
Traditional single stimulus/obvious/ first-hand/nonelaborated	I complete my assignments on time. a) Strongly disagree b) Disagree c) Neither agree nor disagree d) Agree e) Strongly agree	Please use the rating scale to indicate how accurately each item describes you.
Warning of response verification	I complete my assignments on time. a) Strongly disagree b) Disagree c) Neither agree nor disagree d) Agree e) Strongly agree	Please use the rating scale to indicate how accurately each item describes you. When making a response decision, consider that people you know might be asked to verify your responses. Examples of people who might be asked to verify your responses include present/past friends, teachers, and employers as well as family members. Keep in mind that your scores will be considered invalid if the people interviewed indicate that you gave inaccurate information about yourself.
Warning of a lie scale	I complete my assignments on time. a) Strongly disagree b) Disagree c) Neither agree nor disagree d) Agree e) Strongly agree	Please use the rating scale to indicate how accurately each item describes you. When deciding on a response, consider that the inventory includes questions designed to detect false or inaccurate responses. Points will be deducted from your score if you are identified as having provided false or inaccurate responses

Forced choice

For each item, indicate the phrase that **most accurately** describes you as well as the phrase that **least accurately** describes you.

a) I complete my assignments on time.
b) I have difficulty understanding abstract ideas.
c) I know the answers to many questions.
d) I am not easily amused.

Option _____ is most accurate.
Option _____ is least accurate.

Subtle item

Please use the rating scale to indicate how accurately each item describes you.

When my friends don't complete their assignments on time, they usually have a good reason.

Or

It annoys me when others don't complete their assignments on time.

a) Strongly disagree
b) Disagree
c) Neither agree nor disagree
d) Agree
e) Strongly agree

Second-hand item

My high-school teachers would say that I complete my assignments on time.

a) Strongly disagree
b) Disagree
c) Neither agree nor disagree
d) Agree
e) Strongly agree

Please use the rating scale to indicate how accurately each item describes you.

At-work frame of reference item

I complete my assignments on time at work

a) Strongly disagree
b) Disagree
c) Neither agree nor disagree
d) Agree
e) Strongly agree

Please use the rating scale to indicate how accurately each item describes you.

Elaborated

I complete my assignments on time.

a) Strongly disagree
b) Disagree
c) Neither agree nor disagree
d) Agree
e) Strongly agree

Please use the rating scale to indicate how accurately each item describes you. If you select options D or E, please provide a brief 2-3 sentence justification for your rating.

Obvious key

I complete my assignments on time

Please use the rating scale to indicate how accurately each item describes you.

The scoring weight for each response option is shown in parentheses.

a) Strongly disagree (1)
b) Disagree (2)
c) Neither agree nor disagree (3)
d) Agree (4)
e) Strongly agree (5)

Subtle key

I complete my assignments on time

Please use the rating scale to indicate how accurately each item describes you.

The scoring weight for each response option is shown in parentheses.

a) Strongly disagree (1)
b) Disagree (1)
c) Neither agree nor disagree (3)
d) Agree (5)
e) Strongly agree (4)

related to the level of certainty an individual has about the views of the audience to whom he/she is accountable.

When there is no warning of response verification, applicants who fake only need to know the characteristics that define a qualified job candidate. Research has shown that applicants not given a warning have little trouble responding as a qualified job candidate when completing a personality inventory (e.g., Furnham, 1990; Schmit & Ryan, 1993; Vasilopoulos, Reilly, & Leaman, 2000). This suggests a weak relationship between personality scales and cognitive ability test scores. On the other hand, when a warning of response verification by others is given, applicants who fake need to consider characteristics of a qualified candidate was well how they are perceived by all individuals who could be asked to verify their responses. Because applicants experience varying degrees of certainty about how they are perceived by those who could be contacted, they may engage in *preemptive self criticism* in an effort to avoid making responses that others will object to (Tetlock, 1983, 1985). The process of preemptive self-criticism results in an *integratively complex* response decision that involves an increase in the amount of distinct pieces of information considered (referred to as *differentiation*) and a more elaborate definition of the covariation among the pieces of distinct information (referred to as *integration*). Effective differentiation is assumed to be a function of general memory and broad retrieval ability, whereas integration is assumed to be a function of reasoning ability. This manifests itself as stronger correlations between personality scales and cognitive ability test scores.

In general, Vasilopoulos et al. (2005) found support for their hypotheses. In a field study, all personality-cognitive ability score correlations were significantly more positive and larger among applicants who received a warning of response verification by others, ranging from .17 to .33 (for Dependability and Emotional Stability respectively) when a warning was given versus -.09 to .17 (also for Dependability and Emotional Stability) when no warning was given. A similar pattern was found for four of the six personality scales included in a laboratory study, ranging from .14 to .46 (for Emotional Stability and Openness to Experience, respectively) when a warning was given versus -.04 to .23 (for Conscientiousness and Openness to Experience, respectively) when no warning was given. Evidence that providing a warning increased the complexity of faked response decisions came from the results of analyses involving item response latency data. With the exception of the results for the impression management scale in the laboratory study, applicants who received a warning took significantly longer to respond to items on all scales than did applicants who did not receive a warning.

The results reported above demonstrate that the use of a warning of response verification can increase the cognitive load of the personality item response process. However, it is worth noting that the results may not generalize to other selection settings. The applicant sample used by Vasilopoulos et al. (2005) consisted of candidates for a federal law enforcement job that includes a personal background investigation as part of the hiring process. Because applicants knew about the background investigation requirement when the personality test was administered, it seems reasonable that many applicants believed that investigators would seek to verify their responses during the course of the background investigation. More research is needed to see whether the same results are obtained when applicants are candidates for a job that does not include a background investigation as part of the selection process.

Warning of a Lie Scale

Although the studies conducted by Vasilopoulos et al. (2005) are informative, the results cannot necessarily be extrapolated to situations where a warning of a lie scale is used. Having said this, it seems likely to expect that attempts to avoid detection through a lie scale will also increase the cognitive load of the item response process. Research has shown that the inclusion of a warning of a lie scale in the test instructions reduces, but does not eliminate, response inflation (Doll, 1971). Instead, applicants adopt a strategy that involves only endorsing items that are most relevant for the job. We propose that successful attempts to inflate responses without scoring too high on a lie scale require applicants to accurately identify the traits that are (a) most relevant for the job, and (b) measured by the items on the inventory. This knowledge allows applicants who are warned that a lie scale is embedded in the inventory to appropriately select the items on which to inflate their responses.

There are several reasons to expect that cognitive ability is related to the ability to identify both job-relevant traits and the traits measured by items on the inventory. Two lines of research support the idea that the accurate identification of job-relevant traits is a function of cognitive ability. An indirect inference can be drawn from research on performance appraisal showing significant relationships between cognitive ability and the accuracy of job performance ratings (e.g., Borman, 1979; Smither & Reilly, 1987). In this research, accuracy was defined as the correct identification and evaluation of the traits underlying performance. More direct evidence comes from Christiansen, Burns, and Montgomery (2005) who reported a significant correlation between cognitive ability and an implicit job theory measure. Implicit job theories identify job relevant traits as well as the relationships among them.

There is also evidence that cognitive ability is related to the accurate identification of the trait measured by a personality item. Christiansen, Wolcott-Burman, Janovics, Burns, and Quirk (2005) reported a significant relationship between cognitive ability and a measure of *dispositional intelligence*—the ability to accurately infer traits from behaviors. Since personality items are often presented as behaviors, dispositional intelligence should play a role in determining the accuracy with which applicants who fake can identify the traits being assessed by the items on the inventory.

ITEM FORMAT AND THE PERSONALITY-COGNITIVE ABILITY RELATIONSHIP

The majority of proactive attempts to deal with applicant faking have involved manipulations of the item format. In this section, we discuss how using forced-choice items, subtle items, second-hand items and situation-specific items can increase the cognitive load of the item response process and, as a result, the relationship between cognitive ability and personality.

Forced-Choice Items

In recent years, researchers have expressed renewed interest in the use of a forced-choice item format as a way to deal with applicant faking. Briefly, forced-choice items require applicants to choose between two statements that describe equally desirable behaviors or traits. Converse et al. (see chapter 10 of this book) provide a detailed description of the forced-choice item format. An example forced-choice item is provided in Table 12.1. Interest in the forced-choice format was bolstered by the results of research showing that forced-choice personality scales yielded higher criterion-related validity estimates than single-stimulus scales (Christiansen, 2000; Jackson, Wroblewski, & Ashton, 2000). The explanation offered for the higher levels of criterion-related validity of forced-choice scales is that forced-choice items are harder to fake. However, there is also evidence to suggest that the higher criterion-related validity estimates for forced-choice scales reflects the fact that they reflect individual differences in cognitive ability as well as personality. For example, Vasilopoulos, Cucina, Dyomina, Morewitz, and Reilly (in press)[2] demonstrated that using a forced-choice item format to deal with applicant faking also has the unintended consequence of increasing the cognitive load of the response process for applicants who fake. They proposed that faking responses to forced-choice personality items is more complex than faking responses to single-stimulus personality items because faked responses to forced-choice items require the added step of accurately rank ordering the relevance of stems with similar job desirability. The role of cognitive ability in accurately rank ordering job-relevant traits is suggested by the performance appraisal research showing significant relationships between cognitive ability and differential accuracy—the extent to which a rater correctly rank orders performance dimensions in terms of relevance (Hauenstein & Alexander, 1991).

For the most part, the results reported by Vasilopoulos et al. (in press) supported their hypotheses. When participants answered honestly, cognitive ability was unrelated to the Conscientiousness and Openness to Experience scales on the single-stimulus ($r = .02$ and $.09$ respectively) and forced-choice ($r = .03$ and $-.09$ respectively) inventories. However, when participants responded as if they were applying to a college they really wanted to attend, cognitive ability was unrelated to the Conscientiousness and Openness to Experience scales on the single-stimulus inventory ($r = -.05$ and $-.03$ respectively), and positively related to Conscientiousness and Openness to Experience scales on the forced-choice inventory ($r = .36$ for both scales).

One interesting finding was the fact that low cognitive ability participants who completed the forced-choice inventory as an applicant actually scored lower on the Conscientiousness and Openness to Experience scales than low cognitive ability participants who responded honestly ($d = -.12$ and $-.21$, respectively). In contrast, high cognitive ability participants who completed the forced-choice inventory as an applicant scored substantially higher on the Conscientiousness and Openness to Experience scales than the high cognitive ability participants who responded honestly ($d = .76$ and $.78$, respectively). It is worth noting that similar effect sizes were obtained for the single stimulus Conscientiousness and Openness to Experience scales among participants high in cognitive ability ($d = .72$ and $.94$, respectively). Thus, low cognitive ability applicants who choose to fake on a forced-choice inventory may hurt—rather than help—their chances of getting hired, whereas high cognitive ability applicants will be unaffected by the complexity induced by a forced-choice item format.

Also of note was the finding that the forced-choice personality scales obtained in the applicant condition were related to grade point average (GPA) after controlling for cognitive ability. In the applicant condition, GPA was more strongly correlated with the forced-choice Conscientiousness and Openness to Experience scales ($r = .24$ and $.22$, respectively) than with the single stimulus scales ($r = .02$ and $.07$, respectively), a finding consistent with earlier research by Christiansen (2000) and Jackson et al. (2000). Although controlling for cognitive ability reduced the validity of the Conscientiousness and Openness to Experience forced-choice scales (pr = .13 and .12, respectively), the estimates were still larger than those obtained for Conscientiousness and Openness to Experience scales on the single-stimulus inventory (pr = .04 and .08, respectively). Thus, it appears that, while forced-choice scales tap cognitive ability, they also provide information about individual differences in personality.

In a recent study, Christiansen, Burns et al. (2005, Study 3) replicated the findings of Vasilopoulos et al. (in press). An important aspect of this study was that it demonstrated that the effect of cognitive ability on forced-choice scales is mediated by the applicant's implicit job theory. Specifically, they suggested that individuals high in cognitive ability should be able to develop more accurate representations (i.e., implicit job theories) of the personality dimensions associated with successful performance for a specific job. The quality of the implicit job theory will then influence the effectiveness of an individual's faking efforts on the forced-choice measure. Christiansen, Burns et al. administered a measure of implicit job theory, whereby participants indicated which personality dimensions were important for a specific occupation. Scores on the implicit job theory measure for Conscientiousness were modestly correlated (i.e., r = .20) with cognitive ability test scores. A path analysis depicting the mediating role of implicit job theory demonstrated acceptable fit and indicated that implicit job theory is a partial mediator of the relationship between cognitive ability and the increase in forced-choice scores that resulted from instructions to fake.

Obvious Versus Subtle Items

The use of subtle (as opposed to obvious) items is another approach to reducing faking in self-report measures. With subtle items, the construct(s) assessed by an item are less clear to applicants, making it harder for applicants to fake (see Table 12.1 for an example). Subtle items are often identified when an empirical item selection strategy is used (Hough, Eaton, Dunnette, Kamp, & McCloy, 1990). In contrast, the construct(s) measured by an obvious item are very clear to applicants (see Table 12.1 for an example). This makes it easier for applicants to respond in a manner that gives maximal points for the item, thus making it easier for applicants to fake.

Several researchers have recommended the use of subtle items in self-report measures. Two major review articles on biodata mention the use of subtle items. Citing past research by Lautenschlager (1985), Mumford and Stokes (1992) conclude that it is harder for applicants to fake when subtle items are used. Mael (1991) also discusses the resistance of subtle items to faking; however, he mentions some of the negative aspects of items that are not obviously job-relevant to nonpsychologists (e.g., applicants and managers).

A good amount of the research on this topic has focused on comparing measures composed entirely of obvious items to those composed entirely of subtle items. For example, Bornstein, Rossner, Hill, and Stepanian (1994) compared the fakability of an obvious (objective) measure and a subtle (projective) measure of Dependability. In a series of laboratory studies they found that participants were more easily able to identify the construct measured by the obvious measure than that measured by the subtle measure. In addition, participants were able to fake their responses on the obvious measure, but not on the subtle measure. In a third study, participants were told that the trait being measured was either a desirable, undesirable or neutral trait. This manipulation only affected scores on the obvious measure. Taken together, these studies support the hypothesis made by Bornstein et al. that face validity and fakability are inversely related; thus obvious measures are more face valid, but less resistant to faking than subtle measures.

Alliger, Lilienfeld, and Mitchell (1996) examined the fakability of overt and covert integrity tests. Integrity tests (which are used to predict counterproductive behaviors such as theft) can be divided into two types: overt tests (which are obvious measures containing items that inquire about an applicant's history of, and attitudes toward, counterproductive behaviors) and covert tests (which are subtle measures containing personality items purported to relate to integrity). In essence, overt tests measure an applicant's history of, and attitudes toward, counterproductive behaviors, whereas covert tests assess constructs that make individuals prone to engage in counterproductive behaviors. Alliger et al. found that coaching led to enhanced faking on overt but not covert integrity tests. A meta-analysis by Alliger and Dwight (2000) also found that covert integrity tests were less susceptible to faking and coaching than overt integrity tests.

Meta-analytic work by Ones, Viswesvaran, and Korbin (1995) demonstrated that scales composed of subtle items are less susceptible to both faking good and faking bad than scales composed of obvious items. However, research focusing on item-level validity suggests that subtle items are less valid than obvious items. Lefkowitz, Gebbia, Balsam, and Dunn (1999) found some evidence that subtle items (i.e., those appearing less job relevant) had lower validity than obvious items. Johnson (2004) also found that subtle items had lower validity when self-report personality items were used to predict acquaintance personality ratings. These findings are consistent with early research demonstrating that subtle items have lower item-level validities than obvious items (e.g., Duff, 1965, McCall, 1958; Wiener, 1948).

One of the assumptions behind the use of subtle items to decrease faking is that it is more difficult for applicants to identify the dimensions assessed by an item and to determine the response that will maximize their score on an item. We might expect that subtle items are more effective at decreasing faking in low cognitive ability applicants. Since subtle items are more abstract, it seems plausible that high cognitive ability applicants (who should be better at abstract reasoning) might be better able to identify the dimension(s) assessed by subtle items and to determine the relationship between an item and job performance. In contrast, only a small degree of cognitive ability would be required to identify the dimension(s) assessed by an obvious item and the relationship between an obvious item and job performance. Therefore, cognitive ability might not be expected to relate to scores on obvious items under faking conditions.

Although there is some theoretical justification for a relationship between cognitive ability and scores on subtle (but not obvious) measures under faking conditions, the empirical literature is less clear. There is evidence that respondents higher in cognitive ability are better able to fake overt integrity tests and are better able to effectively use coaching material to increase scores on both overt and covert integrity tests (Alliger et al., 1996). When respondents were instructed to fake, there was a significant positive correlation between cognitive ability and scores on an overt integrity test. Alliger et al. reported a significant positive correlation between cognitive ability and scores on a covert integrity test when respondents were coached on how to fake a covert test, but not when coaching was either not provided or was given for an overt test. A similar finding was found for overt test scores when coaching was given for overt tests. Thus, it appears that the relationship between cognitive ability and faking on obvious vs. subtle items may be affected by coaching.

First-Hand Versus Second-Hand Items

Another proposed method of decreasing faking is the use of second-hand (as opposed to first-hand) items. In his taxonomy of biodata items, Mael (1991) distinguishes between first-hand items (which inquire about an applicant's personal self-knowledge, judgments, opinions, and evaluations) and second-hand items (which ask an applicant to indicate what they believe others' knowledge, judgments, opinions, and evaluations would be about them). Examples of first- and second-hand items are provided in Table 12.1. Mael hypothesized that second-hand items are more susceptible to faking than first-hand items. According to Mael, second-hand items add a level of "speculative subjectivity" (p. 775) making the item less objective and hence more fakable. Mael also suggested that second-hand items could have higher validities than first-hand items when applicants answer honestly.

Two published studies have followed-up on Mael's speculations. First, Lefkowitz et al. (1999) observed higher validities for second-hand items than for first-hand items. Second, Graham, McDaniel, Douglas, and Snell (2002) investigated the validity and fakability of first- and second-hand biodata items. They found that second-hand items had higher validities (average item level validity of .13) than first-hand items (average item level validity of .08) when respondents answered honestly. However, both first- and second-hand items were equally fakable (ds of .36 and .34, respectively) and had equally low validities (average item level validities of .05 and .03, respectively) when respondents faked.

In addition to differing in validity, responses to first- and second-hand items might also be differentially related to cognitive ability under faking conditions. Specifically, it might be expected that responses to second-hand items (but not first-hand items) would be related to cognitive ability when applicants fake. The processes involved in responding to second-hand items is somewhat similar to the processes involved when a warning of response verification is given. Both situations ask applicants to make a determination as to how others (e.g., past supervisors, teachers, peers, etc.) would evaluate their personality and behaviors. Individuals high in cognitive ability may be better at retrieving and integrating information from memory regarding the opinions of others on their personality and past behaviors. In addition, cognitive ability may be positively related to the ability to come up with rational reasons and justifications for stating that others would view them positively.

Frame-of Reference: Situation-Specific Item Stems

Some researchers have attempted to increase the criterion-related validity of personality measures by using situation-specific item stems (see Table 12.1 for an example). Schmit, Ryan, Stierwalt, and Powell (1995) proposed that personality is somewhat situationally specific, and that personality inventories designed for a specific situation (e.g., an at-school setting) should demonstrate higher levels of criterion-related validity. They modified personality items to reflect behavior in an academic setting and found that Conscientiousness scale scores demonstrated higher levels of criterion-related validity. One main drawback of this method is the fact that not all personality items can easily be adapted to at-work or at-school settings. Schmit et al. noted the difficulty of using this method for items measuring Openness to Experience, which is conceptually and empirically related to academic and training performance (e.g., Barrick & Mount, 1991; Costa & McCrae, 1992; Cucina & Vasilopoulos, 2005). In addition, Hunthausen, Truxillo, Bauer, and Hammer (2003) noted the possibility that using an at-work frame of reference could make items more transparent and thus, more fakable.

An argument could be made for a relationship between cognitive ability and scores on a personality test with an at-work frame of reference under applicant conditions. Cognitive ability may be related to the recall of specific instances that are consistent with an at-work frame of reference. In our discussion of cognitive ability and warnings of response verification and second hand items, we indicate that cognitive ability is related to the retrieval and integration of personality related information from memory. Thus, high cognitive ability individuals may be better able to retrieve and integrate work-specific personality information than low cognitive ability individuals. On the other had, using a work-specific frame of reference may make a personality inventory more obvious, for which the empirical literature on the relationship with cognitive ability is less clear.

Two research studies on frame of reference effects appear to rule out the possibility that this approach introduces a cognitive load into personality scale scores. In a concurrent-validity study using airline customer service managers, Hunthausen et al. (2003) reported higher validity coefficients when an at-work frame of reference was used with Extraversion and Openness to Experience items. This effect was significant after controlling for cognitive ability. In general, there was little difference in the relationship between Big Five personality scores and cognitive ability under the standard and at-work frame of reference conditions. In a laboratory study, Bing, Whanger, Davison, and VanHook (2004) reported higher correlations for a school specific Conscientiousness scale than a general scale in predicting GPA, under both honest and faking conditions. Faking had a similar effect on the mean and criterion-related validity of Conscientiousness scale scores for both the general and school frame of reference conditions. Finally, the general and school specific scales had similar relationships with cognitive ability in both the honest and faking conditions. The results of both studies indicate that using a work or school specific frame of reference does not introduce a cognitive load under honest or faking conditions. In addition, the school specific items in Bing's study retained their validity under faking conditions, despite being more obvious.

ITEM RESPONSE ELABORATION AND THE PERSONALITY-COGNITIVE ABILITY RELATIONSHIP

In addition to altering the format of noncognitive items, some researchers have suggested using item response elaborations to decrease faking. Specifically, applicants can be instructed that they will have to elaborate on and defend their responses to the self-report items in a follow-up interview or by writing a justification for their responses (see Table 12.1 for an example). A poor elaboration or defense of a desirable response to an item would indicate possible faking. In addition, applicants may engage in less response distortion if they anticipate having to justify their responses. This particular approach is less frequently mentioned in the literature. Schmitt and Kunce (2002) examined the effects of elaboration for biodata items in field study. Note that participants were not warned of an impending elaboration, but instead were asked to elaborate on their answers while they completed the inventory. The biodata scores were lower when participants were asked to elaborate on their ratings. Schmitt and Kunce asked participants to elaborate on only a subset of the items in the inventory. Although the effects were strongest for the items in the inventory that required elaboration, elaboration did lead to slightly lower scores on items that did not require elaboration, indicating a carryover effect. The authors did report slightly higher correlations between elaborated biodata scores and scores from a job knowledge exam, verbal test and an essay; however, these differences were not statistically significant. In a follow-up study, Schmitt, Oswald, Kim, Gillespie, Ramsay, and Yoo (2003) replicated the effects of elaboration on lowering biodata scale scores; however, they did not observe a carryover effect, nor did elaboration affect the criterion-related validity, subgroup differences and correlations with social desirability of the biodata scores. One drawback of this approach is the fact that like verifying responses on a self-report inventory with others, this approach can be resource intensive. For example, an organization might not have the resources to conduct follow-up interviews to verify responses on the self-report inventory. Subsequent interviews might have to solely focus on other questions due to resource constraints.

Another drawback could be the possibility that the use of elaboration once again introduces a cognitive load on personality test scores. This phenomenon could occur for several reasons. First, as described elsewhere in this chapter, we take the position that cognitive ability is related to the ability to retrieve and integrate personality-related information. Cognitive ability is also related to the ability to produce rational reasons and justifications for an inflated personality item response. In addition, cognitive ability could be related to the ability to articulate this information in written elaborations. Finally, individuals rating the elaborations might have difficulty separating an applicant's writing ability from their personality-related information in the written elaborations. Research by MacLane, Martin, Liberman, and Nickels (2004) showed that writing ability and cognitive ability can impact ratings on accomplishment record dimensions. However, research by Schmitt and Kunce (2002) may not support the possible cognitive load of elaborated responses. They found no differences in the relationship between verbal ability and writing ability with elaborated and nonelaborated biodata responses. However, their measure of verbal ability was in essence a multiple-choice writing test containing items covering grammar, punctuation, sentence and paragraph structure, etc. Although this type of test should be indicative of how well a respondent can articulate personality-related information in writing, it may not fully tap the specific cognitive abilities that are involved in the process of inflating item responses when an elaboration is required.

ITEM PLACEMENT AND THE PERSONALITY-COGNITIVE ABILITY RELATIONSHIP

Altering the location of items in an inventory is another proposed method of reducing faking in self-report measures. This method is similar to the use of subtle items, in that it attempts to make the constructs measured in an inventory less obvious to applicants. To explain, many multidimensional personality inventories disperse items measuring different constructs using an alternating or random approach. For example, the first five items in a Big Five measure might measure Neuroticism, Extraversion, Openness to Experience, Agreeableness and Conscientiousness, in that order. This ordering is then repeated throughout the inventory. This approach has been purported to make the dimensions assessed by an inventory less salient to applicants, and thus less fakable (Anastasi, 1976; Mumford & Stokes, 1992). In contrast, items can be grouped according to the dimension they assess (e.g., all of the items assessing Conscientiousness could be administered in a single section). Mumford and Stokes have suggested that the dimensions assessed by an inventory are more salient to applicants when this approach is used.

Although the practice of alternating or randomizing items in an inventory has long been a rule of thumb, there has been little empirical research on this topic, with the notable exception of a study by McFarland, Ryan, and Ellis (2002). They note that the practice of alternating or randomizing items goes against theoretical work in the cognitive psychology literature. Drawing from work by Knowles and Byers (1996) and Tourangeau and Smith (1996), McFarland et al. proposed that grouping items makes the dimensions assessed by the items more salient and transparent. Presumably, knowing which dimensions a set of items taps should make it easier for applicants to retrieve dimension-relevant information from memory. Knowles and Byers suggest that when a respondent has a clearer understanding of the construct being assessed, he or she will be able to conduct more efficient memory searches, and will focus searches on information relevant to the construct of interest. In other words, grouping items makes it easier for applicants to understand the dimensions assessed, and this understanding helps applicants to provide more accurate item responses. Knowles and Byers go further and suggest that grouping items refines a respondent's interpretation of the meaning of a set of items and helps to reduce any alternative (construct-irrelevant) interpretations of an item. They also hypothesize that inventories will be more reliable and valid if the construct(s) they measure are made clear to respondents.

Knowles and Byers (1996) also suggest that as a respondent progresses through an inventory, their response process to the items changes. When responding to the first set of items, respondents are likely to interpret items in multiple ways and will give thought into how best to respond. After responding to many items, the nature of the construct being assessed is clearer and respondents may begin to integrate and summarize information related to the construct being assessed. At this point, respondents may invoke a self-schema, which is used to respond to items. Most likely, it would be easier for respondents to develop this self-schema when items are grouped by dimension, rather than being placed in alternating or random order. When items are not grouped, applicants would have to switch from one dimension to another as they progress through the inventory and would not be able to focus solely on developing a self-schema for one dimension at a time. Presumably, the activation of the self-schema would lead to more accurate and schema congruent item responses. However, it is also possible that respondents recognize the similarity of grouped items and give the same response to each item with little consideration of the actual item content. On a similar note, it is possible that the automatic processing resulting from the use of a self-schema may yield responses that are more consistent across items, but not necessarily more accurate in terms of describing the individual's personality. This would especially be true if one takes the perspective that personality varies to some degree across the different situations that each specific item measures.[3] To continue, it is also possible that the self-schema could be impacted by other factors such as mood and motivated reasoning.

As noted by McFarland et al. (2002) it is important to consider the cognitive processes involved in responding to items. The associative network models are popular models for how concepts, features and propositions are activated and retrieved from memory (Anderson, 1983; Carlston & Smith, 1996; Collins & Loftus, 1975; Fiske & Taylor, 1991; Smith, 1998; Wyer & Srull, 1986). Kunda (1999) provides a good overview of these models in the context of social cognition. According to these models, concepts, features and propositions are represented as nodes and are connected via links to other nodes in a network. For example, suppose a respondent is attempting to answer the question "Get chores done right away" from Goldberg's (1999) IPIP. We might envision a central node for performing chores, which is linked to many related nodes (e.g., specific chores, instances of performing or neglecting chores, personality traits related to performing chores on time). Activating one node increases the chances of activation of connected nodes. Thus, when the respondent activates the central node related to performing chores, other linked nodes are more likely to become activated. If the nodes are activated they are more likely to be brought to the respondent's attention and to be used by the respondent in making a determination of how to respond to the item (under honest conditions). According to the associative network models, nodes tend to stay activated for only a limited time. Thus, if the respondent diverts his or her attention away from the central node, the activation of the other linked nodes will dissipate.

If personality items are grouped by dimension, then the nodes related to the dimension being assessed will continue to be active as the respondent progresses through the inventory. For example, if all of the Conscientiousness items are grouped together, then the portions of the network related to Conscientiousness will continue to remain active throughout the Conscientiousness section of the inventory. This has the potential of increasing the number of nodes that the respondent can potentially tap into when deciding upon a response. Thus, the respondent will have access to a larger pool of information when responding to items.

In contrast, consider the situation where personality items are randomly dispersed throughout an inventory or are in an alternating order (by dimension). It's possible that the nodes related to a dimension lose activity when the respondent moves to items tapping other dimensions. By the time the respondent returns to the dimension in question, the nodes associated with that dimension may have to be reactivated. This has the potential of decreasing the number of nodes that the respondent can use when answering an item.

On the other hand, McFarland et al. (2002) point out that making the dimensions of an inventory more salient to respondents (by grouping items) should make the inventory easier to fake. This argument is built on the literature on the fakability of subtle versus obvious items (discussed above). McFarland et al. conducted a laboratory study to explore this possibility and found that personality inventories can be easier to fake when items are grouped by dimension. This effect was most prevalent for Conscientiousness and Neuroticism.

One potential implication of this line of research that has not been investigated is the possibility that narrow and unidimensional measures of personality may be more susceptible to faking than broad and multidimensional measures. For example, suppose that a practitioner conducts a criterion-related validity study using an inventory containing measures of each of the Big Five dimensions and determines that of the Big Five factors, only Conscientiousness predicts performance. If *only* the items in the Conscientiousness scale are administered to future applicants and they are administered as a group, then the dimension assessed by the inventory will be more salient to applicants. Furthermore, it is possible that applicants would perceive the Conscientiousness items as having more importance if they are given in isolation of the other Big Five dimensions than if they are administered in conjunction with the other Big Five dimensions.

It is possible that cognitive ability moderates the effectiveness of alternating or randomizing items to reduce faking. McFarland et al. (2002) suggest that alternating or randomizing items introduces a cognitive load due to the fact that respondents need to switch to a different dimension with each item. In addition, according to Mumford and Stokes (1992) the dimensions assessed are less salient when items are alternated or randomized. Individuals high in cognitive ability might be better able to recognize the pattern of dimensions present in alternating and randomized items and thus identify the dimensions assessed in an inventory. This effect is likely given the fact that recognizing patterns and sequences is considered a part of cognitive ability. For example, many cognitive ability tests include number series items or may ask applicants to discern patterns in a series of figures and indicate which figure would appear next in the pattern. Although, to our knowledge, this line of research has not included patterns of personality items, it seems plausible that cognitive ability should be related to recognizing patterns in items that are alternated or randomly distributed throughout an inventory. Individuals high in cognitive ability should also have the requisite reasoning abilities for identifying the dimensions present in alternated or randomly distributed items.

Applicants high in cognitive ability might also be better at (a) maintaining the activation of nodes related to each dimension while answering items related to other dimensions and/or (b) switching between the different networks related to each personality dimension. This is consistent with the suggestion made by McFarland et al. (2002) that randomizing items places a greater cognitive load on respondents. In addition to cognitive ability, cognitive complexity and adaptability may play a role as well. Thus, there is a possibility that alternating or randomizing items is more effective at reducing faking for low cognitive ability applicants than for high cognitive ability applicants. That said, many existing personality instruments alternate or randomize items and in large part there has been little relationship between cognitive ability and personality scale scores for traditional single-stimulus items given without warnings, etc. (see meta-analytic research by Ackerman & Heggestad, 1997 and Cortina, Goldstein, Payne, Davison, & Gilliland, 2000).

OPTION-LEVEL KEYING AND THE PERSONALITY-COGNITIVE ABILITY RELATIONSHIP

Another possible approach to reducing applicant faking is the use "subtle keys," whereby the response option that gives maximal points is less clear to applicants, making it harder for applicants to fake (see Table 12.1 for an example). The subtle keys are usually generated using option-level empirical keying. In option-level empirical keying, the scoring weights for the individual response options of an item are a function of the empirical relationship between endorsing a response option and a criterion. Empirical keying was first developed for use in weighted application blanks in the insurance industry. The method was soon extended to biodata instruments, where it has enjoyed most of its use. Empirical keying has also been used to key situational judgment inventories (Bergman, Drasgow, Donovan, & Juraska, 2003; Dalessio, 1994; Krokos, Meade, Cantwell, Pond, & Wilson, 2004; MacLane, Barton, Holloway-Lundy, & Nickels, 2001; Weekley & Jones, 1997). In contrast to option-level scoring, item-level scoring assumes that there is a linear relationship between response options and a criterion; thus items are often scored using a Likert scale where the first option receives one point, the second, two points, and so forth.

Mumford and Stokes (1992) hypothesized that empirically keyed biodata measures would be less prone to faking due to the fact that the scoring keys for these measures are often less transparent and less obvious to applicants than those for rationally keyed measures. Indeed, a laboratory study by Kluger, Reilly, and Russell (1991) suggests that biodata measures scored at the option level may be more resistant to faking than those scored at the item level. With option-level scoring the particular option with the largest weight is not easily apparent to applicants. Thus applicants who attempt to fake may not in actuality inflate their scores. Kluger et al. tempered their findings with the possibility that the presence of faking on an option-keyed inventory may introduce random error into scores on the inventory, thus decreasing validity, even though scores on the inventory aren't inflated by faking.

Cognitive ability might be positively related to the ability of applicants to fake option-keyed noncognitive measures. As mentioned above, there is evidence that high cognitive ability applicants are better able to develop an adopted schema (or implicit job theory) detailing the traits of a successful employee. In the context of forced-choice tests, adopted schema has mainly been used to explain how high cognitive ability applicants identify which personality dimensions are relevant to performance. With option-keyed measures, this concept can be taken one step further, whereby applicants identify the optimal level on each dimension in addition to the relevance of each dimension to performance. Based on this argument, we hypothesize that option-keyed instruments should positively correlate with cognitive ability under faking conditions.

CONCLUSION

The impact of faking on the use of noncognitive measures in applied settings has been—and continues to be—a challenge to personnel psychologists. Frustration with social desirability scales (Goffin & Christiansen, 2003) has spurred interest in proactive approaches to faking designed to make it harder for applicants to successfully inflate their scores. In this chapter, we outlined how several of these proactive approaches to faking can introduce a cognitive load into scale scores and, as a consequence, reduce construct validity.

The introduction of a cognitive load associated with proactive approaches to faking also has implications for the utility of noncognitive measures used to make selection decisions. For example, the use of measures incorporating a proactive approach might increase the adverse impact of the selection process or decrease incremental validity when included in a selection battery with a cognitive ability test. Low adverse impact and significant incremental validity are often cited as reasons for using noncognitive measures in personnel selection.

NOTES

1. Throughout this chapter we assume that a personality dimension is scored so that it is positively related to performance. In other words, we assume that dimensions that might be negatively related to performance (e.g., Neuroticism) have been recoded to create scale scores (e.g., Emotional Sta-

bility) that are positively related to performance. Thus, faking should increase personality scale scores.

2. The results of this research were first reported in Dyomina, Vasilopoulos, Cucina, and Reilly (2003).

3. We thank Richard Griffith for making us aware of these two possibilities.

REFERENCES

Ackerman, P. L., & Heggestad, E. D. (1997). Intelligence, personality, and interests: Evidence of overlapping traits. *Psychological Bulletin, 121,* 219–245.

Alliger, G. M., & Dwight, S. A. (2000). A meta-analytic investigation of the susceptibility of integrity tests to faking and coaching. *Educational and Psychological Measurement, 60*(1), 59-72.

Alliger, G. M., Lilienfeld, S. I., & Mitchell, K. E. (1996). The susceptibility of overt and covert integrity tests to coaching and faking. *Psychological Science, 7,* 32-39.

Anastasi, A. (1976). *Psychological testing* (4th ed.). New York: Macmillan.

Anderson, J. R. (1983). *The architecture of cognition.* Cambridge, MA: Harvard University Press.

Barrick, M. R., & Mount, M. K. (1991). The Big Five personality dimensions and job performance: A meta-analysis. *Personnel Psychology, 44*(1), 1-26.

Barrick, M. R., & Mount, M. K. (1996). Effects of impression management and self-deception on the predictive validity of personality constructs. *Journal of Applied Psychology, 81,* 261-272.

Bergman, M. E., Drasgow, F., Donovan, M. A., & Juraska, S. E. (2003, April). *Scoring of situational judgment tests.* Paper presented at the 18th annual conference of the Society for Industrial and Organizational Psychology, Orlando, FL.

Bing, M. N., Whanger, J. C., Davison, H. K., & VanHook, J. B. (2004). Incremental validity of the frame of reference effect in personality scale scores: A replication and extension. *Journal of Applied Psychology, 89*(1), 150-157.

Borman, W. C. (1979). Individual differences correlates of accuracy in evaluating others' performance effectiveness. *Applied Psychological Measurement, 3,* 103-115.

Bornstein, R. R., Rossner, S. C., Hill, E. L., & Stephanian, M. L. (1994). Face validity and fakability of objective and projective measures of dependency. *Journal of Personality Assessment, 63,* 363-386.

Carlston, D. E., & Smith, E. R. (1996). Principles of mental representation. In E. T. Higgins & A. W. Kruglanski (Eds.), *Social psychology: Handbook of basic principles* (pp. 184-210). New York: Guilford Press.

Christiansen, N. D. (2000, April). *Utilizing forced-choice item formats to enhance criterion-related validity.* Paper presented at the 15th annual conference of the Society for Industrial and Organizational Psychology, New Orleans, LA.

Christiansen, N. D., Burns, G. N., & Montgomery, G. E. (2005). Reconsidering forced-choice item formats for applicant personality assessment. *Human Performance, 18*(3), 267-307.

Christiansen, N. D., Wolcott-Bernham, S., Janovics, J. E., Burns, G. N., & Quirk, S. W. (2005). Good judge revisited: Individual differences in the accuracy of personality judgments. *Human Performance, 18*(2), 123-149.

Collins, A. M., & Loftus, E. F. (1975). A spreading-activation theory of semantic processing. *Psychological Review, 82*, 407-428.

Cortina, J. M., Goldstein, N. B., Payne, S. C., Davison, K., & Gilliland, S. W. (2000). The incremental validity of interview scores over and above cognitive ability and conscientiousness scores. *Personnel Psychology, 53*, 325–352.

Costa, P. T., & McCrae, R. R. (1992). *Revised NEO Personality Inventory (NEO-PI-R) and the NEO Five-Factor Inventory (NEO-FFI).* Odessa, FL: Psychological Assessment Resources.

Cucina, J. M., & Vasilopoulos, N. L. (2005). Nonlinear personality-performance relationships and the spurious moderating effects of traitedness. *Journal of Personality, 73*, 227-260.

Dalessio, A. T. (1994). Predicting insurance agent turnover using a video-based situational judgment test. *Journal of Business and Psychology, 9*, 23-32.

Doll, R. E. (1971). Item susceptibility to attempted faking as related to item characteristics and adopted fake set. *Journal of Psychology, 77*, 9-16.

Duff, F. L. (1965). Item subtlety in personality inventory scales. *Journal of Consulting Psychology, 29*, 565-570.

Dyomina, N. V., Vasilopoulos, N. L., Cucina, J. M., & Reilly, R. R. (2003, April). *Forced-choice personality tests: A measure of personality or "g"?* Paper presented at the 18th annual conference of the Society for Industrial and Organizational Psychology, Orlando, FL.

Fiske, S. T., & Taylor, S. E. (1991). *Social cognition.* New York: Random House.

Furnham, A. (1990). Faking personality questionnaires: Fabricating different profiles for different purposes. *Current Psychology: Research & Reviews, 9*(1), 46-55.

Goffin, R. D., & Christiansen, N. D. (2003). Correcting personality tests for faking: A review of popular personality tests and an initial survey of researchers. *International Journal of Selection and Assessment, 11*, 340-344.

Goldberg, L. R., (1999). A broad-bandwidth, public-domain, personality inventory measuring the lower-level facets of several five-factor models. In I. Mervielde, I. J. Deary, F. De Fruyt, & F. Ostendorf (Eds.), *Personality psychology in Europe* (Vol. 7, pp. 7-28). The Netherlands: Tilburg University Press.

Graham, K. E., McDaniel, M. A., Douglas, E. F., & Snell, A. F. (2002). Biodata validity decay and score inflation with faking: Do item attributes explain variance across items? *Journal of Business and Psychology, 16*(4), 573-592.

Hauenstein, N. M., & Alexander, R. A. (1991). Rating ability in performance judgments: The joint influence of implicit theories and intelligence. *Organizational Behavior and Human Decision Processes, 50*, 300-323.

Hough, L. M. Eaton, N. K., Dunnette, M. D., Kamp, J. D., & McCloy, R. A. (1990). Criterion-related validities of personality constructs and the effect of response distortion on those validities. *Journal of Applied Psychology, 75*, 581-595.

Hunthausen, J. M., Truxillo, D. M., Bauer, T. N., & Hammer, L. B. (2003). A field study of frame-of-reference effects on personality test validity. *Journal of Applied Psychology, 88*(3), 545-551.

Jackson, D. N., Wroblewski, V. R., & Ashton, M. C. (2000). The impact of faking on employment tests: Does forced-choice offer a solution? *Human Performance*, *13*(4), 371-388.

Johnson, J. A. (2004). The impact of item characteristics on item and scale validity. *Multivariate Behavioral Research*, *39*(2), 273-302.

Kluger, A. N., Reilly, R. R., & Russell, C. J. (1991). Faking biodata tests: Are option-keyed instruments more resistant? *Journal of Applied Psychology*, *76*(6), 889-896.

Knowles, E. S., & Byers, B. (1996). Reliability shifts in measurement reactivity: Driven by content engagement or self-engagement? *Journal of Personality and Social Psychology*, *70*(5), 1080-1090.

Krokos, K. J., Meade, A. W., Cantwell, A. R., Pond, S. B., & Wilson, M. A. (2004, April). *Empirical keying of situational judgment tests: Rationale and some examples.* Paper presented at the 19th annual conference of the Society for Industrial and Organizational Psychology, Chicago.

Kunda, Z. (1999). *Social cognition: Making sense of people.* Cambridge, MA: MIT Press.

Lautenschlager, G. J. (1985, March). Controlling response distortion of an empirically-keyed biodata questionnaire. In G. S. Shaffer (Chair), *Twenty years of biodata research.* Symposium presented at the annual meeting of the Southeastern Psychological Association, Atlanta, GA.

Lefkowitz, J., Gebbia, M. I., Balsam, T., & Dunn, L. (1999). Dimensions of biodata items and their relationships to item validity. *Journal of Occupational and Organizational Psychology*, *72*, 331-350.

MacLane, C. N., Barton, M. G., Holloway-Lundy, A. E., & Nickels, B. J. (2001, April). *Keeping score: Empirical vs. expert weights on situational judgment responses.* Paper presented at the 16th Annual Convention of the Society for Industrial and Organizational Society, San Diego, CA.

MacLane, C. N., Martin, N. R., Liberman, B. E., & Nickels, B. J. (2004, April). *Written communication and writing skill: Confusion or combination.* Paper presented at the 19th annual conference of the Society for Industrial and Organizational Psychology, Chicago.

Mael, F. A. (1991). A conceptual rationale for the domain and attributes of biodata items. *Personnel Psychology*, *44*, 763-791.

McCall, R. J. (1958). Face validity in the D scale of the MMPI. *Journal of Clinical Psychology*, *14*, 77-80.

McFarland, L. A., Ryan, A. M., & Ellis, A. (2002). Item placement on a personality measure: Effects on faking behavior and test measurement properties. *Journal of Personality Assessment*, *78*(2), 348-369.

Mumford, M. D., & Stokes, G. S. (1992). Developmental determinants of individual action: Theory and practice in applying background data measures. In M. D. Dunnette & L. M. Hough (Ed.), *Handbook of industrial and organizational psychology* (Vol. 3, pp. 61-138). Palo Alto, CA: Consulting Psychologists Press.

Ones, D. S., & Viswesvaran, C. (1998). The effects of social desirability and faking on personality and integrity assessment for personnel selection. *Human Performance*, *11*, 245-269.

Ones, D. S., Viswesvaran, C., & Korbin, W. P. (1995). *Meta-analyses of fakability estimates: Between-subjects versus within-subjects designs.* Paper presented at the 10th annual conference of the Society for Industrial and Organizational Psychology, Orlando, FL.

Ones, D. S., Viswesvaran, C., & Reiss, A. (1996). Role of social desirability in personality testing for personnel selection: The red herring. *Journal of Applied Psychology, 81,* 660-679.

Schmit, M. J., & Ryan, A. M. (1993). The Big Five in personnel selection: Factor structure in applicant and nonapplicant populations. *Journal of Applied Psychology, 78*(6), 966-974.

Schmit, M. J., Ryan, A. M., Stierwalt, S. L., & Powell, A. B. (1995). Frame-of-reference effects on personality scale scores and criterion-related validity. *Journal of Applied Psychology, 80,* 607-620.

Schmitt, N., & Kunce, C. (2002). The effects of required elaboration of answers to biodata questions. *Personnel Psychology, 55,* 569-587.

Schmitt, N., Oswald, F. L., Kim, B. H., Gillespie, M. A., Ramsay, L. J., & Yoo, T. (2003). Impact of elaboration on social desirability and the validity of biodata measures. *Journal of Applied Psychology, 88,* 979-988.

Smith, E. R. (1998). Mental representations and memory. In D. Gilbert, S. T. Fiske, & G. Lindzey (Eds.), *Handbook of social psychology* (Vol. 1, 4th ed., pp. 391-445). New York: McGraw-Hill.

Smither, J. W., & Reilly, R. R. (1987). True intercorrelation among job components, time delay in rating, and rater intelligence as determinants of accuracy in performance ratings. *Organizational Behavior and Human Decision Processes, 40*(3), 369-369.

Snell, A. F., & McDaniel, M. A. (1998, April). *Faking: Getting data to answer the right questions.* Paper presented at the 13th annual conference of the Society for Industrial and Organizational Psychology, Dallas, TX.

Snell, A. F., Sydell, E. J., & Lueke, S. B. (1999). Towards a theory of applicant faking: Integrating studies of deception. *Human Resource Management Review, 9*(2), 219-242.

Tetlock, P. E. (1983). Accountability and complexity of thought. *Journal of Personality and Social Psychology, 45,* 118–126.

Tetlock, P. E. (1985). Accountability: The neglected social context of judgment and choice. In B. M. Staw & L. Cummings (Eds.), *Research in organizational behavior* (Vol. 1, pp. 297-332). Greenwich, CT: JAI Press.

Tourangeau, R., & Smith, T. W. (1996). Asking sensitive questions: The impact of data collection mode, question format, and question context. *Public Opinion Quarterly, 60*(2), 275-304.

Vasilopoulos, N. L., Cucina, J. M., & McElreath, J. M. (2005). Do warnings of response verification moderate the relationship between personality and cognitive ability? *Journal of Applied Psychology, 90*(2), 306-322.

Vasilopoulos, N. L., Cucina, J. M., Dyomina, N. V., Morewitz, C. L., & Reilly, R. R. (in press). Forced-choice personality tests: A measure of personality and cognitive ability? *Human Performance.*

Vasilopoulos, N. L., Reilly, R. R., & Leaman, J. A., (2000). The influence of job familiarity and impression management on self-report measure response latencies and scale scores. *Journal of Applied Psychology, 85,* 50-64.

Weekley, J. A., & Jones, C. (1997). Video-based situational testing. *Personnel Psychology, 50,* 25-49.

Wiener, D. N. (1948). Subtle and obvious keys for the Minnesota Multiphasic Personality Inventory. *Journal of Consulting Psychology, 12,* 164-170.

Wyer, R. S., & Srull, T. K. (1986). Human cognition in its social context. *Psychological Review, 93,* 322-359.

CHAPTER 13

LET'S GO FAKING!

Culture and Response Distortion in International Employment Testing

Richard Frei, Yukiko Yoshita, and Joshua Isaacson

As more international companies incorporate personality-based inventories into their selection batteries, human resources (HR) practitioners have begun to take a closer look at how response distortion differs across cultures. While there has been considerable research comparing personality constructs and translated scales between countries (e.g., Paunonen & Ashton, 1998), far fewer studies have examined applicant response distortion with non-U.S. samples, and only a handful have looked at how cultural issues may influence the criterion and construct validity of these selection tests. The impact of such research would be far-reaching. It could provide industrial/organizational (I/O) psychologists with a better understanding of how cultural issues influence the selection process beyond Schneider's (1987) Attraction/Selection/Attrition (ASA) model. Likewise, personality researchers could gain a better understanding of culture-specific personality traits and how context and situation influence self-reporting in other cultures. HR practitioners

A Closer Examination of Applicant Faking Behavior, 333–355
Copyright © 2006 by Information Age Publishing
All rights of reproduction in any form reserved.

could use such knowledge to create culture-specific cutoff scores or weighting schemes for selection battery components.

It is, therefore, disheartening that industrial/organizational psychologists have taken a behavior as cognitively complex and multifaceted as response distortion and reduced it to a simplistic concept known as *faking*. Unlike comparable terms such as social desirability and impression management, which imply at least some knowledge of social norms or a propensity to positively self-present to people of influence, faking has no redeeming qualities. Proponents of the "faking matters" argument characterize the behavior as deliberate and amoral misrepresentation for the sole greedy purpose of obtaining employment. Since personality-based selection tests often tap into respondents' feelings and thoughts and are largely unverifiable, I/O psychologists rarely use the term *lying* to describe this type of deception (whereas it is common to say that an applicant *lied* on a resume or *lied* during an interview). Regardless, it is clear that the term faking is synonymous with lying in noncognitive employment testing.

Over the past 10 years, the debate over faking has gotten both more narrowly focused and more acrimonious, for a number of reasons. Whereas past generations of researchers who have tackled this issue focused on methods to minimize deliberate faking (e.g., forced choice response formats, subtle items, warnings), the current batch of faking research has focused primarily on between-group comparisons of applicant and nonapplicant samples or respondents using different instructional sets meant to simulate an applicant setting (e.g., "fake good" vs. "answer honestly"). This has led to a series of statistical slap fights, with researchers using more elaborate methods of structural equation modeling and item response theory to highlight or minimize these differences. These studies tended to minimize messy situational or cultural factors that could possibly influence response distortion in favor of complex statistical tests that were of little practical use and told us nothing about the nature of response distortion in employment testing (but were very publishable, nonetheless).

Researchers in this field have been more likely to stake out the extreme positions that faking either: (a) destroys criterion validity or applicant rank ordering in a top-down selection system or (b) is just another form of positive self-enhancement that has negligible or no effects on validity. In other words, the vast middle ground, in which research on understanding response distortion in selection as a complex human behavior that may have many influences beyond job-relevant positive self-enhancing or outright lying to procure employment, remains largely fallow (for a notable exception, see Snell, Sydell, & Lueke, 1999). This polarization of the faking debate, like the word faking itself, seems oddly American.

As the faking debate spreads beyond the United States, it seems like the perfect opportunity to look at the potential influence of culture on response distortion. Culture consists of a complex set of variables, often difficult to measure and operationalize. Definitions of culture vary, but have certain commonalities. Matsumoto (2000) defines culture as a dynamic and yet relatively stable system of shared rules, both explicit and implicit, that are established by groups and communicated across generations, which can include attitudes, values, beliefs, norms, and behaviors. Fiske (2002) provides a definition of culture that is very similar to that of Edgar Schein's (1992) view of organizational culture; "A socially transmitted or socially constructed constellation consisting of such things as practices, competencies, schemas, symbols, values, norms, institutions, goals, constitutive rules, artifacts, and modifications of the physical environment" (p. 84).

When using noncognitive selection tests in international settings, differences in culture could manifest themselves in a number of ways. Norms and cutoff scores created with U.S. samples may not be applicable in another country. When the applicant pool consists of respondents from multiple countries, cultural differences in response distortion may influence the rank ordering of applicants in a top-down selection system. Cultures may differ in terms of the methods and very nature of performance appraisal (e.g., the importance of employee lateness or conflict avoidance in giving feedback), thus influencing the criterion-related validity of personality tests. Cultural differences may also influence applicants' perceptions of the self, employer expectations about responding, or even the interpretation of the (outwardly) simple instructions "Please respond honestly."

Cross-cultural differences on noncognitive selection test scores could also reflect real group-level differences on personality traits. Allik and McCrae (2004) refer to this as the reverse causality hypothesis. It is based on the trait perspective that personality traits are largely influenced by genetics (as opposed to shared environment) and are stable throughout adulthood, mostly impervious to life events. Therefore, the influence of culture on personality should be negligible. Instead, proponents of reverse causality argue that culture is a social adaptation based on the aggregate personality traits of its members. McCrae (2000, 2002) correlated mean Big Five scores (as measured by the NEO-PI-R) from 36 global samples with Hofstede's dimensions of national culture (power distance, individualism, masculinity, uncertainty avoidance). He identified several positive, significant relationships between Big Five factors and national culture, such as individualism/extraversion and uncertainty avoidance/neuroticism. Hofstede and McCrae (2004) re-examined the data and offered radically different explanations for these relationships. While Hofstede argued that national cultures affect the way people respond to personality tests (e.g., what behaviors are deemed socially desirable in a given culture), McCrae hypothesized that differences in national culture were merely a reflection of true group-level difference in personality traits. Thus, those cultures that had people who were higher in neuroticism would create value systems and associated institutions that maximized uncertainty avoidance.

The goal of this chapter is to highlight cultural variables that might influence applicant faking and summarize the few published studies on response distortion in international employment testing. Using Japan as an example, we offer ideas and suggestions as to *why* response distortion might differ between cultures. It is the authors' hope that this chapter will not only guide those who wish to further explore possible cultural differences in response distortion, but also inspire researchers who study faking in this country to delve further into the very nature of response distortion beyond simple between-groups differences.

INTERNATIONAL FAKING STUDIES

The authors could only locate five published studies on response distortion in employment testing that were conducted outside of the United States. Like faking research in the United States, the idea that culture may influence faking behavior is mostly ignored in these studies. Only one study (Sandal & Endresen, 2002) directly tested for cross-cultural differences in faking, between the United States and Norway. Of the remaining four, two (Goeters, Timmermann, & Maschke, 1993; and Steffens, 2004) compared the fakability of different personality tests and two (Peeters & Lievens, 2005; Law, Mobley, & Wong, 2002) looked at within-group differences in faking under various conditions.

Sandal and Endresen (2002) administered the Good Impression scale from the California Personality Inventory (CPI) to Norwegian students under "straight take" and "fake good" instructions and compared their scores with normative data using U.S. students from the CPI Manual (Gough & Bradley, 1996). They hypothesized that Norwegian students would score lower on the scale than the American normative sample because the high feminine Norwegian culture (Hofstede, 1980) may apply normative pressure against self-promotion. They found that, compared to the U.S. normative samples, Norwegian students did score significantly lower than the U.S. samples in both the "straight take" and "fake good" conditions on the CPI scale. They also compared the student sample with a sample of 494 Norwegian job applicants, who took the entire CPI as part of selection battery. The mean for the scale was two points higher for the applicant sample than for the "fake good" sample. While this difference is not significant, it is interesting to note that, using Gough's (1987) cutoff score of 30 or above as indication of faking, an equal number of applicants and "fake good" students scored at or above the cutoff. Further, the standard deviation for the "fake good" group was nearly 1.5 points higher than that of the applicant group. This indicates that perhaps not everyone has the ability to fake good when told to do so. Finally, while the authors provided normative data for U.S. students under honest and fake good instructional sets, they failed to provide the normative data for applicants in the United States. Such information would be useful in determining whether the mean differences between U.S. and Norwegian students under the honest and fake good instructions also hold true for U.S. and Norwegian applicants as well.

Goeters et al. (1993) compared the fakability of the Personality Research Form (PRF) to the Temperament Structure Scales (TSS) in a sample of applicants for pilot training with a major European airline. The TSS is a personality scale designed specifically for pilots. The pilots were given the inventories twice, once before they applied for the job, and once after they applied for the position. Their results indicated that the TSS was much less susceptible to faking than the PRF. However, the authors admitted that "due to its susceptibility to faking tendencies, the PRF manual warns against the administration of the test for selection (p. 127)." The authors make no mention of any possible differences in faking among applicants from different European countries.

Steffens (2004) compared the conscientiousness scale of the NEO-FFI with an Implicit Association Test (a test of automatic cognition) designed to measure the same construct. Steffens argued that implicit measures have been shown to be more robust against deception than more explicit measures. She was also interested in whether the IAT would be more susceptible to faking if respondents took the test more than once. She assigned participants to either fake good or fake bad conditions. In the fake good condition, respondents were asked to pretend that they were applying for a job where the employer was most interested in conscientious work. In the fake bad condition, the respondents (in a wonderful example of cultural influence) were given the following instructions,

> The not conscientious instruction asked them to imagine they were renting a room in an apartment-sharing community. People in the community were totally cool and would not rent to bourgeois people ... the least conscientious person will be selected. (p. 168)

Correlations between the conscientiousness scale of the NEO-FFI scales and the implicit measure of the construct were of a medium order of magnitude. The IAT was much more difficult to fake than the NEO-FFI, although subsequent practice with implicit measures reduced its robustness to faking.

Peeters and Lievens (2005) looked at the influence of faking on the predictiveness of Bess and Mullins' (2002) situational judgment test with college students in Belgium. Students were administered a battery of tests, including the Advanced Progressive Matrices and the authorized Flemish version of the NEO-FFI. They were also given a translated version of the SJT, with half the subjects instructed to answer as if they were taking part in a college admissions exam and to get the highest score possible, while the other half were instructed to answer as honestly as possible. When students were told to fake, they were able to raise their scores on the SJT. The correlation between the SJT and GPA was significantly higher in the "honest" condition ($r = .33$) than the "fake" condition ($r = .09$), which the authors interpreted as the negative impact of faking on the criterion-related validity of the SJT. Further, the authors found that, in a selection battery which included tests of cognitive ability and the Big Five, the SJT added incremental validity to the prediction of GPA in the honest condition, but not in the faking condition. Peeters and Lievens concluded that "when students fake, and they probably do this in a selection context, our SJT is no longer a valid predictor" (p. 82). Again, there was no mention that Belgian respondents may in any way differ from other cultures with regards to faking.

Law et al. (2002) used a unique methodology to differentiate between social desirability and faking of a biodata inventory in a sample of graduating actively job-seeking college seniors in China. Over 1 thousand students were sent a packet of surveys, including the biodata inventory, the Personal Characteristics Survey (PCI; Barrick, Mount, & Strauss, 1993), and an ethics/attitude toward lying scale designed by the authors, and told that the career center at the university was collecting information about graduating seniors. Approximately 40% of the students returned the surveys. The authors argued that students who were predisposed to respond in a socially desirable way would have done so on this first administration.

They then separated the sample into three subgroups and sent them the same biodata survey, with one of three instructional sets: (1) Research only (students were told that their results would be used for research purposes only), (2) job applicant (students were told that employers would have access to the biodata survey results) and, (3) best presentation (students were told that employers would have access to the biodata survey results, so they should try to present themselves as favorably as possible). 164 students returned both sets of surveys and respondents were approximately evenly distributed across the three instructional sets. Using a repeated measures design, the authors controlled for socially desirable responding by partialing out the effect of the Time 1 biodata scores.

The authors hypothesized that, consistent with previous faking literature, students would score highest under the "best impression" instructional set and lowest on the "research only" instructional set. Further, they hypothesized that students who scored lowest on the biodata scale at Time 1 would engage in the most faking at Time 2. Both of these hypotheses were supported. They also found that students who scored lower on the ethics scale had a higher propensity to fake. However, the authors also noted that, as compared to American norms for the scale, Chinese students scored significantly lower. They attributed this to two potential cultural factors First, the biodata scoring key was developed on applicants who had preselected themselves into the company's selection process. Because of the company has a reputation as being very selective, it is probable that the normative sample was qualitatively different than the current sample. They also hypothesized that Hong Kong participants did not engage in as much self-promotion, under any condition, as compared to their Western counterparts. The authors recommend that true cross-cultural studies be conducted to explore these differences.

The aforementioned studies on international applicant faking are similar to American studies in that they minimize the importance of cultural and situational factors, as epitomized by studies in which respondents are given instructions to fake or answer like an applicant. The entire job application process, including pressures to succeed, fear of failure or getting caught telling a lie, and the intrinsic and extrinsic motivation associated with any given job, is reduced to a one or two sentence instructional set, with little experimental or mundane realism. Researchers on both sides of the faking argument agree that respondents can raise their scores on noncognitive selection tests when told to do so. But whether this type of "instructed faking" actually reflects the true complexity of applicant response distortion has been a main criticism offered by personality testing proponents (e.g., Rosse, Stecher, Miller, & Levin, 1998), who dismiss these studies as artificial. However, when it comes to better understanding these complexities, the very cultural and situational factors that make faking such a multifaceted behavior, researchers have remained largely on the sidelines.

CULTURE AND RESPONSE DISTORTION IN JAPAN: AN EXAMPLE

In order to generate discussion and debate regarding the future of international faking research, and to highlight the potential influences of culture on faking, we turn to Japan as an example. Personality tests and other noncognitive selection tools are popular in Japan, and yet very little research has been conducted on the influence of response distortion on such tests (we could find no published studies that directly addressed applicant response distortion in Japan). While the cross-cultural equivalence of many U.S.-developed selection scales has been established in Western cultures, their utility in Japanese employment testing has not been clearly demonstrated. While the United States and Japan have a very active business relationship, few Westerners actually understand the intricacies and complexities of Japanese culture and find doing business with the Japanese strange and frustrating. It is not surprising that famed Japanese historical novelist Ryootaroo Shiba defined culture as

> something that is shared and treasured by a group of people who feel comfortable within that cultural boundary as if they were wrapped up in a soft warm cover no matter how strange and ridiculous it may look to outsiders. (Ikeda, 2004, p. 43)

Surprisingly, some Japanese organizations rely almost *solely* on personality measures to select employees, while others use personality measures as screening tools, as supplementary material for conducting interviews, or in the placement of the employees within the organization (SPI note no kai, 2003). Takeda (2003) provided a detailed description of Japanese organizations' personnel selection practices for new university graduates. Since almost all Japanese organizations offer extensive on-the-job training, selection batteries place stronger emphasis on interpersonal skills, such as the ability to communicate and collaborate with superior and colleagues (Shimada, 2005). Recruiters look for applicants who will fit in with the culture of the corporation; those who will absorb the goals of the organization and be content to build a career within the organization (Price, 2004). There is less emphasis on evaluating professional knowledge (it is assumed that the prestige of an applicant's university is an adequate proxy variable for task-related KSAs) and more emphasis on selecting applicants who can practice cooperation and harmony (Takeda, 2003).

This emphasis on an applicant's ability to work seamlessly with a group is also reflected in the traditional Japanese selection procedure of recruiting applicants for a particular organization rather than a specific job. In many ways, it is similar to the U.S. military, which recruits young adults to enlist and then places them in specific units based on their results on a battery of placement tests. Japanese organizations often have vague department structures and employees may be asked to move from one function to another depending on the flow of work. This is why Japanese workers describe themselves as "salary men."

Social Desirable Responding

A review of the literature regarding response distortion in Japan indicates that Japanese are more self-effacing, more modest, and more critical in self-descriptions than their Western counterparts. In a meta-analysis of cross-cultural enhancement studies, Heine and Hamamura (in press) found that East Asians showed significantly weaker self-enhancement than Western samples in 79 out of the 81 studies with a significant average effect size ($d = .83$). In a recent special edition of *Journal of Cross-Cultural Psychology* dedicated to self-enhancement in Japan, every author concurred that the Japanese have weaker self-enhancement motivations than North Americans, although they differed as to why such motivations were weaker (Heine, 2003). Japanese are also especially self-critical in situations that they generate themselves (Kitayama, Markus, Matsumoto, & Norasakkunkit, 1997). Heine and Lehman (1995a) found no difference between Canadian and Japanese students on social desirability as measured by the Balanced Inventory of Desirable Responding (BIDR). Stening and Everett (1984) found that, when responding using a Likert-type scale, Japanese managers were more likely to choose the midpoint than their American counterparts.

Researchers often turn to Hofstede's (1980) classification of Japan as a collectivist culture as a possible cause for this lack of self-enhancement (although a recent reanalysis of Hofstede's classification suggests that Japan may be a more individualist culture than once thought; Bond, 2002). They posit that countries high on individualism may have a greater propensity to fake because these countries emphasize the promotion of individual achievement and self-interest over group ties (Hofstede, 1983). Kurman and Sriram (1997) argue that the essence of collectivism is the tendency to subordinate personal needs to in-group needs. Therefore, self-enhancement is less important in collectivist countries. Conversely, other researchers argue that countries high on collectivism may be more likely to fake. In his multiple studies of the individualist/collectivist dimension, Triandis (Triandis, 1995; Triandis & Suh, 2002) argued that members of collectivist countries tended to engage in more deception, lying, and face-saving behavior than those living in individualist countries.

Bedford and Hwang (2003) suggested that the emotion of guilt and shame are pervasive in daily life. They help to maintain sense of personal identity and function as mechanisms of social control. Both guilt and shame are associated with religion and ethics, and different cultures consider and experience them differently (Ho, Fu, & Ng, 2004). Benedict (1946) in her influential book *The Chrysanthemum and the Sword* depicted Japan as a "shame culture." Japanese are more concerned with bringing shame on others than any internal guilt associated with negative behavior. How might this influence response distortion in Japan? Respondents might be less inclined to endorse negative items, as admission of these traits would reflect poorly on the group. On the other hand, research shows that Japanese demonstrate a substantial self-criticism bias which works as a motivator for self-improvement. We know that Americans engage in socially desirable responding, even on anonymous surveys, if they feel guilty about their behavior (e.g., smokers will report smoking less cigarettes a day to make themselves feel better about their smoking). Would Japanese applicants, when confronting a question about their own negative behavior, respond hypercritically, making their behavior sound actually *worse* than it is as a way of showing modesty and the desire to improve?

In an attempt to explain the discrepancies in cross-cultural social desirability research, Lalwani, Shavitt, Johnson, and Zhang (2004) looked at the differences between individualist and collectivist cultures in terms of socially desirable responding on the BIDR. The authors found that while respondents from both individualist and collectivist cultures engaged in socially desirable responding, they differed in their self-presentation strategy. Respondents from individualist cultures were more like to engage in self-deceptive enhancement, characterized by a belief in one's competence to be self-reliant and independent. Respondents from collectivist cultures were more likely to engage in impression management, characterized by image protection, relationship maintenance, and a willingness to dissemble to save face.

However, Heine (2003) argued that "when [Japanese] individuals are motivated to compete with others (*soto*) and aspire to achieve dominance over others, self-enhancing motivations become more functional" (p. 597). Evidence indicates that, as a result of recent economic downturns, competition for employment in Japan is fiercer than ever. The concept of lifetime employment (*shushin koyo*), which was once the cornerstone of the Japanese economic system, has been slowly falling out of favor. Ono (2005) estimates that only approximately 20% of Japanese adults have lifetime employment, with women and younger workers largely left out of the system. Additionally, the number of *freeters* (people between the age of 15 and 34 who lack full time employment and are often underemployed in part-time or temporary jobs) has doubled in the past 10 years to nearly 2 million (Kosugi, 2005). The introduction of *restra* (a reduction in work force, comparable to American restructuring) as a method of making organizations more cost-effective has further tightened the job market (Kikuno, 2003). Thus, the competition for full-time positions that offer the opportunity of lifetime employment has become even more intense; therefore, one might hypothesize that the faking of noncognitive selection tests in Japan might become more prevalent. Snell et al. (1999) provide an excellent interactional model of applicant faking that takes into account both dispositional and motivational facets such as financial incentive that might be a good starting point for researchers studing response distortion in Japanese employment testing.

Given the importance of modesty and humility in Japanese culture, there has been considerable discussion regarding whether Japanese might actually *feign* modesty when responding to questionnaires (Heine & Lehman, 1995a, 1995b; Heine, Lehman, Marcus, & Kitayama, 1999; Kitayama et al., 1997; Marcus & Kitayama, 1991). In other words, it is possible that, in applicant settings, Japanese may (by Western standards) *purposely* fake bad, making themselves look more average to highlight their modesty. In Western cultures, people tend to measure their success normatively rather than ipsatively and usually view themselves as above average, a la the Lake Wobegon Effect. However, Marcus and Kitayama suggest that the Japanese use a different self-system that emphasizes "beauty in modesty." In Japan, to describe oneself as "the average man" (*average* meaning similar to everyone else, as opposed to the Western meaning of mediocrity) is thought to be socially desirable. Barnlund (1989) argued that, since the group is the measure of all things in the Japanese workplace, too frequent or too strong an assertion of the self may threaten the group's solidarity. Respondents in Western cultures emphasize their superiority over others; however those in Eastern cultures tend to focus on their similarity to others (Marcus & Kitayama; Triandis, 1998).

Overall, the research on socially desirable responding in Japan is mixed. While Japanese exhibit self-criticism and modesty biases, they also show a strong tendency towards impression management and self-presentation. Would Japanese job applicants attempt to appear modest or inflate their self-evaluations on personality tests used for selection? The answer is that a desirable job applicant knows *when* it is appropriate to appear modest and *when* to present a positive public face.

The Importance of Context and Situation in Personality

Research on attribution theory in Asian cultures is consistent with the notion that people's behavior is viewed more as a function of situational pressures than the product of dispositional factors (Miller, 1984; Morris & Peng, 1994; Weisz, Rothbaum, & Blackburn, 1984). Many public situations in Japan require modest self-presentations (Kitayama, Masuda, & Lehman, 1998) and individuals may stand to lose the respect of their peers if they do not present themselves in a modest, self-effacing manner (e.g., Bond, Leung, & Wan, 1982). In a series of four experiments that looked at a variety of Asian cultures, Sanchez-Burks, Lee, Choi, Nisbett, Zhao, and Koo (2003) found that Americans paid less attention to relational cues in work settings compared with nonwork settings, while respondents in Eastern cultures were more attuned to situational cues at work and, thus, were more likely to engage in indirect communication at work. The authors attributed these differences to Protestant Relational Ideology (PRI), the widely-held American belief that relational concerns are considered inappropriate in the workplace and are attended to less than in social settings.

Situations in which a person behaves one way (*tatemae*) but truly feels something quite different (*honne*) are more prevalent in Japan than in Western cultures. *Tatemae* means to prop up or to present a frontal or upfront view (but certainly not in the Western sense of "being up front about something"). Politicians often engage in this behavior; saying something in a speech that they clearly do not mean, but think that their constituents want or expect to hear. In any business situation, including employment selection, one is always in public. *Honne* literally translates to mean "real bone," "from the bone" or "the real truth." *Honne* is supposed to be "kept for oneself," not for others to know, and rarely expressed in public. This is especially true if the attitude is negative. Since the expression of *tatemae* and *honne* is clearly situation-specific, personality test items that ask about general behavior without providing a situation or context may be confusing or even insulting to Japanese applicants. Robie, Schmitt, Ryan, and Zickar's (2000) research on narrowing item context specificity on personality-based selection tests (e.g., making items more work-specific) may be useful in adapting Western-developed scales to Japan selection system.

The ability to read a situation and then manipulate tatemae to fit the situation is part of the larger concept of amae. *Amae* is an indigenous psychological concept first proposed by Doi (1971) that permeates all aspects of life and behavior in Japan. Doi explains that it is the noun form of *amaeru*, which means "to depend and presume upon another's benevolence," typified by "helplessness and the desire to be loved." The origin of amae lies in the relationship between a child and its mother. Once a developing infant realizes that that his mother exists independently of himself, he gradually sees his mother as indispensable and craves close contact (Doi, 1971). The mother responds by lavishing attention on the child. While there is no direct translation for *amae* in English, the rather obscure word "mollycoddle" may help us to understand the nature of this relationship. People who want to be mollycoddled do not articulate their desire for attention, but hope by their actions to elicit indulgence from another without the use of language. As soon as they put their desire into language they are putting themselves on an equal footing, as another separate desiring individual. It is through *amae* that Japanese children learn the importance of indirect communication, developing situation-specific behavior, and knowing when to exhibit it. The ability to manipulate *tatemae* is a practiced art and, thus, is viewed as a sign of maturity (Doi, 1986). Amae is the prototype on which all human relationships in Japan are patterned, especially (though not exclusively) when one person is senior to another (Doi, 1971). Gibney (1975) stated, "He may be your father or your older brother or ... your section head at the office ... the amae syndrome is pervasive in Japanese life" (p. 119).

Nisbett, Peng, Choi, and Norenzayan (2001) proposed a theoretical model to explain individualist/collectivist differences in situation sensitivity using ancient Greek and Chinese society and philosophy. The Euro-American civilization was greatly influenced by Greek civilization, while Japanese civilization traces its origins to ancient Chinese civilization. In ancient Greek society, power was perceived to be held by the individual. Greek philosophy focused on categories and rules that would help them understand objects, independent of context. Life was filled with choices and absent of constraints, and ancient Greeks, in some small way, had some personal control over their environment. Chinese society, on the other hand, emphasized that individuals were part of a larger system, characterized by close, complex and well-structured relationships with other group members. They were concerned with relationships among objects and events. Under these conditions, people learned to be hypersensitive to small changes in social situations and how to adapt their behavior to fit these situations (Masuda & Nisbett, 2001).

This sensitivity to situation certainly plays a role in the Japanese perception of what behavior constitutes a lie. This discrepancy between internal attitudes and outward behavior arouses less dissonance among Japanese than it does for Westerners (cf. Heine & Lehman, 1997b; Kashima, Siegal, Tanaka, & Kashima, 1992). In general, Westerners (and especially Americans) attribute lying to dispositional factors and consider it a sign of a defective character. In the United States, the phrase "the truth is subjective" is often met with derision by many who believe that truth is an absolute, rather than a socially constructed phenomenon. One might argue that the idea of absolute truth is a reflection of our Christian roots, in which the term "Truth" is associated, or sometimes even synonymous, with Jesus Christ (e.g., "I am the way, the truth, and the light"). It manifests itself in our legal system, where witnesses swear to tell the truth, the whole truth, and nothing but the truth, so help them God. It appears in our national folklore in the form of a young George Washington who, when confronted with a chopped-down cherry tree (later proven to be apocryphal), boldly told his father, "I cannot tell a lie."

Japanese values are more dependent on the situation (Creighton, 1990). It is said that *tatemae* is the social grease that keeps collectivist cultures moving. The gradients of responses are a well-rehearsed part of public face in Japan. Thus, lying does not have the intensely negative and moral stigma that it does in the United States. In Japan workplace, two common sayings, "A lie is something expedient" and "You have to lie to get ahead in the world" reflect the utility of lying. This is not to say that Japanese organizations are untrustworthy; rather, they prefer to hire people who know *when* it is appropriate to deceive as part of *tatemae*.

Seiter, Bruschke, and Bai (2002) looked at the degree to which deception was perceived to be a socially acceptable form of communication for American and Chinese students. The authors hypothesized that the cultural background of the student (Chinese vs. American), the motivation for the lie (e.g., benefit self, avoid conflict), and the relationship to the target of the lie (e.g., spouse, boss, stranger) would influence the perceived acceptability of the lie in written scenarios. Chinese students rated lies associated with benefiting others, affiliation, privacy protection, conflict avoidance, and impression management as more acceptable than U.S. students. There was also an interaction between motivation and target; in other words, the acceptability of lying to avoid conflict depends on whom the deception is directed towards. The authors noted that Chinese acceptability of deception should not be viewed pejoratively, but instead should be taken to mean that Chinese prioritize values differently than Americans. Cross-cultural differences in perceived acceptability may also be due to an overall depression of acceptability scores in the American sample. They posited that individualist cultures view ethical judgments in terms of absolutes, which may have minimized the distinction between different motivations for American respondents.

The Education System

Colleges in Japan put great emphasis on preparing students for the job application process. Part of that process is learning how to develop and maintain relationships and hone public presentation skills. In a recent survey conducted by the Japanese Ministry of Internal Affairs and Communications (1998), college students and recent graduates from Japan and nine other countries were asked to rate the goals of a higher education. The results were striking. Students from every country outside of Japan indicated that the primary goal of higher education was "to gain common and basic knowledge," while recent college graduates rated "to gain professional knowledge and specialty" as the primary goal. However, both Japanese students and college graduates indicated that the primary goal of higher education was "to develop relationship and friendship with other people." Japanese organizations classify and interview new graduates according to reference groups, based on the prestige of his or her university, department, or club. Poole (2003) argued that clubs serve an important function in Japanese higher education, bringing students together and giving them practice at developing and maintaining close relationships. Large universities may have as many as several hundred clubs. Because of this emphasis on character and personality by prospective employers, participation in clubs is an important factor in a company's selection process. Student leaders of clubs, especially sports club leaders, are highly favored.

Many Japanese organizations recruit and hire new graduates only once a year. While colleges and universities in the United States may offer workshops on interview skills and resume writing, preparation for employment testing is actively taught in every major university in Japan. Students are put through a series of simulated interviews to hone public presentation skills and given tips on how to answer common employment test questions. Preparation for employment testing begins in the third year of college and, by the beginning of the fourth year many students already have an offer of employment. The Japanese are accustomed to this type of rigorous testing; it is ingrained into the education system at a very early age. Poole (2003) called this an escalator educational system, the idea that, in order to get a job with a leading organization, one must attend a leading university. In order to be accepted by a leading university, one must have attended a leading high school, junior high school, and even elementary school, all of which require the extremely competitive entrance examinations.

For those who do not attend university (or do not get hired by the organization of their choice during the traditional hiring period), there are numerous books, private schools, and seminars to better prepare applicants for employment testing. Several leading Japanese employment testing prep books suggest that there are "right" answers to personality-based selection tests and even go as far as to explain what each major organization looks for and how to respond to these exams to get placed in the position of your choice (SPI note no kai, 2003). In the United States, researchers who look at faking assume that the vast majority of job applicants have had neither access to the test nor instruction on how to respond favorably. Researchers who study faking in Japan many not have that luxury.

Snell et al. (1999) argued that academic dishonesty may be a possible correlate of applicant faking. Nonis and Swift (2001) found that students who engaged in dishonest acts in college classes were more likely to engage in dishonest acts in the workplace. If people are disposed cheat and lie in an academic setting, they may also be willing and motivated to cheat and lie in other test-taking situations, such as employment selection. Therefore, cross-cultural differences in academic dishonesty may provide an interesting (yet albeit incomplete) starting point for looking at possible cross-cultural differences in applicant faking. For example, de Lambert, Ellen, and Taylor (2003) looked at academic cheating at tertiary institutions in New Zealand. They gave students a list of possible cheating scenarios and asked them if they thought the scenario was indicative of cheating. Students were grouped according to ethnicity (Asian, European, and New Zealand/Maori/ Pakeha). Results indicated that, for every scenario, a higher proportion of Asian students viewed the scenarios as not examples of cheating. Diekhoff, LaBeff, Shinohara, and Yasukawa (1999) compared self-reported cheating between samples of American and Japanese college students. They found that Japanese reported a higher rate of cheating on exams, a greater tendency to justify their cheating behavior, and a greater passivity in their reactions to the observed cheating of others. Among cheaters of both nationalities, Japanese students rated social stigma and fear of punishment as less effective deterrents than American students did. Among noncheaters, guilt was the most effective deterrent.

Cross-cultural differences between the United States and Japan on the acceptability and prevalence of academic dishonesty may be an artifact of the Japanese employment selection system. Organizations pay more attention to the prestige of an applicant's university or how well he or she will fit into the overall organizational culture and less attention to GPA when hiring new employees. Therefore, academic dishonesty may be viewed as less important and more situational to Japanese students, as it seems to be with other Eastern cultures. However, this lax view towards academic dishonesty should not be confused with a lack of respect for teachers or professors. In Seiter et al. (2002) study of the perceived acceptability of lying, Chinese students rated lying as more acceptable than American students for every target, *except* teacher.

CONCLUSION

Fiske (2002) argues that, instead of trying to describe culture in terms of blanket dimensions such as individualism/collectivism, we should look at aspects of culture that have pervasive consequences for everyday life and psychological processes. These aspects include subsistence and economic systems, religion, marriage, sex and food, and institutions such as schools and businesses. Our examination of Japan gives researchers a starting point from which to develop theories of applicant faking in Japan, and, hopefully, some new ideas of how response distortion can be influenced by culture.

For example, knowledge of basic indigenous psychological concepts associated with the parent/child relationship such as *amae*, *tatemae*, and *honne* may help us better understand the importance of modesty, self-criticism, public presentation skills, and overall impression management in Japan. Such knowledge may be useful in developing hypotheses regarding cross-cultural differences in applicant faking. If Japanese selection systems emphasize organizational fit more than U.S. systems, Japanese applicants may be more likely to engage in impression management on items related to cooperation, teamwork, and conflict avoidance. Or, combined with the extensive preparation for employment selection testing in Japanese universities, faking in Japan may be better operationalized as an applicant's attempt to create a profile that indicates fit with a specific organization (especially the most prestigious and desirable organizations), rather than the inflation of personality traits that applicants perceive as relevant, important, and more desirable. The education system could further potentially influence applicant faking by acculturating Japanese to standardized testing at a very early age and providing college students with practice at employment tests. The changing economic conditions (as epitomized by dwindling lifelong employment and the growing ranks of freeters) may change the very nature or employment selection in Japan, as organizations move away (even if slightly) from selecting employees based on fit, and towards a more KSA-based approach. Even religion plays a role in Japanese perceptions of the acceptability of lying and subjective nature of truth.

Allik and McCrae (2004) has proposed the reverse causality hypothesis; cross-cultural differences on personality test scores reflect actual group-level differences on Big Five personality traits which, in turn, shape the national culture. To test the reverse causality hypothesis, Robie, Brown, and Bly (2005) compared managers from the United States and Japan on a measure of the Big Five. Using Hofstede's categorization of United States and Japan on the four dimensions of national culture and McCrae's (2002) data on personality and culture, Robie et al. made a number of hypotheses regarding differences between U.S. and Japanese managers. Given that the United States is relatively high on individualism and low on uncertainty avoidance, they hypothesized that U.S. managers would score higher on extraversion (which McCrae said was positively correlated with individualism), as well as agreeableness and emotional stability (which McCrae said were negatively correlated with uncertainty avoidance) than their Japanese counterparts. The authors compared 3,458 U.S. managers' and 410 Japanese managers' scores on the Global Personality Inventory (Schmit, Kihm, and Robie, 2000), a measure of the Big Five. The managers had taken the test for various unspecified selection, promotion, and development purposes. While U.S. managers did score, on average, one standard deviation higher on extraversion and agreeableness and nearly two standard deviations higher on emotional stability, they also scored about one standard deviation higher on conscientiousness and openness to experience. U.S. managers rated themselves higher on *all* traits, not just those predicted by Hofstede and McCrae's personality trait/national culture dimension link.

As further evidence of cross-cultural differences in responding between the United States and Japan, each manager's predicted job performance was rated on a 5-point scale by an "assessor." About 54% of the managers had been assessed for selection or promotion purposes, the remaining 46% for development purposes. As the U.S. managers rated themselves higher than their Japanese counterparts, U.S. assessors rated the U.S. managers nearly one standard deviation higher than the Japanese assessors rated the Japanese managers on predicted job performance. Robie et al. (2005) concluded that mean cross-cultural differences in personality dimensions and predicted job performance were more likely related to some form of systematic response bias than a function of reverse causality.

Given that Japanese responding is more influenced by situational factors, we contend that combining personality tests scores used for selection, promotion, and development may be masking situation-specific responding, particularly in the Japanese sample. For example, the use of self-criticism to motivate self-improvement in Japan may severely suppress Big Five scores in the Japanese development respondents. Robie et al. (2005) reported that nearly 50% of the sample had provided data in the context of employee development (as opposed to selection or promotion, which made up the other fifty percent). The authors did not further break down these percentages by country. If the Japanese sample (already considerably smaller than the U.S. sample) consisted of a higher proportion of development-influenced respondents, the hypercritical self-reporting that the Japanese use to motivate themselves in development settings may skew the national means.

In order to get a better understanding of how culture plays a role in applicant faking, industrial/organizational psychologists and HR practitioners could do as Fiske (2002) suggests; immerse themselves within the culture before conducting research. Fiske apparently has little confidence that psychologists will do this, at one point sarcastically saying (as if he was a psychologist) "Realistically, psychologists have to publish six articles a year, and you can't do that if you're hanging out in some village far away from your lab." Fiske suggests more cross-cultural research involving multiple researchers from different cultures who may have a better understanding of how locals may react to a noncognitive selection test. However, Fiske warns that people immersed within a culture sometimes do not have a good understanding of how culture influences their day-to-day behavior and only an outside eye can truly identify differences between cultures. As Fiske notes, that is why deTouquiville's critique of early American democracy remains a staple on many college campuses; the indispensable and unique perspective of the outsider.

Applicant response distortion is a complex behavior that cannot be captured simply by looking at social desirability or deliberate deception. The possible influence of culture on international employment testing will hopefully give researchers a new perspective from which to view the faking debate. We recommend that industrial psychologists take a page from organizational psychologists and systematically analyze a culture's organizations, institutions, relationships, and other artifacts to determine how culture and social context influence faking behavior in applicant settings.

AUTHOR'S NOTE

Address all correspondence to: Richard Frei, Department of Behavioral Sciences, Community College of Philadelphia, 1700 Spring Garden Street, Philadelphia, PA 19130 (rfrei@ccp.edu)

REFERENCES

Allik, J., & McCrae, R. (2004). Toward a geography of personality traits: Patterns of profiles across 36 countries. *Journal of Cross Cultural Psychology, 35*(1), 13-28.

Barnlund, D. (1989). *Communicate style of Japanese and Americans: Images and realities.* Belmont, CA: Wadsworth.

Barrick, M. R., Mount, M. K., & Strauss, J. P. (1993). Conscientiousness and performance of sales representatives: Test of the mediating effects of goal setting. *Journal of Applied Psychology, 78,* 715–722.

Bedford, O., & Hwang, K. K. (2003). Guilt and shame in Chinese culture: A cross-cultural framework from the perspective of morality and identity. *Journal for the Theory of Social Behaviour, 33,* 127-144.

Benedict, R. (1946). *The chrysanthemum and the sword.* Boston: Houghton Mifflin.

Bess, T. L., & Mullins, M. E. (2002, April). *Exploring the dimensionality of situational judgment: Task and contextual knowledge.* Paper presented at the 17th annual conference of the Society for Industrial and Organizational Psychology, Toronto, Ontario, Canada.

Bond, M. H. (2002). Reclaiming the individual from Hofstede's ecological analysis- A 20-year odyssey: Comment on Oyserman et al. (2002). *Psychological Bulletin, 128,* 73-77.

Bond, M. H., Leung, K., & Wan, K. C. (1982). How does cultural collectivism operate? *Journal of Cross Cultural Psychology, 13,* 186-200.

Creighton, M. R. (1990). Revising shame and guilt cultures: A forty-year pilgrimage. *Ethos, 18,* 279-307.

de Lambert, K., Ellen, N., & Taylor, L. (2003). Prevalence of Academic dishonesty in tertiary institutions: The New Zealand Story, Working Paper, Christchurch College of Education.

Diekhoff, G. M., LaBeff, E. E., Shinohara, K., & Yasukawa, H. (1999). College cheating in Japan and the United States. *Research in Higher Education, 40,* 343-353.

Doi, K. (1971). *Amae no kozo..* Tokyo, Japan: Kobundou.

Doi, T. (1986). *The anatomy of Self.* Tokyo, Japan: Kodansha.

Fiske, A. P. (2002). Using individualism and collectivism to compare cultures—A critique of the validity and measurement of the constructs: Comment on Oyserman et al. (2002). *Psychological Bulletin, 128,* 78-88.

Gibney, F. (1975). *Japan: The fragile superpower.* New York: Norton.

Goeters, K., Timmerman, B., & Maschke, P. (1993). The construction of personality questionnaires for selection of aviation personnel. *International Journal of Aviation Research, 3,* 123-141.

Gough, H. G. (1987). *CPI Administrator's Guide.* Palo Alto, CA: Consulting Psychologist Press.

Gough, H. G., & Bradley, P. (1996). *CPI Manual* (3rd ed.). Palo Alto, CA: Consulting Psychologists Press.

Heine, S. J., & Hamamura, T. (2005). *In search of East Asian self-enhancement.* Unpublished manuscript.

Heine, S. J., & Lehman, D. R. (1995a). Socially desirable among Canadian and Japanese students. *Journal of Social Psychology, 135,* 777-779.

Heine S. J., & Lehman, D. R. (1995b). Cultural variation in unrealistic optimism: Does the West feel more invulnerable than the East? *Journal of Personality and Social Psychology, 68,* 595-607.

Heine, S. J., Lehman, D. R., Marcus, H. R., & Kitayama, S. (1999). Is there a universal need for positive self-regard? *Psychological Review, 106,* 766-794.

Heine, S. J. (2003). Making sense of East Asian self-enhancement. *Journal of Cross-Cultural Psychology, 34,* 596-602.

Ho, D. Y. F., Fu, W., & Ng, S. M. (2004). Guilt, shame, and embarrassment: Revelations of face and self. *Culture & Psychology, 10,* 64-84.

Hofstede, G. (1980). *Culture's consequences: International differences in work-related values.* Beverly Hills, CA: Sage.

Hofstede, G. (1983). The cultural relativity of organizational practices and theories. *Journal of International Business Studies, 14,* 75-89.

Hofstede, G., & McCrae, R. (2004). Personality and culture revisited: Linking traits and dimensions of culture. *Cross-Cultural Research, 38*(1), 52-88.

Ikeda, S. (2004, June). *Teaching Japanese culture and language*. Paper presented at the Biennial Conference of the Asian Studies Association of Australia, Canberra.

Kashima, Y., Siegel, M., Tanaka, K., & Kashima, E. S. (1992). Do people believe behaviors are consistent with attitudes? Toward a cultural psychology of attribution process. *British Journal of Social Psychology, 331*, 111-124.

Kikuno, K. (2003). The new look of Japanese restructuring "Restra." *21seiki shakai design kenkyu, 2*, 17-27.

Kitayama, S., Marcus, H. R., Matsumoto, H., & Norasakkunkit, V. (1997). Individual and collective processes in the construction of the self: Self-enhancement in the United States and self-criticism in Japan. *Journal of Personality and Social Psychology, 72*, 1245-1267.

Kitayama, S., Masuda, T., & Lehman, D. R. (1998). *Cultural psychology of social inference: The correspondence bias largely vanishes in Japan*. Unpublished manuscript, Kyoto University, Japan.

Kosugi, R. (2005) Considering the responses to freeter and jobless youth issues. In The Japan Institute for Labor Policy and Training (Ed.), *Labor situation in Japan and analysis: Detailed exposition 2005/2006* (pp. 2-9). Tokyo, Japan: The Japan Institute for Labor Policy and Training.

Kurman, J., & Sriram, N. (1997). Self-enhancement, generality of self-evaluation, and affectivity in Israel and Singapore. *Journal of Cross-Cultural Psychology, 28*, 421-441.

Lalwani, A., Shavitt, S., Johnson, T., & Zhang, J. (2004, October). *The relation between cultural orientation and social desirable responding*. Paper presented at the Sheth Foundation/Sudman Symposium on Cross-Cultural Survey Research, University of Illinois at Urbana-Champaign.

Law, K. S., Mobley, W. H., & Wong, C. (2002). Impression management and faking in biodata scores among Chinese job-seekers. *Asia Pacific Journal of Management, 19*, 541-556.

McCrae, R. (2000). Personality traits and culture: New perspectives on some classic issues [Special Issue]. *American Behavioral Scientist, 44*(1), 2000.

McCrae, R. (2002). NEO-PI-R data from 36 cultures: Further intercultural comparisons. In R. R. McCrae & J. Allik (Eds.), *The Five-Factor Model across cultures* (pp. 105-126). New York: Kluwer Academic/Plenum.

Marcus, H., & Kitayama, S. (1991). Culture and the self: Implications for cognition, emotion, and motivation. *Psychological Review, 98*, 224-253.

Masuda, T., & Nisbett, R. E. (2001). Attending holistically versus analytically: Comparing the context sensitivity of Japanese and American. *Journal of Personality and Social Psychology, 81*, 922-934.

Matsumoto, D. (2000). *Culture and psychology* (2nd ed.). Pacific Grove, CA: Brooks Cole.

Miller, J. G. (1984). Culture and the development of everyday social explanation. *Journal of Personality and Social Psychology, 46*, 961-978.

Ministry of Internal Affairs and Communications (1998). *Sekai no seinann tono hikaku karamita nihon no seinann- dai6kai sekai seinen ishikichousa houkokusho* [A Summary Report of The Sixth World Youth Survey]. Youth Affairs Administration Management and Coordination Agency: Tokyo: Author

Morris, M. W., & Peng, K. (1994). Culture and Cause: American and Chinese attributions for physical and social events. *Journal of Personality and Social Psychology, 67,* 949-971.

Nisbett, R. E., Peng, K., Choi, I., & Norenzayan, A. (2001). Culture and systems of thought: Holistic versus analytic cognition. *Psychological Review, 108,* 291-310.

Nonis, S., & Swift, C. O. (2001). An examination of the relationship between academic dishonesty and workplace dishonesty: A multicampus investigation. *Journal of Education in Business, 77,* 69-77.

Ones, D., Viswesvaran, C., & Reiss, A. (1996). The role of social desirability in personality testing for personnel selection: The red herring. *Journal of Applied Psychology, 81,* 660-679.

Ono, J. (2005, September). *Lifetime employment in Japan: Concepts and measurement.* Paper presented at the NBER Conferences, Tokyo, Japan.

Paunonen, S. V., & Ashton. M. C. (1998). The structured assessment of personality across cultures. *Journal of Cross-Cultural Psychology, 29,* 150-170.

Paulhus, D. L. (1998). *Manual for the Balanced Inventory of Desirable Responding: Version 7.* Toronto/Buffalo: Multi-Health Systems.

Peeters, H., & Lievens, F. (2005). Situational judgment tests and their predictiveness of college students' success: The influence of faking. *Educational and Psychological Measurement, 65,* 70-89.

Price, A. (2004). *Human resource management in a business context.* London: Thompson Learnin.

Poole, G. (2003). Higher education reform in Japan: Amano Ikuo and "The University in Crisis." *International Education Jouranl, 4*(3), 149-176.

Robie, C., Brown, D., & Bly, P. (2005). The big five in the USA and Japan. *Journal of Management Development, 24*(8), 720-737.

Robie, C., Schmit, M. J., Ryan, A. M., & Zickar, M. J. (2000). Effects of item context specificity on the measurement equivalence of a personality inventory. *Organizational Research Methods, 3,* 348-365.

Rosse, J. G., Stecher, M. D., Miller, J. L., & Levin, R. A. (1998). The impact of response distortion on preemployment personality testing and hiring decisions. *Journal of Applied Psychology, 83,* 634-644.

Sanchez-Burke, J., Lee, F., Choi, I., Nisbett, R., Zhao, S., & Koo, J. (2003). Conversing across cultures: East-West communication styles in work and nonwork contexts. *Journal of Personality and Social Psychology, 85*(2), 363-372

Sandal, G. M., & Endresen, I. M. (2002). The sensitivity of the CPI good impression scale for detecting "faking good" among Norwegian students and job applicants. *International Journal of Selection and Assessment, 10,* 304-311.

Schein, E. H. (1992). *Organizational culture and leadership.* San Francisco: Jossey Bass.

Seiter, J. S., Bruschke, J., & Bai, C. (2002). The acceptability of deception as a function of perceiver's culture, deceiver's intention, and deceiver-deceived relationship. *Western Journal of Communication, 66,* 158-180.

Schmit, M. J., Kihm, J. A., & Robie, C. (2000). Development of a global measure of personality. *Personnel Psychology, 53,* 153-193.

Schneider, B. (1987). The people make the place. *Personnel Psychology, 40,* 437-453.

Snell, A., Sydell, E., & Lueke, S. (1999). Towards a theory of applicant faking: Integrating studies of perception. *Human Resource Management Review.*, *9*(2), 219-242.

SPI note no kai (Ed.). (2003). Kono gyoukai, kigyou de kono saiyou tesutoga tukawareteiru!: shuyou saiyou tesutono maruhi monndairei to <sokkaihou> wo denjyu, 2005nen ban ["The personnel selection test" being used by this industry and this organization: Initiating into (giving instruction of) major "personnel selection test" secret item examples and how to quickly solve them, 2005 ed.]. Tokyo, Japan: Yosensha.

Steffens, M. C. (2004). Is the implicit association test immune to faking. *Experimental Psychology*, *5*, 165-179.

Stening, B. W., & Everett, J. E. (1984). Response styles in a cross-cultural managerial study. *Journal of Social Psychology*, *122*, 151-156.

Takeda, K. (2003). Nihonkigyou niyoru daisotsushano sennbatu saiyouwo maguru genjyou to kadai [The current issues and problem regarding Japanese organization personnel selection practice for University graduate]. *Bulletin of Toyohashi Sozo College*, *7*, 99-110.

Triandis, H. C. (1995). *Individualism and collectivism*. Boulder, CO: Westview Press.

Triandis, H. C. (1998). The self and social behavior in differing cultural context. *Psychological Review*, *96*, 506-520.

Triandis, H. C., & Suh, E. M. (2002). Cultural influences on personality. *Annual Review of Personality*, *53*, 133-160.

Weisz, J. R., Rothbaum, F. M., & Blackburn, T. C. (1984). Standing out and standing in: Then psychology of control in America and Japan. *American Psychologist*, *39*, 955-969.

CHAPTER 14

WHAT DO WE KNOW AND WHERE DO WE GO?

Practical Directions for Faking Research

Ann Marie Ryan and Anthony S. Boyce

It would be wrong of us to presume that we know what motivates the authors in this volume to pursue research on faking on personality tests in selection contexts. However, we would venture to guess that the "voice of the common man" plays a role for many of the authors, just as it does for us. We hear via e-mail or phone from the manager who wants to do away with his organization's personality testing because "people can fake" or from the human resources (HR) professional who is frustrated because she cannot convince top management of the utility of personality testing. We have encounters with the friend/family member/acquaintance who complains about the unfairness of using a test others will "lie" on and gain an unfair advantage. We recognize that there is a public sentiment that faking is a problem when it comes to using personality tests in selection (Hoffman, 2000; Song, 2005). Thus, we are probably not inferring too much if we were to say that the authors in this volume are pursuing this research topic for the same reason that many psychologists pursue a

A Closer Examination of Applicant Faking Behavior, 357–371
Copyright © 2006 by Information Age Publishing
All rights of reproduction in any form reserved.

research stream—to determine the veracity of the folk wisdom. Psychologists, as social scientists, are often addressing the question of whether the received wisdom regarding human behavior is indeed correct (Siegfried, 1994), and thus research on faking is really an investigation into a lay theory of response to personality tests in selection contexts.

As is often true of lay theories of human behavior, broad generalities make up the components of the theory: People want jobs, therefore they will try to make themselves look good, and that means these types of tests will pick the liars rather than the best applicants. The lay theorist simply states the relationships he/she believes to exist as truisms. As scientists, faking researchers are obligated to present a conceptual rationale for relationships (e.g., what motivates faking), marshal empirical evidence to support views (e.g., do people fake), and consider alternatives (e.g., positive self-presentation enhances the validity of personality tests). Hence, this volume reviews the substantial body of work testing this lay theory of response to personality tests in selection contexts using the standards of science. Overall, the authors carefully present evidence supporting or refuting basic conclusions and cautiously note what can and cannot be concluded based on the evidence.

But if we were to take a step backward and remove our "scientist hats," how would we speak to the lay theorist? One conclusion of many researchers in the personnel testing arena has been that it does not matter to what extent positive self-presentation or even intentional distortion is occurring because personality tests still enhance our ability to predict who will be the best performers. The answer to the lay theorist might be that of Salgado and deFruyt (2005) "[these measures] can confidently be used for selection purposes" (p. 194). As Peterson and Griffith (in this volume) note, some researchers contend that the lack of straightforward findings on the effects of faking on criterion-related validity of personality measures indicates that faking research is at a dead end.

However, most of the chapters in this volume, while respecting this body of evidence, suggest there is merit in the lay theorist view that faking is a problem. That is, the conclusion of the majority of the authors in this volume is that positive self-presentation bias warrants further investigation, that faking matters. Further, it does not appear that the perspective "faking doesn't matter" has been widely embraced by hiring managers. In practice, the lay theory regarding faking contributes to decisions not to adopt personality tests in selection contexts and to complaints regarding their use, and the lay theorist remains relatively oblivious to the notion that there is any debate over whether faking on personality testing used in selection is a problem. One might question whether research suggesting faking does not make a difference falls into the category of what Anderson

(2005) termed as "robust research failing to influence professional practice."

Given the strong convictions of the lay theorist, why the equivocality among researchers? Why do some view this as a key problem to investigate and others in the selection area consider it a "dead end?" Given that research on this lay theory of responding to personality testing commenced in the 1930s (Zickar & Gibby, in this volume), why is there not greater consensus among researchers?

Looking across the chapters in this volume, some clear themes emerge as to what researchers in this domain see as needed changes to approaching this topic. Our aim in this chapter is to summarize how the directions suggested by authors in this volume can help overcome the equivocality regarding the value of research in this area. Specifically, we discuss five major themes that emerge from the chapters in this volume:

1. Develop broad models of response to personality tests.
2. Clarify the conceptualization of faking and ensure appropriate operationalizations.
3. Redefine what "matters" means.
4. Design interventions to be applicant supportive rather than adversarial.
5. Provide concrete advice for the manager interested in using personality tests.

In our view, following these prescriptions for future research will enable researchers to slice through the equivocality and come up with a sharper picture of the evidence that can then be taken to both the skeptical scientist and the lay theorist, whether in the end it supports or challenges their views of faking.

Develop Broad Models of Response to Personality Tests

Several chapters suggest that researchers in this area need to work from models of responses to test items rather than models of faking, and/or to place faking into the broader question of how to improve testing (Griffith, Malm, English, Yoshita, & Gujar; Johnson & Hogan; Snell & Fluckinger; Tett, Anderson, Ho, Yanf, Huang, & Hanvongse; Pace & Borman). Research niches can become grounds for inbreeding—those in them can end up speaking mostly to each other and thinking alike. While we would not consider faking research of quite that ilk, the authors in this volume recognize the dangers of focusing on very narrow questions on faking, including having less influence on the course of research and practice in the broader domain of selection. These chapters suggest that faking should be addressed within the larger question of what affects the ability and motivation to respond to items in an accurate manner. By shifting the question to "what affects responding to personality test items" rather than "what leads to faking behavior," greater insights can be gained and the ultimate goal, improvement of the use of personality testing in selection contexts, can be attained.

How do the chapters in this volume propose responding be modeled? Griffith et al. focus on understanding why favorable self-presentation occurs at both conscious and unconscious levels and under varying situational conditions. Snell and Fluckinger focus on a response model from a conditional reasoning framework. Tett et al. propose an interactionist model of response distortion. Johnson and Hogan suggest a model of item responding that holds across test-takers regardless of the degree or type of deception that occurs. What links these models is that they all recognize that responding to an item is multiply determined by both individual characteristics and situational factors and that interventions to address response distortion can occur via multiple mechanisms.

Operating from the basis of models of item responding rather than models of faking may lead to different approaches to interventions to prevent or detect faking. For example, Pace and Borman argue that understanding the process of responding to personality test items would be a more effective approach to understanding which warnings work, and why, rather than a scattered, trial-and-error method. Snell and Fluckinger provide a number of examples of how considering justification mechanisms might lead to different research investigations.

Broader models might provide better answers to unaddressed questions regarding faking, such as those raised by Frei, Yoshita, and Isaacson (in this volume) regarding culture. For example, does culture affect response distortion because of its broader role in influencing expectations applicants have regarding how selection should proceed (Steiner & Gilliland, 2001)? Or is culture's role a direct influence on norms regarding how to respond, including norms regarding the acceptability of distortion (Heine & Lehman, 1997; Markus & Kitayama, 1991)? By looking more deeply at the literature on culture and responding rather than framing the question as one of culture and faking, more meaningful advances in understanding can be made.

While the chapters in this volume suggest a broad focus on models of responding to personality items, we would urge a consideration of even wider circles of reach. That is, rather than thinking in terms of responding to personality items, should the focus be on responding to self-description items in general? A wider focus would encompass other types of selection measures (e.g., biodata, integrity tests) for which propositions regarding self-presentation and responding may or may not hold. Additionally, a broader focus might include examining the roles of various self-presentation constructs (e.g., looking at both impression management and self-deception, as suggested by Tett and his colleagues). In sum, a key theme of this volume is the need to stop researching "faking" and instead research item responding.

Clarify the Conceptualization of Faking and Ensure Appropriate Operationalizations

Many of the chapters in this volume noted the historical evolution and discussed distinctions among terms such as faking, impression management, and self-deception (Griffith et al.; Mesmer-Magnus & Viswesvaran; Tett et al.; Zickar & Gibby). Yet questions still remain as to what the construct of faking is, and whether a given study is actually capturing the behavior of interest. In undergraduate research methods courses we teach that it is important to attend to what a concept is labeled vis à vis other concepts (precision of conceptual definitions) and to how you measure that concept (quality of operationalization). Yet, faking research as a domain has inconsistency in conceptualization and operationalization, and, much worse, is the tendency to ignore these inconsistencies in interpreting and applying findings.

Tett et al. raise the question "Is faking an identifiable construct?" and proceed to note that it is important to see faking as distinct from other forms of socially desirable responding in both honesty and deliberateness. They note that the way one defines the construct of faking influences not only what results one might find in a research study, but also what questions one is likely to pursue. Griffith et al. note that stereotyping fakers as all operating under the same motivations, having the same abilities, and achieving the same outcomes for the same wrong purpose is a gross oversimplification of how individuals self-present on personality tests. They suggest that future research needs to consider ways to break down "faking variance" in terms of both individual and situational determinants. Such breakdowns may suggest that some of what might be construed as faking behavior might best be labeled as something else, as suggested by Snell and Fluckinger in their chapter on applicant responding. In returning to the previous theme, the construct of faking needs to be discussed within the nomological net of item response behaviors.

Even with a similar conceptual definition, researchers can make different choices as to operationalizations. Mesmer-Magnus and Viswesvaran summarize how different data collection strategies lead to conceptualizing faking differently (i.e., as an individual difference versus situationally induced), describe a number of different operationalizations of faking (e.g., comparisons across groups, comparisons of individuals under different instructional sets), and note the pros and cons associated with these methodological choices. While research strategies will have some tradeoffs in terms of maximizing internal, construct and external validity (Shadish, Cook, & Campbell, 2002), the overall body of evidence in a given domain should employ a diversity of strategies so that it cannot be assaulted on these grounds. Unfortunately, as a number of the chapters in this volume attest (Burns & Christiansen; Griffith et al.), reliance on a single approach to address a question (e.g., partialling of social desirability scales, comparing incumbent and applicant means) may lead to a different answer than if other operationalizations are examined. It is imperative that faking researchers acknowledge how the particular methodological choices in a given study constrain the generalizability of findings; consideration needs to be given to multiple replications under various operationalizations before conclusions are drawn regarding the presence or absence of faking effects.

Redefining "Matters"

As noted in the introduction to this chapter, some researchers have dismissed the faking in selection contexts issue as one that is not important (e.g., Schmitt & Oswald, in press). The dismissals can take the form of "it doesn't affect the validity of the test so it doesn't matter," or, as Snell and Fluckinger observed, "personality tests do not have great validity so they should not be used regardless of faking's role." Is a moratorium on this research warranted?

Certainly, before reaching a conclusion that faking does not matter, there is a need to tackle the theme just discussed (i.e., conceptualization and operationalization differences). A further task is to give attention to defining what "matters" means. While there is research on faking effects on criterion-related validity of personality tests predicting task performance, on construct validity, and on selection decisions, the conception of "matters" still has had a narrow focus. Even within those areas receiving greater investigation, there is equivocality regarding what one should conclude. We briefly discuss each stream of research on faking effects.

Criterion-Related Validity

Quite a substantial amount of research has been devoted to the effect of faking on criterion-related validity (see Peterson & Griffith chapter for a review). While some researchers conclude there is no effect (e.g.,

Johnson & Hogan; Mesmer-Magnus & Viswesvaran; Tett et al.; Ones, Viswesvaran, & Reiss, 1996), other researchers argue that criterion-related validity can be attenuated by faking (e.g., Burns & Christiansen; Mueller-Hanson, Heggestad, & Thornton, 2003; Pace & Borman). The Peterson and Griffith chapter describes the research as not definitive but mixed, largely for two methodological reasons. First, studies that have generally shown little effect on criterion-related validity have tended to examine this question using social desirability measures as proxies for faking, a potentially questionable practice (see also Burns & Christiansen; Griffith et al.; Smith & Ellingson, 2002). Alternatively, studies indicating that faking can negatively impact criterion-related validity have tended to use laboratory studies with "fake good" or "incentive" instructional sets (e.g., Dunnette et al, 1962; Mueller-Hanson et al., 2003). The external validity of such studies is also questionable due to problems associated with equalizing respondent motivation to engage in faking and the use of student samples (Mesmer-Magnus & Viswesvaran; Peterson & Griffith).

So what can be done to address these limitations and attempt to unravel when and if faking impacts criterion-related validity in *real applicant* settings? Many of the authors in this volume suggest the need for field research involving actual applicants and direct assessments of faking, rather than the use of social desirability scales or other proxies, to shed light onto this question (Burns & Christiansen; Griffith et al.; Peterson & Griffith; Tett et al.). The "gold standard" in this regard would involve a design where applicants take a personality test as part of the selection process, the personality test does not enter into the selection decision, the same applicants are tested at a later period under confidential conditions as incumbents, and detailed job performance data is obtained for each individual. This type of design allows for the direct assessment of faking as the difference, in algebraic or latent growth modeling terms, between applicant and incumbent scores. This difference could then be examined in relation to individuals' job performance to assess the impact of faking on criterion-related validity with actual applicants in real applicant settings. It may also be useful to include impression management and self-deception scales to elucidate the construct validity of these measures (Burns & Christiansen; Griffith et al.).

In addition to methodological issues, the authors in this volume point out that "what is the criterion?" has not been thoroughly attended to when examining the impact of faking on criterion-related validity. As Peterson and Griffith note, summaries of how personality relates to other work behaviors exist: Johnson (2003) describes potential relationships to organizational citizenship behavior and adaptive performance (see also Borman, Penner, Allen, & Motowidlo, 2001; Hurtz & Donovan, 2000), Cullen and Sackett (2003) note links to counterproductive behavior, and

Stewart (2003) suggests possible connections to team-level outcomes. As any psychology student can tell you, one should not talk about "the validity of the test" but the inferences drawn from the test for a given purpose. In discussing whether faking affects the criterion-related validity of personality tests, researchers should clarify that the evidence thus far has focused predominately on overall performance and/or task performance, not other criteria. Peterson and Griffith argue that faking researchers examining criterion-related validity need to expand the criterion domain. This is important as it is possible that while faking may have little effect on the validity of tests with certain task performance dimensions as the criterion, faking may actually enhance the validity of tests when social behaviors such as teamwork or customer service behavior serve as the criterion. For similar reasons, it is also important to examine the influence of faking on criterion-related validity for different jobs as faking may enhance, or have minimal impact on, validity for those jobs in which impression management skills contribute to successful job performance (e.g., sales jobs) while attenuating validity for jobs in which impression management skills are unrelated to performance (e.g., engineering jobs). Provision of different types of warnings (Pace & Borman) or testing applicants under conditions that influence different faking-related justification mechanisms (Snell & Fluckinger) may also help to elucidate the *when* and *if* of faking effects.

Researchers need to be more proactive and persuasive in pursuing the applicant samples and organizational settings necessary to employ "gold standard" designs with multiple criteria. Given organizations' interests in having the most valid selection tests possible and managers' lay theories about applicants and faking, it seems that gaining access to these samples and settings should not be an insurmountable obstacle. Such research is necessary before any consensus can be reached regarding faking effects on criterion-related validity.

Construct Validity

From a purely pragmatic perspective, one can be less concerned with the construct validity of selection tools as long as criterion-related validity is demonstrated. That is, if the best job performers (or least likely to turnover or whatever criterion is of focus) are selected, perhaps it does not matter exactly what was assessed to make that determination. However, to better understand positive self-presentation, faking, and related conceptualizations, construct validity of selection tools definitely matters.

Tett and his colleagues summarize the mixed research on the effects of response distortion on construct validity, concluding that there is evidence in both factor analytic and item response theory studies of negative repercussions of faking on construct validity. Further, one recent study suggests that cognitive ability may be related to faking and thus test scores saturated with response distortion may reflect, in part, cognitive ability (Dilchert & Ones, 2005).

Tett et al. allude to a "validity paradox" in that effects are observed with regard to construct validity but not, arguably, criterion-related validity. This validity paradox cannot help but be compared to that observed in relation to interviews and assessment centers where criterion-related validity is observed, construct validity is often lacking, and researchers are left wondering what exactly is being measured (Arthur, Woehr, & Maldegen, 2000; Van Iddekinge, Raymark, & Roth, 2005). As suggested by Tett and his colleagues, research comparing the convergent and discriminant validity of different personality assessment methods, such as clinical interviews or significant other reports, to self-reported applicant personality may help to clarify exactly what is being measured, and the relative influence of response distortion. In sum, there is a growing body of research suggesting that faking may matter in terms of effects on construct validity.

Who is Selected

Higgs, Papper, and Carr (2000) noted that the definition of what makes a selection process successful encompasses much more than empirical validity. So to should attention regarding faking relate to more than its effect on validity. As researchers in this volume point out (e.g., Burns & Christiansen; Mesmer-Magnus & Viswesvaran; Tett et al.) the effects of faking may relate to the quality of selection decisions in terms of which individuals are selected (Mueller-Hanson et al., 2003; Rosse, Stecher, Miller, & Levin, 1998). Some may argue that switches in rank order that have no strong effect on overall validity should not be a concern because it is no different than misordering that occurs from unreliability of measures. To us, any source of error needs to be a concern of researchers, and improving reliability of assessment and reducing systematic bias of any kind is a worthy goal. Indeed, accepting error in decision making from uncontrollable variance (e.g., the unreliability in measurement due to test-taker fatigue, etc.) is more palatable than accepting systematic and potentially reducible error (e.g., deliberate response distortion, discrimination in hiring decisions, etc.). Hence, research on changes in selection decisions as a function of response distortion may lead to the conclusion that faking matters.

Fairness Perceptions

Even if one does not evaluate findings regarding rank order changes as particularly troublesome in terms of effects on the quality of hires, one should certainly see them as troubling from a standpoint of fairness in hiring processes. If test-takers view personality tests as less fair because they are fakable, as suggested by research linking fairness perceptions to self-reported faking (McFarland, 2003), this is a concern because perceptions of unfairness may lead to applicant self-selection out of the hiring process, turning down of job offers, badmouthing the organization, and so forth. (Hausknecht, Day, & Thomas, 2004). Perceptions of unfairness can have consequences for individuals as well, affecting, for example, job satisfaction and self-efficacy (e.g., Colquitt, Conlon, Wesson, Porter, & Ng, 2001; Truxillo, Bauer, & Sanchez, 2001). Even if such perceptions are ill-founded and the use of such tools is fair from a psychometric standpoint, perceptions should matter to researchers because they affect test-taker behavior and attitudes.

Marketability

Some of the other criteria mentioned by Higgs et al (2000) as important to the success of selection systems, such as face validity and marketability, should also enter into researchers' definitions of "what matters" in relation to faking and personality testing. That is, personality tests that do not meet the commonsense criteria of a manager or HR representative (i.e., we don't want something people can distort) will not be adopted and used. Marketability involves the ability to clearly communicate why the test is being used and why it is valuable to the organization. In a practical sense, an HR manager must weigh the tradeoffs between a gain in the quality of hires to the time and energy required to effectively implement the hiring process (Jayne & Rauschenberger, 2000). Researchers and practitioners need to be cognizant of how faking (or beliefs regarding fakability) affects the ability to easily implement a personality test—how much time and energy an organization is willing to invest to explain the validity of tests to managers, to handle any applicant complaints, and so forth. The types of evidence and messages that managers perceive to be persuasive in regard to the use of personality tests are useful and practical areas for future research.

Quality of Tools

Ryan and Tippins (2004) noted that gaps exist between research and practice with regard to selection tool use for lots of reasons, including that specific tools can be less useful than one would assume based on average validities if they are poorly developed. A critical challenge to convincing managers that personality testing has value is there are many poorly developed instruments being marketed for use in selection contexts. One point of education regarding personality testing surrounds what well-developed instruments might do versus poorly developed ones. Peterson and Griffith note that researchers should not "hastily conclude that faking is not a concern" but rather continue to work to improve measurement and understanding of personality constructs.

In sum, researchers should broaden the definition of "matters" to encompass all those things that matter to successful implementation of a selection tool: criterion-related validity in relation to multiple criteria, construct validity, changes in selection decisions, fairness perceptions, and manager perceptions of face validity and marketability.

Design Interventions to be Applicant Supportive Rather Than Adversarial

Johnson and Hogan suggest that positive self-presentation needs to be construed as normal and positive behavior, rather than as dubious or deviant. Following from this, there is a need to take a positive tact toward design of interventions to increase the accuracy of personality assessment. One theme that underlies the chapters is that researchers ought to be working on ways for applicants to adequately express their true natures, rather than as police officers trying to catch offenders. Johnson and Hogan echo this theme strongly and suggest that rather than focusing on discouraging positive self-presentation, thinking of ways to get applicants to present their best self accurately would be a more appropriate approach.

Research on interventions to reduce or mitigate faking has reached consensus in a number of ways: warnings work (Pace & Borman), social desirability response scales to correct scores do not (Burns & Christiansen) and format adaptations like forced choice methodologies have limits and potential (Converse et al.). Future research on interventions can build on this knowledge base by focusing on ways to encourage accurate self-presentation rather than just to discourage false self-presentation. For example, Pace and Borman suggest that warnings focused on raising test-taker awareness that faking is "against their best interests" can promote accurate self-presentation. Additionally, Snell and Fluckinger suggest that contextual cues raising awareness of the idea that faking is the same as lying may encourage more accurate self reports. Interventions that support applicants in their quest to find the right job may take the field further than interventions focused on catching fakers.

A related theme that cross-cuts the chapters on interventions is that interventions can occur at multiple points in the response model—to limit the ability to distort (e.g., through forced choice methods), to limit the motivation to distort (e.g., warnings), or to catch those who may have distorted (e.g., social desirability measures). Burns and Christiansen provide a cogent summary of whether the "intervention after the fact" approach has merit, and their conclusion is that this is not an effective strategy. Rather, research and practice should focus on enhancing the ability of test-takers to accurately self-describe, both through test design and situational interventions, and on enhancing the motivation to respond accurately.

We also have to consider what interventions are practical. For example, warnings are a low cost intervention. Converse et al. make it clear that to get to a psychometrically useful forced choice format one has to do a tremendous amount of work (the unfolding technique) and one might encounter applicant dislike, suggesting this is not a viable solution. Converse et al. raise the specific question: while further research may enable the development of useful forced choice tools, is this the best focus of research energy for those seeking practical solutions? Taking that question to a broader level, if the aim is to promote accurate self-presentation, what strategies are both effective and practical?

Provide Concrete Advice for the Manager Interested in Using Personality Tests

Having read through this volume that provides an overview of the research on faking and personality assessment in selection contexts, what advice should be given to hiring managers? A cohesive message has yet to emerge regarding whether one would recommend using personality tests on a widespread basis, or how one would detect or mitigate undesirable response distortion, or whether the gains in incremental validity over other selection tools are offset by the time and effort needed to communicate around the issue of distortion. For example, in addition to the conflicting conclusions regarding the influence of faking on criterion-related validity discussed above, authors in this volume provide conflicting suggestions on the use of social desirability scales. That is, some authors in this volume recommend the use of social desirability scales to detect fakers (e.g., Johnson & Hogan; Mesmer-Magnus & Viswesvaran) while others rail against this practice suggesting that these scales do not represent valid operationalizations of faking (e.g., Burns & Christiansen; Griffith et al; Tett et al.). Clearly, even among faking researchers there is substantial divergence in opinions regarding how to approach the use of personality testing in selection contexts. In the face of this lack of agreement, if one truly believes that faking matters, what advice should be given to organizations considering whether and how to implement personality tests?

Despite the fact that there is no complete answer to this question, limited advice is available for those organizations considering implementation. First, predictive validity studies should be used to confirm that even in the presence of applicant faking, the specific personality test being considered has criterion-related validity for the intended purposes. As we noted earlier, there are many poorly developed and useless tests available, and efforts should be made to better educate managers on how to differentiate among them. The value of personality testing may also be better established via a focus on multiple criteria and not solely task performance. Second, warnings of some sort should be utilized to deter faking as they appear to be effective and low cost. These should be coupled with instructions that emphasize the value of accurate self-presentation for applicants. Third, personality tests should be used to select-out rather than to select-in and as part of a comprehensive process, not as a sole selection device. Fourth, careful attention should be paid to how best to present personality testing to various stakeholder groups so that perceptions are well managed. Finally, organizations should be encouraged to pursue "gold standard" research to assess any effects of response distortion on the multiple outcomes described (e.g., validity, fairness perceptions, selection decisions, marketability). More systematic and high-quality research can assist the individual organization as well as the field as a whole in determining how best to implement personality tests in selection contexts.

In the end, there are still many directions in which to invest research energy. Zickar and Gibby indicate how far research has come but how much has yet to be addressed. Their chapter suggests that researchers have focused on the antecedents of faking and the preemption of faking behavior, but the focus on consequences is a bit more recent. The historical review suggests researchers in this area have not wavered from the view that faking matters, yet the broader field may not be in complete agreement on that point. We would advocate a continued focus on the practical utility questions: Can we create tools that are more accurate assessments of personality? Can we create situations that elicit accurate self-presentations? Can we create tests that are more predictive of important workplace criteria including but also going beyond task performance? Can we create tools that are viewed positively by hiring managers and test-takers? Addressing these questions – none of which include the word "faking"—is the direction likely to have the greatest theoretical and practical impact for faking research.

AUTHOR'S NOTE

Address correspondence to: Ann Marie Ryan, 333 Psychology Building, Michigan State University, East Lansing, MI 48824, ryanan@msu.edu, 517-353-8855.

REFERENCES

Anderson, N. (2005). Relationships between practice and research in personnel selection: Does the left hand know what the right is doing? In A. Evers, N. Anderson, & O. Voskuijl (Eds.), *The Blackwell Handbook of Personnel selection* (pp. 1-24). Oxford, England: Blackwell.

Arthur, W., Woehr, D. J., & Maldegen, R. (2000). Convergent and discriminant validity of assessment center dimensions: A conceptual and empirical reexamination of the assessment center construct-related validity paradox. *Journal of Management, 26*(4), 813-835.

Borman, W. C., Penner, L. A., Allen, T. D., & Motowidlo, S. J. (2001). Personality predictors of citizenship performance. *International Journal of Selection and Assessment, 9*, 52-69.

Colquitt, J. A., Conlon, D. E., Wesson, M. J., Porter, C. O., & Ng, K. Y. (2001). Justice at the millennium: A meta-analytic review of 25 years of organizational justice research. *Journal of Applied Psychology, 86*(3), 425-445.

Cullen, M. J., & Sackett, P. R. (2003). Personality and counterproductive work behavior. In M. R. Barrick and A. M. Ryan (Eds.), *Personality and work: Reconsidering the role of personality in organizations* (pp. 150-182). San Francisco: Jossey-Bass.

Dilchert, S., & Ones, D. S. (2005). *Race differences in social desirability scores partly due to g.* Paper presented at the 20th annual conference of the Society for Industrial and Organizational Psychology, Los Angeles.

Dunnette, M. D., McCartney, J., Carlson, H. C., & Kirchner, W. K. (1962). A study of faking behavior on a forced-choice self-description checklist. *Personnel Psychology, 15*(2), 13-24.

Hausknecht, J. P., Day, D. V., & Thomas, S. C. (2004). Applicant reactions to selection procedures: An updated model and meta-analysis. *Personnel Psychology, 57*, 639-683.

Heine, S. J., & Lehman, D. R. (1997). The cultural construction of self-enhancement: An examination of group-serving biases. *Journal of Personality and Social Psychology, 72*, 1268-1283.

Higgs, A. C., Papper, E. M., & Carr, L. S. (2000). Integrating selection with other organizational processes and systems. In J. F. Kehoe (Ed.), *Managing selection in changing organizations: Human resource strategies* (pp. 73-122). San Francisco: Jossey-Bass.

Hoffman, E. (2000). *Ace the corporate personality test.* New York: McGraw-Hill.

Hurtz, G. M., & Donovan, J. J. (2000). Personality and job performance: The Big Five revisited. *Journal of Applied Psychology, 85*(6), 869-879.

Jayne, M. E. A., & Rauschenberger, J. M. (2000). Demonstrating the value of selection in organizations. In J. F. Kehoe (Ed.), *Managing selection in changing organizations: Human resource strategies* (pp. 123-157). San Francisco: Jossey-Bass.

Johnson, J. W. (2003). Toward a better understanding of the relationship between personality and individual job performance. In M. R. Barrick & A. M. Ryan (Eds.), *Personality and work: Reconsidering the role of personality in organizations* (pp. 83-120). San Francisco: Jossey-Bass.

Markus, H. R., & Kitayama, S. (1991). Culture and the self: Implications for cognition, emotion, and motivation. *Psychological Review, 98,* 224-253.

McFarland, L. A. (2003). Warning against faking on a personality test: Effects on applicant reactions and personality test scores. *International Journal of Selection and Assessment, 11,* 265-276.

Mueller-Hanson, R., Heggestad, E. D., & Thornton, G. C., III. (2003). Faking and selection: Considering the use of personality from select-in and select-out perspectives. *Journal of Applied Psychology, 88*(2), 348-355.

Ones, D. S., Viswesvaran, C., & Reiss, A. D. (1996). Role of social desirability in personality testing for personnel selection: The red herring. *Journal of Applied Psychology, 81*(6), 660-679.

Rosse, J. G., Stecher, M. D., Miller, J. L., & Levin, R. A. (1998). The impact of response distortion on preemployment personality testing and hiring decisions. *Journal of Applied Psychology, 83*(4), 634-644.

Ryan, A. M., & Tippins, N.T. (2004). Attracting and selection: What psychological research tells us. *Human Resource Management, 43,* 305-318.

Salgado, J. F., & deFruyt, F. (2005). Personality in personnel selection. In A. Evers, N. Anderson, & O. Voskuijl (Eds.), *The Blackwell Handbook of Personnel selection* (pp. 174-198). Oxford, England: Blackwell.

Schmitt, N., & Oswald, F. L. (in press). The impact of corrections for faking on the validity of noncognitive measures in selection settings. *Journal of Applied Psychology.*

Shadish, W. R., Cook, T. D., & Campbell, D. T. (2002). *Experimental and Quasi-Experimental Designs for Generalized Causal Inference.* Boston: Houghton Mifflin.

Siegfried, J. (1994). The status of common sense in psychology. Westport, CT: Ablex.

Smith, D. B., & Ellingson, J. E. (2002). Substance versus style: A new look at social desirability in motivating contexts. *Journal of Applied Psychology, 87*(2), 211-219.

Song, K. M. (2005, June 22). Faking your type to "pass" a personality test. *Seattle Times,* p. F1.

Steiner, D. D., & Gilliland, S. W. (2001). Procedural justice in personnel selection: International and cross-cultural perspectives. *International Journal of Selection and Assessment, 9,* 124-137.

Stewart, G. L. (2003). Toward an understanding of the multilevel role of personality in teams. In M. R. Barrick & A. M. Ryan (Eds.), *Personality and work: Reconsidering the role of personality in organizations* (pp. 183-204). San Francisco: Jossey-Bass.

Truxillo, D. M., Bauer, T. N., & Sanchez, R. J. (2001). Multiple dimensions of procedural justice: Longitudinal effects on selection system fairness and test-taking self-efficacy. *International Journal of Selection and Assessment, 9,* 336-349.

Van Iddekinge, C. H., Raymark, P. H., & Roth, P. L. (2005). Assessing personality with a structured employment interview: Construct-related validity and susceptibility to response inflation. *Journal of Applied Psychology, 90*(3), 536-552.

ABOUT THE AUTHORS

Michael G. Anderson is a doctoral student in the industrial/organizational (I/O) psychology program at the University of Tulsa under the mentorship of Dr. Robert Tett. He received his master's degree from the University of Northern Iowa in 2003. His research interests include personality and job performance, small group research, and performance appraisal. Michael is currently a researcher at Hogan Assessment Systems in Tulsa, OK, and is preparing his dissertation on the role of personality in groups and teams, using meta-analysis.

Dr. Walter C. Borman received his PhD in industrial/organizational psychology from the University of California (Berkeley). He is currently CEO of Personnel Decisions Research Institutes (PDRI) and is professor of industrial/organizational psychology at the University of South Florida. He is a Fellow of the Society for Industrial and Organizational Psychology, and in 1994-95 served as president of the society. Borman has written more than 300 books, book chapters, journal articles, and conference papers. He recently coedited the I/O volume of the *Handbook of Psychology*, and, with two PDRI colleagues, wrote the personnel selection chapter for the *Annual Review of Psychology*. He also has served on the editorial boards of several journals in the I/O field, including the *Journal of Applied Psychology, Personnel Psychology, Human Performance*, and the *International Journal of Selection and Assessment*. He was the recipient of SIOP's Distinguished Scientific Contributions Award for 2003.

Anthony S. Boyce is currently a PhD candidate at Michigan State University. His primary research areas include applicant faking, the stigmatization of temporary workers, and applicant reactions. His interests in applicant faking center on the impact of faking on the construct and criterion-related validity of personality tests and the resulting selection decisions. His work on the stigmatizing treatment of temporary workers focuses on the affective and behavioral consequences for these workers as well as the more macro organizational consequences. His current research efforts on applicant reactions focus on the impact of individual differences in cultural values in applicants' perceptions of the fairness of different selection procedures. Additionally, Anthony has been involved in the development, validation, and implementation of various types of selection tests for a number of Fortune 500 companies.

Gary N. Burns is finishing his doctorate in industrial and organizational psychology at Central Michigan University, focusing on different methods of examining the nature of faking and its influence on the selection process. Currently he is examining different methods of computationally modeling faking in an attempt to bridge the gap between laboratory and field experiments.

Hilary B. Butera is a research associate in the Research and Development Department at Caliper. Ms. Butera is the lead statistician on all internal studies focused on the psychometric properties of the Caliper assessments. She has published articles relating to the psychometric evaluation of personality assessments.

Dr. Neil D. Christiansen has been doing research in the area of applicant distortion of personality measures for over a decade, focusing on the detection and prevention of faking. In particular, he has critically studied the use of social desirability scales in terms of their potential to facilitate hiring decisions and their limitations as indicators of applicant faking. His publications in this area have appeared in Personnel Psychology, Human Performance, and the International Journal of Selection and Assessment.

Dr. Patrick D. Converse (PhD, 2005, Michigan State University) is an assistant professor at the Florida Institute of Technology. In addition to the use of personality measures in selection contexts, his research interests include cognitive ability and the ability requirements of occupations and self-regulation.

Jeffrey M. Cucina is a doctoral student in industrial/organizational psychology at the George Washington University. He is actively involved in research on personnel selection issues, including studies focusing on personality testing. In particular, he has conducted research on the construct validity and spurious moderating effects of traitedness and the presence of a cognitive load in personality test scores when warnings against faking and forced choice instruments are used. His publications appear in the *Journal of Personality,* the *Journal of Business and Psychology,* the *Journal of Applied Psychology,* the *International Journal of Selection and Assessment* and *Human Performance.*

Dr. Andrew English is ThoughtLink's director of research and the Florida Institute of Technology's 3T (teams, training, and technology) research supervisor. He has extensive experience in the development of online training systems. Andrew recently developed an online tutorial on using models, simulations and games for training, and exercising for the Department of Homeland Security. His research interests span team training and modern technology to personality assessment and theories of individual differences. His strengths include: conducting needs analysis, leading focus groups, measurement of psychological constructs, and the design and evaluation of training systems. He has produced several reports under contract with the Department of Homeland Security and Department of Defense on using models, simulations and games for training and exercising, and is published in *Small Group Research*.

Christopher D. Fluckinger is currently a graduate student in the industrial/organizational psychology program at the University of Akron and completed his undergraduate work at Bowling Green State University. His primary research focus includes understanding the reasons that people use, both explicitly and implicitly, to justify the selection of responses on self-report inventories. Other areas of interest revolve around improving usability in survey design and implementation.

Dr. Richard Frei has worked as a college professor, researcher, and consultant for nearly 20 years. His publications and conference presentations cover a wide variety of applied psychology topics, including cross-cultural differences in perception of time, employment testing, and technology in the workplace. As a consultant, Richard specializes in program evaluation and team processes, and he works with colleges and universities on developing strategies to increase minority and international student retention. He is also a faculty advocate for the MyPsychLab interactive psychology Web Site sponsored by Allyn and Bacon. Richard currently teaches in the Behavioral Science department at the Community College of Philadelphia. He has previously taught psychology and management courses at Temple University, Florida Institute of Technology, and Kent State University.

Dr. David C. Funder does research in personality and social psychology that addresses personality assessment, person-situation interactions, and the accuracy of personality judgment. The work has focused on individual differences in impulse control, resilience, conscientiousness, subjective well-being, extraversion, and many other traits, emphasizing their implications for behavior and important life outcomes. The author of two books, his publications have also included articles in *Psychological Review*, *Psychological Bulletin*, *Behavioral and Brain Sciences*, the *Journal of Personality and Social Psychology*, and the *Journal of Experimental Social Psychology*.

Dr. Robert E. Gibby works for Procter & Gamble's organizational research unit. His research in the area of applicant faking on personality assessments, including his dissertation, has focused on trying to better define faking and nonfaking groups of respondents through the use of advanced psychometric techniques (IRT, MM-IRT, SEM) as opposed to using predefined groups (e.g., applicants vs. incumbents) or social desirability scales. His background and experience in selection includes work on the development of a number of different assessments for P&G (pre-screening, personality, biodata, situational judgment, cognitive ability, and behavioral interviews), delivered through varied technologies (CAT, forms casting). Dr. Gibby has also performed historical research on the early use of personality inventories in industry.

Dr. Richard Griffith is an associate professor in the I/O psychology program at the Florida Institute of Technology. His research examines personality issues in the workplace, specifically the faking of personality measures used in the selection context. His publications have examined the effect of faking on the measurement qualities of an instrument as well as the impact on rank order. His recent work has focused on developing and testing a theoretical model of applicant faking behavior, and examining the job performance of those individuals identified as fakers.

Abhishek Gujar received his master's degree in I/O psychology from the Florida Institute of Technology. His master's thesis was on the effect of self-deceptive enhancement on personality measures used in personnel selection. Abhishek has conducted research in the areas of personality, response distortion on noncognitive measures used in employment selection, frame of reference effects, and cognitive factors effecting individuals' perception of the self. Specifically, he has studied how unconscious components invloved in response distortion may impact the interpretation of non-cognitive measures He has presented his research at poster sessions and symposia at the Society for Industrial Organizational Psychology,

American Psychological Association, and Industrial Organizational Psychology and Organizational Behavior annual conferences.

Apivat Hanvongse is a master's student in I/O psychology at the University of Tulsa on a Fulbright Scholarship. His research interests include performance appraisal, emotional regulation, self-efficacy, positive psychology, and social entrepreneurship. He likes to travel to distant places in search of new experiences.

Dr. Cynthia Hedricks is a senior vice president at Caliper, a global human resources assessment and consulting firm in Princeton. Dr. Hedricks leads the research and development team in new product development, psychometric analysis of all Caliper assessments, and validation studies on the relationships between Caliper personality traits, cognitive measures, and work performance in the applied setting. Most recently, she directs a cross-cultural research program on the personality traits that distinguish women corporate leaders. She regularly presents the results of Caliper's global studies to academic and corporate audiences.

Chia-Lin (Rose) Ho is a first-year doctoral student in the I/O psychology program at the University of Tulsa. She received her master's degree from the University of North Carolina at Charlotte in 2005. Her thesis topic involved the development of a Taiwanese workplace personality measure, using a combined emic-etic approach. Her research interests more broadly include individual differences, cross-cultural studies, and diversity issues at work. She is currently examining the structure of personality across cultures, individual differences in the importance of person-job and person-organization fit, and the influence of cultural values on the propensity to trust and organizational trust.

Dr. Robert Hogan, president of Hogan Assessment Systems, is an international authority on personality assessment, leadership, and organizational effectiveness. Dr. Hogan was McFarlin Professor and Chair of the Department of Psychology at the University of Tulsa for 14 years. Prior to that, he was professor of psychology and social relations at The Johns Hopkins University. He has received a number of research and teaching awards, and is the editor of the *Handbook of Personality Psychology* and author of the Hogan Personality Inventory. Dr. Hogan received his PhD from the University of California, Berkeley, specializing in personality assessment. Dr. Hogan is the author of more than 300 journal articles, chapters, and books. He is widely credited with demonstrating how careful attention to personality factors can influence organizational effectiveness in a variety of areas—ranging from organizational climate and leadership to selection and effective team performance. Dr. Hogan is a fellow of the American Psychological Association and the Society for Industrial/Organizational Psychology.

Lei (Jason) Huang is a secind-year doctoral student in the I/O psychology program at the University of Tulsa. His research interests include the prediction of job performance, individual differences in computer-based training, and cross-cultural issues in I/O psychology. Coming from China, Lei sees it as his long-term goal to promote I/O psychology in China through teaching and research.

Anna Imus is currently a doctoral candidate at Michigan State University where she is pursuing a degree in industrial and organizational psychology. She graduated with honors in psychology at George Mason University in 2003. Her research interests, broadly speaking, relate to various aspects of the selection process. Specifically, she is working on papers related to the stigmatization of perceived preferential selection, developing alternative predictors of college performance, faking on personality-based selection tests, and job search behavior. She has presented her work at various national conferences and was recently published in the *Blackwell Handbook of Personnel Selection*. In her spare time, Anna enjoys playing with her two cats and cross-country skiing.

Joshua Isaacson is currently a MS graduate student at the Florida Institute of Technology. He is actively conducting research in the areas of applicant response distortion, group/team training, selection, and issues dealing with technology in the workforce. He has presented research at national conferences and written several papers in the area of applicant faking. The Naval Air Warfare Center (NAVAIR), Training Systems Division, currently employs Joshua where he is helping revise the noncognitive component of the Aviation Selection Test Battery (ASTB). NAVAIR specializes in state of the art innovative approaches to personality scale development in order to reduce faking and therefore improve the validity of such measures.

Dr. John A. Johnson completed his PhD in personality psychology at The Johns Hopkins University in 1981. Since then, he has been continuously employed at the Pennsylvania State University, where he is currently professor of psychology. In 1990-91 he was an Alexander von Humboldt-Stiftung Research Fellow at the University of Bielefeld in Germany. His research has been aimed at improving the validity of self-report personality scales in personnel selection by testing alternative models of the cognitive, motivational, and social dynamics underlying the act of responding to individual personality items. He has also studied methods for improving the validity and pragmatic utility of Web-based narrative personality reports. He has published on these topics in *European Journal of Personality, Human Performance, Journal of Personality and Social Psychology, Journal of Research in Personality, Multivariate Behavioral Research*, and *Personnel Psychology*. Dr. Johnson has advised both government agencies and commercial organizations about personnel selection.

Tina Malm is currently working on her PhD in industrial/organizational psychology under the supervision of Dr. Richard Griffith at the Florida Institute of Technology. Her main research interest within the area of applicant response distortion focuses on the prediction of faking behavior by individual difference variables. She also conducts research in the prevention of faking behavior and the identification of reliable types and levels of response distortion.

Dr. Michael A. McDaniel received his PhD in industrial/organizational psychology at George Washington University in 1986. He is a professor of management at Virginia Commonwealth University and president of Work Skills First, Inc., a human resource consulting firm. Dr. McDaniel is nationally recognized for his research and practice in personnel selection system development and validation. Dr. McDaniel has published in several major journals including the *Academy of Management Journal*, the *Journal of Applied Psychology, Human Performance*, and *Personnel Psychology*. Dr. McDaniel is a member of the Academy of Management and a Fellow of the Society of Industrial and Organizational Psychology, the American Psychological Association, and the American Psychological Society. His current research interests include applicant faking, situational judgment tests, applications of meta-analysis to I/O psychology, publication bias, and demographic differences in the workplace.

Dr. Jessica Mesmer-Magnus (PhD, 2005, Florida International University) is an assistant professor of management with the Cameron School of Business at the University of North Carolina at Wilmington. Her research interests include work/family conflict/balance, organizational training program design, whistleblowing/ counterproductive behavior, and business ethics. She has published in the *Journal of Vocational Behavior, Journal of Business Ethics* and the *Journal of Labor Research*. She is certified as a senior professional in human resources and has worked as a human resources manager for a U.S.-based national consulting firm.

Dr. Frederick L. Oswald (PhD, 1999, University of Minnesota) is an associate professor at Michigan State University. His research focuses on personnel selection in employment and academic admissions settings, with an emphasis on developing and refining personality measures and other noncognitive measures. His publications have appeared in the *Journal of Applied Psychology, Personnel Psychology,* and the *International Journal of Selection and Assessment.* He has been known to engage in impression management tactics by not admitting in his biography that some of the manuscripts he submits to journals get rejected outright.

Victoria L. Pace is a PhD student at the University of South Florida and a research associate at Personnel Decisions Research Institutes, Inc. (PDRI). Her research and applied work focus on development and validation of selection and other assessment measures, especially those measuring personality. In order to enhance the accuracy of organizational decisions, she continues to explore innovative methods to address factors such as response distortion that may limit the usefulness of these measures.

Mitchell H. Peterson is pursuing his MS and PhD degrees in I/O psychology at the Florida Institute of Technology. Much of his early work has focused on personality measurement in the selection process. More specifically, his research has examined motivational influences on applicant faking behavior, in addition to the effects of faking on the criterion-related validity of personality measures. His work has appeared in book chapters, as well as being presented at professional conferences. Some of his more recent work has investigated the combination of ability and personality measures and the impact of this combination on hiring decisions. Mitchell has also been involved in conducting organizational climate assessments for the United States Department of Defense.

Radha Roy is a senior research associate in the Research and Development Department at Caliper. She is the lead statistician on all client-focused studies that investigate the relationships between Caliper personality traits, cognitive measures, and work performance. Ms. Roy has coauthored scientific presentations on the results of Caliper's validation studies in the applied setting.

Dr. Ann Marie Ryan is a professor of organizational psychology at Michigan State University. She received her PhD in 1987 from the University of Illinois at Chicago. Ann Marie has been recognized as a prolific researcher in the areas of fairness in employee selection and employee attitude surveying. She is a fellow of APA and APS and a past president of Division 14, the Society for Industrial and Organizational Psychology. Ann Marie currently serves as editor for *Personnel Psychology,* a leading I/O journal.

Dr. Andrea F. Snell is currently an associate professor of industrial/organizational psychology at the University of Akron. In addition to her long-standing interests in applied research methodology and statistical modeling, she has investigated the utility of biodata and personality measures as predictors of occupational variables. Along with her students at the University of Akron, she has been searching for mechanisms that will improve the predictive validity of these measures when they are used for applicant selection. With her collaborators, she has most recently been investigating situational effects on applicant responding with the hopes of understanding, and possibly mitigating, the problems associated with intentional response distortion. Her publications have appeared in the *Journal of Applied Psychology, Personnel Psychology, Organizational Research Methods*, and *Journal of Vocational Behavior.*

Dr. Robert P. Tett is an associate professor of psychology at the University of Tulsa, where he teaches courses in personnel selection, leadership, meta-analysis, and univariate and multivariate statistics. He received his doctorate from the University of Western Ontario in 1995 under the mentorship of Douglas N. Jackson, PhD. His current research interests include personality complementarity within teams, the meaning of leadership, and emotional intelligence as a multidimensional trait domain. Dr. Tett has published in outlets including the *Journal of Applied Psychology, Personnel Psychology, Journal of Organizational Behavior, Human Performance, Journal of Personality*, and *Personality and Social Psychology Bulletin*. He is senior author of the recently published self-report *Survey of Emotional Intelligence* (SEI) and a workplace counterpart (SEIW). He is a devoted father of three, enjoys playing hockey, and plays piano any chance he gets.

Dr. Nicholas L. Vasilopoulos is an associate professor and director of the industrial and organizational psychology doctoral program at The George Washington University. Much of his research focuses on identifying factors that influence the validity of self-report measures used for personnel selection. Specifically, he has explored whether (a) self-report item response latencies can be used to detect applicant faking, (b) changes in test characteristics (e.g., including a warning of verification or using forced-choice item format) impact construct and criterion-related validity, and (c) nonlinear relationships exist between personality and performance. His publications in the area have appeared in the *Journal of Applied Psychology, Journal of Personality*, and *Human Performance*.

Dr. Chockalingam Viswesvaran (PhD, University of Iowa) is a professor of psychology at Florida International University. His research interests include business ethics, personnel selection, and human resource management. He has published in *Journal of Applied Psychology, Organizational Behavior and Human Decision Processes*, and *Psychological Bulletin*. He has served on five editorial boards and as an associate editor of the *International Journal of Selection and Assessment*. He is an elected fellow of the Society for Industrial and Organizational Psychology as well as Divisions 5 (Measurement) and 14 (Industrial-Organizational Psychology) of the American Psychological Association.

Tae Seok Yang graduated from Sungkyunkwan University, South Korea with a bachelor of business administration in 2002 and is now a third-year doctoral student in I/O psychology at the University of Tulsa. He has been working with Dr. Kurt Kraiger on research regarding the effects of seductive details on training effectiveness. He is also working with Dr. Robert Tett on a study of emotional intelligence. He is currently a research and teaching assistant for the TU I/O psychology program and a representative for I/O PhD students in the psychology graduate student association at TU.

Yukiko Yoshita received her MS in industrial/organizational psychology from the Florida Institute of Technology in 2002, and is currently pursuing her PhD Her research interests range from applicant distortion on personality measures, with a focus on the modeling of faking and effect of applicant distortion; to cross-cultural issues in I/O psychology in Japan and the United States. She has presented papers in several conferences and has published research in *Personnel Review*.

Dr. Michael J. Zickar has been researching the effects of faking on the psychometric properties of personality tests for over a decade. He is an associate professor of psychology at Bowling Green State University (BGSU) and is the program Chair of BGSU's doctoral program. In addition, he is the historian of the Society for Industrial-Organizational Psychology. His research on faking has appeared in the *Journal of Applied Psychology, Organizational Research Methods*, and *Applied Psychological Measurement*.